Environmental Geochemistry and Health
with special reference to developing countries

Geological Society Special Publications
Series Editor A. J. FLEET

GEOLOGICAL SOCIETY SPECIAL PUBLICATION NO. 113

Environmental Geochemistry and Health

with special reference to developing countries

EDITED BY

J. D. APPLETON
British Geological Survey
Nottingham, UK

R. FUGE
Institute of Earth Studies
University of Wales
Aberystwyth, UK

and

G. J. H. McCALL
Department of Earth Studies
Liverpool University, UK

1996
Published by
The Geological Society
London

THE GEOLOGICAL SOCIETY

The Society was founded in 1807 as The Geological Society of London and is the oldest geological society in the world. It received its Royal Charter in 1825 for the purpose of 'investigating the mineral structure of the Earth'. The Society is Britain's national society for geology with a membership of around 8000. It has countrywide coverage and approximately 1000 members reside overseas. The Society is responsible for all aspects of the geological sciences including professional matters. The Society has its own publishing house, which produces the Society's international journals, books and maps, and which acts as the European distributor for publications of the American Association of Petroleum Geologists, SEPM and the Geological Society of America.

Fellowship is open to those holding a recognized honours degree in geology or cognate subject and who have at least two years' relevant postgraduate experience, or who have not less than six years' relevant experience in geology or a cognate subject. A Fellow who has not less than five years' relevant postgraduate experience in the practice of geology may apply for validation and, subject to approval, may be able to use the designatory letters C Geol (Chartered Geologist).

Further information about the Society is available from the Membership Manager, The Geological Society, Burlington House, Piccadilly, London W1V 0JU, UK. The Society is a Registered Charity, No. 210161.

Published by The Geological Society from:
The Geological Society Publishing House
Unit 7, Brassmill Enterprise Centre
Brassmill Lane
Bath BA1 3JN
UK
(*Orders:* Tel. 01225 445046
Fax 01225 442836)

First published 1996

The publishers make no representation, express or implied, with regard to the accuracy of the information contained in this book and cannot accept any legal responsibility for any errors or omissions that may be made.

© The Geological Society 1996. All rights reserved. No reproduction, copy or transmission of this publication may be made without written permission. No paragraph of this publication may be reproduced, copied or transmitted save with the provisions of the Copyright Licensing Agency, 90 Tottenham Court Road, London W1P 9HE. Users registered with the Copyright Clearance Center, 27 Congress Street, Salem, MA 01970, USA: the item-fee code for this publication is 0305-8719/96/$7.00.

British Library Cataloguing in Publication Data
A catalogue record for this book is available from the British Library.

ISBN 1-897799-64-0
ISSN 0305-8719

Typeset by Bath Typesetting, Bath, UK

Printed by The Alden Press, Osney Mead, Oxford, UK

Distributors

USA
AAPG Bookstore
PO Box 979
Tulsa
OK 74101-0979
USA
(*Orders:* Tel. (918) 584-2555
Fax (918) 584-2652)

Australia
Australian Mineral Foundation
63 Conyngham Street
Glenside
South Australia 5065
Australia
(*Orders:* Tel. (08) 379-0444
Fax (08) 379-4634)

India
Affiliated East-West Press PVT Ltd
G-1/16 Ansari Road
New Delhi 110 002
India
(*Orders:* Tel. (11) 327-9113
Fax (11) 326-0538)

Japan
Kanda Book Trading Co.
Tanikawa Building
3-2 Kanda Surugadai
Chiyoda-Ku
Tokyo 101
Japan
(*Orders:* Tel. (03) 3255-3497
Fax (03) 3255-3495)

Contents

Preface	vii
MILLS, C. F. Geochemical aspects of the aetiology of trace element related diseases	1
PLANT, J. A., BALDOCK, J. W. & SMITH, B. The role of geochemistry in environmental and epidemiological studies in developing countries: a review	7
FORDYCE, F. M., MASARA, D. & APPLETON, J. D. Stream sediment, soil and forage chemistry as indicators of cattle mineral status in northeast Zimbabwe	23
JUMBA, I. O., SUTTLE, N. F., HUNTER, E. A. & WANDIGA, S. O. Effects of botanical composition, soil origin and composition on mineral concentrations in dry season pastures in western Kenya	39
MASKALL, J. & THORNTON, I. The distribution of trace and major elements in Kenyan soil profiles and implications for wildlife nutrition	47
BOWELL, R. J., WARREN, A. & REDMOND, I. Formation of cave salts and utilization by elephants in the Mount Elgon Region, Kenya	63
SELINUS, O. FRANK, A. & GALGAN, V. Biogeochemistry and metal biology	81
EDMUNDS, W. M. & SMEDLEY, P. L. Groundwater geochemistry and health: an overview	91
BOWELL, R. J., McELDOWNEY, S., WARREN, A., MATHEW, B. & BWANKUZO, M. Biogeochemical factors affecting groundwater quality in central Tanzania	107
DISSANAYAKE, C. B. Water quality and dental health in the Dry Zone of Sri Lanka	131
SMITH, B., BREWARD, N., CRAWFORD, M. B., GALIMAKA, D., MUSHIRI, S. M. & REEDER, S. The environmental geochemistry of aluminium in tropical terrains and its implications to health	141
THORNTON, I. Sources and pathways of arsenic in the geochemical environment: health implications	153
SMEDLEY, P. L., EDMUNDS, W. M. & PELIG-BA, K. B. Mobility of arsenic in groundwater in the Obuasi gold-mining area of Ghana: some implications for human health	163
HELIOS RYBICKA, E. Environmental impact of mining and smelting industries in Poland	183
NICHOLSON, K. Lacustrine sediment geochemistry as a tool in retrospective environmental impact assessment of mining and urban development in tropical environments: examples from Papua New Guinea	195
FUGE, R. Geochemistry of iodine in relation to iodine deficiency diseases	201
DISSANAYAKE, C. B. & CHANDRAJITH, R. L. R. Iodine in the environment and endemic goitre in Sri Lanka	213
STEWART, A. G. & PHAROAH, P. O. D. Clinical and epidemiological correlates of iodine deficiency disorders	223
NAIR, M. G., MAXWELL, S. M. & BRABIN, B. J. The protective role of trace elements in preventing aflatoxin induced damage: a review	231
NICHOLSON, R. A., ROBERTS, P. D. & BAXTER, P. J. Preliminary studies of acid and gas contamination at Poas volcano, Costa Rica	239
OLIVER, M. A. Kriging: a method of estimation for environmental and rare disease data	245

HARVEY, R., POWELL, J. J. & THOMPSON, R. P. H. A review of the geochemical factors linked to Podoconiosis 255

Index 261

Preface

Environmental geochemistry has long been known to influence human and animal health, the links being recognized possibly as long ago as the 4th century AD by the Chinese. While early recognized problems such as endemic goitre were clearly due to natural geochemical variations within the environment, more recently anthropogenic perturbations of the environment have been shown to exert a strong influence on human and animal well being.

Although some health problems related to environmental geochemistry manifest themselves in the developed world, they are more keenly felt in the developing countries due to the added stress of such factors as poverty and malnutrition. In addition, while most of the population of the developed world have diets which include food sources from geographically diverse regions, in the developing countries this is frequently not so. Many people in developing countries are dependent on very localized sources of food and water and any geochemical anomaly (enrichment or depletion) within these local environments will have a marked influence on the well-being of the inhabitants.

The latest World Health Organisation figures indicate that more than 800 million people in the developing countries are at risk from iodine deficiency, which is the cause of goitre and cretinism, a severe form of mental retardation. 350 million people in developing countries are estimated to suffer from severe iron deficiency and recent studies have provided evidence that cancer and heart disease are related to selenium deficiencies. High natural environmental concentrations of fluorine cause mottling of teeth and crippling bone and joint deformities in both humans and cattle. Excessive dietary intakes of aluminium, arsenic and lead derived from both natural sources and as a result of industrial pollution are also associated with ill-health in humans. Concern over this issue is increasing as a result of rapid economic and population growth in the developing world. Multidisciplinary studies involving epidemiologists and biochemists as well as environmental and nutritional specialists are essential if the impact of trace element levels on ill-health in developing countries is to be properly addressed and remedies implemented.

The main aim of this volume is to discuss the application of geochemistry to the study of human and animal health problems in developing countries. It stems from a conference held at the Geological Society, Burlington House, London, UK on 20–21 October 1993 which was organized by the Joint Association of Geoscientists for International Development (JAGID) and the Society for Environmental Geochemistry and Health (SEGH). The conference provided a forum for the exchange of ideas, information and experience between geochemists, nutritionists, medical and veterinary researchers and brought together researchers from UK, Poland and Sweden as well as a range of developing countries including India, Kenya, Tanzania, Uganda and Sri Lanka. Of the 34 papers presented at the meeting, 20 are published in this volume while two invited contributions have been added. Abstracts of the other 14 papers are published in *Environmental Geochemistry and Health in Developing Countries, October 1993: Abstracts Volume*. This is available (price code I) from BGS Publications (Tel: 0115 936 3241).

Subjects covered in the volume reflect the breadth of the topic under discussion, ranging through animal and human health issues related to soil, plant, water and volcanic gas chemistry. The role of natural and anthropogenic influences on the environment are covered as are the roles of geochemical mapping, monitoring and baseline identification.

The first two papers in the volume set the scene, with considerations of the aetiology of geochemically related nutritional diseases (**Mills**) and the role of geochemistry in environmental and epidemiological studies in developing countries (**Plant et al.**). The next group of five papers concern animal nutrition in Zimbabwe (**Fordyce et al.**), Kenya (**Jumba et al**; **Maskall & Thornton**) and the Kenya–Uganda border area (**Bowell et al.**). Finally in this section **Selinus et al.** present a Swedish view of environmental monitoring using aquatic mosses, roots of aquatic plants and organ tissues from the moose.

Health aspects of groundwater chemistry are reviewed by **Edmunds & Smedley** followed by contributions on biogeochemical factors affecting groundwater quality in Tanzania (**Bowell et al.**), water quality and dental fluorine in Sri Lanka (**Dissanayake**), and the geochemistry of aluminium and its potential toxicity in Uganda (**Smith et al.**). The environmental behaviour of arsenic is reviewed by **Thornton**, while **Smedley et al.** consider the health implications of this element in groundwater in a gold-mining area in Ghana. The population theme is continued with a consideration of the impacts of mining and smelting on the Polish environment (**Helios Rybicka**) and on the use of lake sediments to assess mining and urban contamination in Papua New Guinea (**Nicholson**).

Iodine with its long recognized link to health is the subject of three papers, beginning with a consideration of its geochemistry (**Fuge**). Papers on the aetiology of endemic goitre in Sri Lanka (**Dissanayake & Chandrajith**) and on the epidemiology of iodine deficiency disorders (**Stewart & Pharoah**) consider the role of iodine and other factors in these diseases.

The role of trace elements in preventing aflatoxin induced cell damage and disease and the potential influence of geographical variations in trace element levels on disease patterns is discussed by **Nair et al. Nicholson et al.** consider the negative environmental impacts of acidic volcanic gas emissions in Costa Rica on crops, buildings and the health of domestic livestock and people. The penultimate paper demonstrates how a geostatistical method (kriging) can be used to investigate the links between the geographical distribution of disease and environmental factors (**Oliver**), while the final paper is a review of geochemical factors which may influence the distribution of Podoconiosis, or non-filarial elephantiasis, in tropical Africa (**Harvey et al.**).

The editors are particularly grateful to JAGID, SEGH, the British Council (Kenya) and the Overseas Development Administration who provided support which facilitated the attendance of some of the participants from developing countries. We wish to express our thanks to the authors for their patience and collaboration and to the many reviewers whose suggestions and comments helped to ensure that the papers published here maintain the consistently high standards of Geological Society Special Publications. The considerable efforts of the production editor, David Ogden, and other staff at the Geological Society Publishing House, Bath, are much appreciated.

Don Appleton
Ron Fuge
Joe McCall

March 1996

Geochemical aspects of the aetiology of trace element related diseases

C. F. MILLS

Rowett Research Institute, Bucksburn, Aberdeen AB2 9SB, UK

Abstract: Deficiencies, excesses or imbalances in the supply of inorganic elements from dietary sources can have an important influence on animal and human health and susceptibility to disease. Many such situations arise from anomalies in the inorganic element composition of food chains. These, in turn, are frequently attributable to the composition of the geochemical environment as modified by the influence of soil composition and botanical or cultural variables upon the inorganic composition of the diet.

Appreciation of the epidemiological importance of such factors in the aetiology of nutritional diseases can contribute, very significantly, to their detection and effective control.

Although the agricultural relevance of geochemical data is firmly established and widely appreciated, investigation of its value in the context of anticipation and control of major human nutritional diseases in the third world has yet to be undertaken adequately.

Until the middle of the 19th century, convictions that geographically related features of the natural environment were associated with local differences in susceptibility to ill health of human communities and their livestock were based, almost entirely, upon anecdote. Nevertheless, many formed the basis of community legislation for land use, animal management or cultural practices, and are now supported by firm evidence derived from our growing understanding of the significance of inorganic elements for human and animal health. The traditional restriction of common grazing rights to maintain animal health on the 'machair' lands of the Outer Hebrides is one such example; the distribution of communities issuing such legislation closely reflects the distribution of cobalt deficient soils. Many such established relationships reflect, directly or indirectly, anomalies in the geochemical environment that influence the flow of inorganic elements into and through the food chain of human subjects and the animals and crops upon which they depend.

Recent developments in geochemical survey techniques are providing an increasing volume of data which can make a significant contribution to understanding the significance and likely causes of those inorganic element-related diseases which are attributable either to deficiencies or excesses of inorganic elements in diets or drinking water. The growth of information on this topic is particularly welcome (Aggett *et al.* 1990). Thus the detection and control of such diseases is often complicated by the fact that signs of a general malaise rather than the appearance of diagnostically specific signs are frequently the most typical manifestations of excessive or deficient element supply. This, and the frequency with which antagonistic interactions arising from inorganic element imbalances are involved in the aetiology of disorders attributable to deficiencies or toxicities (Mills 1995) illustrate why multi-element geochemical survey data can provide information invaluable to the nutritional epidemiologist (Aggett & Mills 1996). This brief contribution will endeavour to illustrate the potential of geochemical studies in this context, while also emphasizing the need for consideration of those multi-disciplinary aspects that describe the influence of variables modifying relationships between the chemical composition of the environment and its influence on susceptibility to nutritional disease.

The influence of geochemical and soil compositional variables

Experience with nutritional disorders in domesticated animals points to many situations in which soil geochemical data can contribute to the identification of problems of health or productivity related to trace element supply. Areas with high intrinsic risks of deficiencies of copper, selenium, cobalt, iodine and of toxicities of fluorine, lead and selenium have thus been identified (SAC/SARI 1982; Mills 1984). Such investigations have also produced evidence that changes in soil and crop management can markedly modify trace element uptake by food crops and thus change trace element disease patterns in livestock. High soil pH, whether arising from calcareous soil parent materials or induced by liming, restricts the supply of available zinc, iron and cobalt but can promote crop accumulation of molybdenum, selenium and fluorine. Intermittent flooding or irrigation of certain soil types can greatly enhance uptakes of molybdenum and selenium while concentrations of iron, manganese and aluminium may increase by several orders of magnitude following the development of acid conditions in soils derived from some parent materials.

Although the value of such information for the detection of microelement-related disorders in animals is now well recognized, such information is less frequently exploited in the context of its relevance to human health (Shaper 1984). However, its potential is becoming evident from recent studies. These include studies of the pathogenesis of the drastic skeletal lesions of *genu valgum* associated with excessive uptakes of fluorine in India. Relationships between a low selenium status of soils and animals and the distribution of the cardiomyopathy of Keshan disease of children and, conversely, of relationships between high soil selenium and selenosis in adults have been demonstrated from Chinese studies. Epidemiological studies leave no doubt as to the significance of relationships between iodine-responsive diseases in human communities and the distribution of iodine deficiency in soils, crops or water supplies. These and other examples are considered in more detail by Mills (1987), Shaper (1984) and Hetzel & Maberly (1986).

Epidemiologically relevant data are emerging from geochemical surveys in several parts of the world. International surveys of regional differences in soil and crop composition are also being promoted by the Food & Agriculture Organisation (Silinpää 1982, 1990). However, the stimulus to many such studies is usually the need to resolve causes of crop failure, or in the instance of geochemical surveys, to search for mineral resources. There is urgent need to emphasize to those financing such studies, that the data they produce are often of value for identifying areas in which health may be prejudiced by imbalances in mineral supply. Thus, in the context of iodine deficiency disorders, the need for such information has been largely ignored since data on the geochemical or soil distribution of iodine are rarely of major interest to the mineral prospector or the plant nutritionist to whom iodine is a 'non-essential' element. Data on the geographical distribution of areas in which 'environmental' iodine is low would clearly assist decision making in according priorities to intervention programmes for the control of a human disease for which it is estimated that at least 1 000 million people are at risk (Hetzel & Maberly 1986).

Soil and geochemical factors influencing trace element uptake

Variables modifying the trace element content of foods or of drinking water can influence, profoundly, the risks that soil and geochemical compositional anomalies may be in terms of pathological consequences to animals or human subjects. However, variety in the selection of food sources can have a major influence on whether such diseases develop (SAC/SARI 1982). Thus, intensification of cropping usually implies that the variety of species available for consumption by animals decreases markedly; the extent of protection afforded by individual differences in the extent of plant species uptake of individual elements is thus lost. Correspondingly, urban human communities usually obtain their food, and sometimes their water supplies, from a variety of geographical sources. They are thus less likely to be exposed to compositional anomalies resulting from the soil or its parent material than are those of rural communities relying heavily on locally grown crops as constituents of diets containing a limited number of food items.

Examples of geochemical and soil variables likely to influence significantly the trace element intake of rural populations are given in Table 1. The examples given are not exhaustive but they do illustrate the following points:

- The geochemical composition of rocks from which some soils are derived can influence, directly or indirectly, both the balance of elements within those soils and the trace element content of food crops grown on them. Regional differences of chromium, copper, iodine, iron, selenium and zinc, and excesses of arsenic, cadmium, fluoride, lead and selenium arise from such causes.
- Differences in soil moisture and acidity or alkalinity arising either from natural causes or from cultivational or irrigation practices, industrialization or urbanization, can markedly affect the uptake of specific elements into edible crops and vegetation (SAC/SARI 1982). Irrigation with alkaline groundwaters, especially of certain types of shales and of some mineralized granites, greatly increases molybdenum and selenium uptake. Iron-rich irrigation waters restrict the uptake of selenium into the food chain. Extensive water leaching of some acid arenaceous soils low in organic matter increases the magnitude of iodine losses and reduces the intrinsic availability of soil selenium and zinc. High soil acidity strongly potentiates crop uptake of aluminium, iron and manganese.
- The impact of geochemical and soil vari-

Table 1. *Typical geochemical and soil features associated with inorganic element anomalies causing nutritional diseases in man and domesticated livestock*

Syndrome	Environmental anomaly	Species affected*
Deficiencies		
Low cobalt	Soils intrinsically low in Co, e.g. extensively leached, acid arenaceous soils or with Co immobilized with Fe/Mn hydroxide complexes.	R
Low phosphorus	High Fe/Al parent materials with low pH and highly organic soils.	R
Low selenium	Soils intrinsically low in Se, e.g. leached arenaceous soils particularly when low in organic and argillaceous fractions. Fixation of Se in soils high in Fe.	M, F
Low zinc	Calcareous parent materials and derived soils especially when adventitious soil present in diets high in cereals and legumes. Arid arenaceous soils.	M, F
Toxicities		
High arsenic	Waters from some hydrothermal sources or soils derived from detritus of mineral ore (especially Au) workings. Well waters or irrigation waters from sandstones high in arsenopyrite.	
High fluoride	Waters from some aquifers especially from rhyolite-rich rocks, black shales or coals; soils from F-containing residues of mineral or industrial deposits. Aggravated by high evaporative losses.	M, F
High molybdenum	Mo from molybdeniferous shales or local mineralization especially if drainage poor and soil pH > 6.5 (a significant cause of secondary Cu deficiency).	R
High selenium	Bioaccumulation of Se in organic-rich soil horizons; accumulation by high evaporative losses of high pH ground waters	M, F

* M, human subjects; F, farm livestock, general; R, ruminant livestock, specific.
Data sources: SAC/SARI (1982); WHO (1996).

ables on human trace element intake depends markedly on the type of crops that are dietary staples. In general, the trace element composition of leguminous crops, pulses and cruciferous crops fluctuates widely as such variables change. In contrast, the trace element content of cereal grains, although frequently lower, is less readily affected. It should be noted, however, that the selenium content and to a lesser extent, the zinc content of cereal grains are markedly influenced by soil conditions and type.

The influence of geochemical and soil conditions on trace element supply is considered in greater detail elsewhere. Reports from the Food & Agricultural Organisation (Silinpää 1982, 1990) describing a wide range of studies in different countries illustrate the value of assessing the intrinsic properties of indigenous soils as trace element sources for a region. Disappointingly little is known about the geographical distribution of abnormalities influencing dietary supplies of iodine and selenium but, apart from these, extensive information banks have been developed for many countries which define the mineral characteristics of rocks and soils in the context of their influences on the inorganic composition of staple crops. Although correspondingly extensive databanks on dietary element intakes in the developing countries are now being assembled (Parr et al. 1992), virtually no effort has been made to quantify relationships between the inorganic composition of soils and diets. The need to examine such relationships more fully in investigations of the causes of deficiencies of copper, iodine, selenium, zinc and possibly of boron, chromium and fluorine and of toxicoses due to fluoride, iron, molybdenum and selenium is urgent.

Those responsible for the development of large-scale irrigation schemes where introducing modifying soil-cultural practices should be strongly encouraged to request information from such sources, before such policies are imposed upon third world communities, on the possible consequences of these developments on trace element flux from parent rocks, through soils, to edible crops. Failure to take this precaution has already caused extensive problems from excesses of fluoride, arsenic and selenium in some communities. It has also caused crop failures and decreased animal productivity from the inadvertent potentiation of trace element related disorders caused by deficiencies of copper, cobalt, manganese, selenium and zinc or by excesses of fluorine, iodine, iron and selenium and, in animals, of molybdenum.

The impact of such changes can differ according to the predominant species of plant that constitutes the staple crop. Plants have a wide tolerance to many trace elements and show marked species variability in the extent to which these elements are accumulated. The variability of trace element content of the specific crops also differs markedly for the different elements; copper has been found to vary 4-fold, zinc 7-fold, boron 21-fold, and molybdenum up to 46-fold depending on soil and crop management techniques (Aggett & Mills 1996). Maize concentrates selenium better than rice; recycling of maize residues as a fertilizer in crops growing in a high selenium soil has been incriminated in instances of selenium intoxication in livestock and human communities in some regions of China (WHO 1996).

Water composition and health

The risks of appearance of trace element related effects on health can be influenced substantially by changes in the sources of drinking water or groundwater. The effects of irrigation, soil water economy and solar evaporation on the entry of elements into the food chain through rural populations are most clearly evident in communities where food choice, particularly for the young, is largely restricted to a few locally grown staples and water is drawn from a single source.

Aquifers in fluoride-bearing rocks or the irrigation of fluoride-rich soils can increase the intake of this element through drinking water or food crops sufficiently to induce major problems of community health as a consequence of the adverse effects of fluoride intoxication on skeletal development. The increased consumption of water in arid environments may be sufficient to increase the risks of fluorosis whenever the fluorine content of water exceeds about $2\,mg\,l^{-1}$.

Conversely, the leaching of iodine from acid mineralized soils is an important determinant of the geographical distribution of iodine deficiency. Although it is suspected that soils low in clay mineral fractions and in organic matter appear to be associated with the increased risks of development of iodine deficiency in animal and human populations, such relationships are insufficiently well characterized to be of value in predicting risks of disease in specific areas. The distribution of selenium deficiency may be influenced, similarly, by the leaching of mineralized soils by rain water and recent work (Arthur & Beckett 1994) highlighting the metabolic interdependence of these two elements suggests that concurrent deficiencies of both selenium and iodine would be particularly hazardous in the induction of iodine deficiency disorders.

Contamination of groundwater or well water with arsenic derived initially from the oxidation of arsenopyrite has now been incriminated as a cause of chronic arsenic toxicity in human communities of at least five continents. Many such cases involve biological concentration of geologically derived arsenic during its passage through food chains (for review see Chatterjee *et al.* 1995).

Geophagia

The involuntary or sometimes the deliberate eating of earth is a widespread practice both in the animal kingdom and in many human communities of the third world. It is a form of pica likely to have pathological implications when it impairs the intestinal uptake of trace elements such as iron and zinc. However, it is also suggested, although not proven, that geophagia is sometimes an evolved adaptive attempt to compensate for mineral deficits or imbalances. Differing properties of various clays clearly affect their ability to meet this need. The eating of clays with high cation exchange capacity can release many elements under acid digestive conditions that can significantly supplement the intakes of available copper, iron, manganese and zinc of some subsistence communities. They may also release macro minerals, the need for which may be the underlying stimulus to geophagia. In contrast, consumption of some inadequately characterized clay minerals in the Middle East is believed to be the cause of growth retardation, anaemia and delayed puberty (WHO 1996). However, deliberate or adventitious contamination of foods with calcareous soils may sufficiently elevate the intake of calcium in some Middle Eastern communities to affect adversely zinc utilization from vegetable-based diets. The consumption of iron-rich laterites or iron-rich waters draining therefrom must be expected to influence adversely the utilization of copper, zinc and possibly of selenium in other regions.

Conclusions

The duration and extent of exposure of animals or human subjects to conditions which modify appreciably the flow of inorganic elements through the food chain can have a profound effect upon the pathological consequences of such exposure. For all species, variety in the selection of food sources is the greatest protection against adverse effects. Nomadic social habits and miscellany in the selection of vegetational species for food can restrict to tolerable limits the magnitude of changes in tissue trace element content caused by anomalies originating from changes in geochemical or soil composition of the environment. In contrast, deliberate or inadvertent restriction of opportunities for food selection as a consequence, for example, of climatic disasters or the introduction by irrigation or cropping intensification can amplify the effects of local anomalies in soil or parent rock composition.

Awareness of these effects and of the factors which modify their consequences can be of enormous assistance to epidemiologists seeking the causes of diseases. These, with the virtual exceptions of the goitre of iodine deficiency and, in sheep, the neurological manifestations of copper deficiency, give rise to ill health unaccompanied by clinical manifestations that are sufficiently specific to permit unequivocal diagnosis.

The potential contribution of the geochemist and soil chemist to the detection and control of diseases arising from such nutritional causes has been undervalued (Mills 1987; Aggett & Mills 1996). Quite understandably, the spectrum of elements covered by most geochemical surveys has usually been determined by mineralogical rather than by biological considerations. It is to be hoped, firstly, that biological aspects will be given greater emphasis and, secondly, that the need for multidisciplinary consideration of geochemical and soil data will be appreciated. Relationships are complex and disappointment will follow if modifying effects of interactive variables are not taken into account in the interpretation of such studies.

The author is indebted particularly to the Hamdard Press, Karachi and World Health Organisation, Geneva for permission to quote extensively from reports he has published previously on this topic.

References

AGGETT, P. J. & MILLS, C. F. 1996. Detection and anticipation of the risks of development of trace element-related disorders. *In: Trace Elements in Human Nutrition and Health*. World Health Organisation, Geneva, 289–308.

——, ——, MORRISON, A., PLANT, J. A., SIMPSON, P. R., *et al.* 1990. A study of environmental geochemistry and health in North East Scotland. *In:* THORNTON, I. (ed.) *Geochemistry & Health*. Monograph Series: Environmental Geochemistry and Health. Science Reviews, Northwood, 81–91.

ARTHUR, J. R. & BECKETT, G. J. 1994. New metabolic roles for selenium. *Proceedings of the Nutrition Society*, **53**, 616–624.

CHATTERJEE, A., DAS, D., MANDAL, B. K., CHOWDHURY, T. R., SAMANTA, G. & CHAKRABORTI, D. 1995. Arsenic in ground water in six districts of West Bengal, India: The biggest arsenic calamity in the world. *Analyst*, **120**, 643–657.

HETZEL, B. S. & MABERLY, G. F. 1986. Iodine *In:* MERTZ, W. (ed.) *Trace Elements in Human and Animal Nutrition*. 5th ed., Vol. 2, Academic Press, Orlando, 139–208.

MILLS, C. F. 1984. Geochemistry and animal health. *In:* BOWIE, S. H. U. & THORNTON, I. (eds) *Environmental geochemistry and health*. Reidel, Dordrecht, 59–95.

—— 1987. The detection of trace element problems in the developing countries. *In:* SAID, M., RAHMAN, M. A. & D'SILVA, L. A. (eds) *Elements in health and disease*. Hamdard University Press, Karachi, 74–89.

—— 1995. Trace element bioavailability and interactions. *In: Trace Elements in Human Nutrition and Health*. World Health Organisation, Geneva, 22–46.

PARR, R. M., CRAWLEY, H., ABDULLA, M., IYENGAR, G. V. & KAMPULAINEN, J. 1992. *Human dietary intakes of the trace elements: a global literature survey, mainly for the period 1970–1991*. International Atomic Energy Agency, Vienna.

SAC/SARI. 1982. *Trace element deficiency in ruminants: report of a study group*. Scottish Agricultural Colleges/Scottish Agricultural Research Institutes, Edinburgh.

SHAPER, A. G. 1984. Geochemistry & Human Health. *In:* BOWIE, S. H. U. & THORNTON, I. (eds) *Environmental geochemistry and health*. Reidel, Dordrecht, 97–119.

SILINPÄÄ, M. 1982. *Micronutrients and the nutrient status of soils: a global study*. FAO Soils Bulletin, **48**, Food & Agriculture Organisation, Rome.

—— 1990. *Micronutrients at the country level*. FAO Soils Bulletin, **63**, Food & Agriculture Organisation, Rome.

WHO 1996. *Trace elements in human nutrition and health*. World Health Organisation, Geneva.

The role of geochemistry in environmental and epidemiological studies in developing countries: a review

J. A. PLANT, J. W. BALDOCK & B. SMITH

British Geological Survey, Keyworth, Nottingham NG12 5GG, UK

Abstract: Concern over the effects of chemicals in the environment on the health of man and animals is growing as rapid economic and population growth extends such problems as land degradation, pollution and urbanization from industrialized nations to the developing world.

In this paper we review the principal socio-economic and environmental pressures on developing countries before discussing the role of geochemistry in: (1) preparing high resolution baseline data to identify potential hazards; (2) understanding the pathways of chemical elements from rocks and soils to man and animals; and (3) developing amelioration strategies to reduce the impacts of inappropriate land use, power generation and mining. The particular geochemical problems of tropical terrains are discussed and some case histories from the international work of the British Geological Survey (BGS), funded by the Overseas Development Administration, are described.

It is recommended that developing nations prepare modern geochemical maps, ideally to the standards set out in International Geological Correlation Programme Project 360 World Geochemical Baseline, and that aid agencies should fund integrated environmental geochemical surveys as being of primary importance, especially for health studies and land use planning; particular attention should be paid to the environmental impact of urbanization. Further understanding of chemical and mineralogical speciation is required to improve the interpretation of geochemical data for environmental purposes.

Multidisciplinary studies, involving epidemiologists, biochemists and nutritional specialists, are essential if natural and anthropogenic impacts are to be properly assessed and practical amelioration measures implemented.

There is a growing awareness of the relationships between animal and human health and the distribution of chemical substances in the environment, first demonstrated by Webb (1964). In industrialized countries, concern continues to focus on anthropogenic accumulations of potentially harmful elements (PHEs) such as As, Cd, Hg and Pb, and on organic compounds such as DDT, PCBs and dioxins. Some of these chemicals may be classified as carcinogens, neurotoxins or irritants; others may cause reproductive failure or birth defects (WHO 1988). In the case of domestic livestock, conditions caused by deficiencies in one or more of the essential trace-elements Co, Cu, Zn, Se and I are also well documented (Mertz 1986). However, the extension of such links to humans in developed countries remains controversial, since diets are diverse and generally considered to provide adequate trace-element levels.

In recent years the availability of regional geochemical data for Britain, Canada, Scandinavia and many other developed countries has demonstrated that in addition to pollution related to man's activities, large areas have high concentrations of heavy or radioactive elements which occur naturally (BGS atlas series 1978–95; Appleton 1992). Extensive regions have also been shown to have levels of essential trace elements well below those recommended for soils and pasture (Darnley *et al.* 1995). Excellent epidemiological data are available for some developed countries and certain associations between environmental geochemistry, diet and degenerative disease have been suggested (e.g. Martyn *et al.* 1989). Attempts to link the occurrence of degenerative diseases, such as cancer and heart disease to diet may, however, be jeopardized by lack of knowledge of the chemistry of dietary components as well as by the wide range of confounding factors. Variations in the trace element status of most crops reflect that of the soil in which they are grown, but in developed countries the effect on humans is masked by the use of food from different areas.

By contrast, people in many developing countries, particularly those living on subsistence agriculture, obtain much of their food from local sources and problems such as land degradation, pollution and increasing urbanization may be particularly intense. Equally, trace-element deficiencies or toxicities may be much more critical for human and animal health than in developed countries. Studies of the relationship between environmental geochemistry and health are therefore likely to be of more immediate value in developing than developed countries, although the results could have worldwide significance. Unfortunately, modern geochemical data are rarely available for developing countries, or may be inadequate for environmental purposes, having been collected principally for mineral exploration.

In this paper we first briefly review the socio-economic threats to the environment of developing countries, with particular reference to the special problems of tropical terrains. The basic requirements for preparing baseline geochemical data are described with reference to International Geological Correlation Programme Project (IGCP) 259/360 guidelines (Darnley et al. 1995) and the geochemical factors affecting the pathways of chemical elements from soils and water to plants, animals and man are discussed. New methods of processing geochemical data to indicate speciation and hence bioavailability are also suggested, and some specific geochemical case studies carried out by the British Geological Survey (BGS) in developing countries are briefly outlined. Such research can help not only to identify potential environmental problems and develop amelioration strategies, but also to act as a basis for the development of appropriate policy and legislative frameworks. It also has the potential to cast new light on several types of degenerative disease which affect man, animals and crops.

Socio-economic pressures on the environment of developing countries

The annual increase in world population is approaching 100 million per year (UN 1992), approximately 90% of which is in the poorest countries where one-quarter of the population live in 'absolute poverty' (having an annual income below US$ 450) as defined by the World Bank (1992). According to McMichael (1993) population pressure and poverty have adverse inter-related environmental impacts. For example, the effects of large-scale coal burning in industrializing developing countries, of increasing ricefield methane emission and of widespread deforestation throughout the third world, can be attributed to population pressure. Problems associated with the use of marginal farmlands, soil erosion from cash cropping, uncontrolled industrial and urban pollution and the accumulation of potentially harmful wastes are particularly intense in some developing countries (McMichael 1993).

Each year around six million hectares of the world's agricultural land are lost through soil degradation (UN 1992), which is perhaps one of the most fundamental problems. Soil is a complex mixture of rock and mineral detritus, organic matter, microbes (bacteria, yeasts and fungi) and invertebrates. The removal of protective plant cover, overgrazing or overcropping leading to the depletion of essential nutrients, the addition of toxic chemicals, salination (in some cases as a result of irrigation) and acidification from power generation and industrialization, can cause all soil organisms to die and soil to turn to inert mineral dust (desertification).

Economic and population pressures on rural economies cause migration, especially to urban centres. In developing countries the growth of very large cities (with populations greater than five million) is of particular concern; on current trends the population of some eight to ten cities will reach 15–25 million by the end of the century (UN 1991; McMichael 1993). Urbanization in turn leads to further adverse environmental effects, such as contamination of surface water and aquifers through poor sanitation. Pollution of air, water and soil from vehicles, power plants and factories is at its most extreme in cities, and may result in increased incidence of childhood diseases associated with heavy metal poisoning (WHO 1988; US Geological Survey 1984). However, despite the growing need, geochemical studies of major cities in developing countries are currently lacking.

Mining has also caused contamination, including cyanide, mercury and arsenic releases from gold mining, and radioactive element pollution from tin mining. Although large international mining companies now generally work to high environmental standards, mineral working by uncontrolled and disorganized groups (especially for gold) continues to cause environmental problems in developing countries, as in the recent incidents of mercury pollution in Brazil (Stigliani & Salomons 1993).

The surface environment in developing countries

The socio-economic pressures on the environment in developing countries are frequently compounded by the nature of their surface environment, which can be particularly susceptible to degradation and pollution, especially in climatic regions classified as equatorial, tropical or sub-tropical (Köppen 1936). Climates have

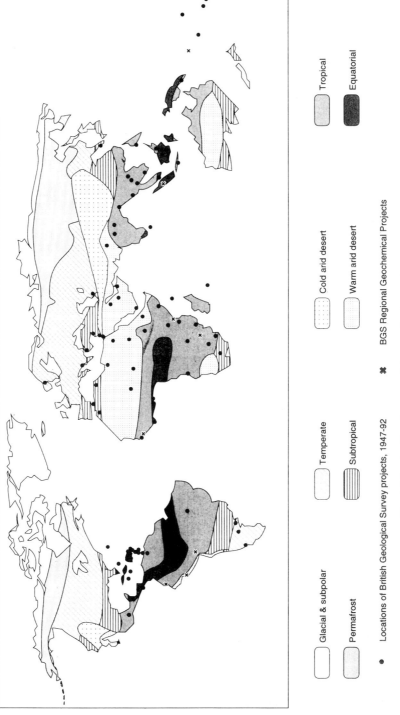

Fig. 1. Morphoclimatic zones of the world (after Budel 1982).

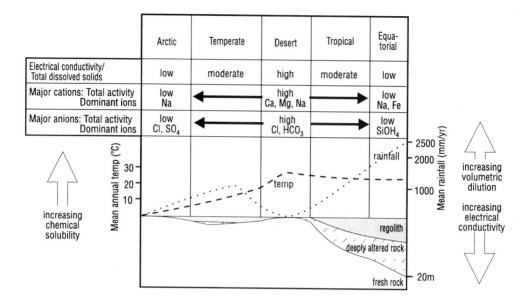

Fig. 2. Generalized relationships between regolith, thickness of the weathering zone, climatic factors and water chemistry (modified after Pedro 1985).

changed frequently and in some cases profoundly in the geological past, so that some areas have been affected by a succession of different weathering and dispersion processes (Fig. 1). At lower latitudes and particularly in continental areas of low relief, the regolith may be an expression of the cumulative effects of subaerial weathering during many millions of years (Butt & Zeegers 1992). Some soils have been deeply leached under various climatic conditions, including long periods of high rainfall and temperature to which they have been repeatedly subjected from Tertiary times or even as far back as the late Proterozoic (Daniels 1975; Butt 1989). Tropical and semi-tropical soils, partly because of their age and the number of weathering cycles to which they have been subjected, tend to be poorer, thinner and more fragile than those in temperate regions (Murdoch 1980). This is particularly the case in shield areas of Africa and Latin America where soils have developed on crystalline basement and are generally older. In contrast, those of southeast Asia developed on alluvium or volcanic rocks are generally younger and more fertile (McMichael 1993).

In tropical countries laterites, which have been forming continuously for at least 100 million years (since the Jurassic) in parts of Africa, India, South America and Asia (King 1957; Michel 1973), are common. Weathering profiles may extend to depths of over 150 m, with weathering fronts at great depth (Trescases 1992), so that the fresh supply of chemical elements from rocks into the biosphere is limited. Laterites, which include Fe- and Al-rich tropical weathering products such as bauxites, ferricretes and Fe- and Al-duricrusts, have markedly homogeneous mineralogical and chemical compositions which usually bear little relationship to underlying bedrock (Nahon & Tardy 1992). They have well defined chemical profiles but, depending on the climatic and tectonic history, the land surface may be highly complex with the profiles truncated or buried by later detritus (Butt & Zeegers 1992).

In such conditions, intense oxidation, which generally extends to the weathering front, results in a lack of organic matter (which is stored in the biota rather than soils) and of other major elements such as phosphorus, nitrogen and potassium, and also in the formation of stable insoluble minerals such as clays (especially kaolinite) which markedly increase the Al/Si ratio. Generally, such soils are deficient in soluble cations (Na^+ and Ca^{2+}) and anions (Cl^-, PO_4^{3-} and NO_3^-), but have high levels of resistant oxides of Fe, Al, Ti and Mn. They are typically kaolinitic or ferralitic, with local calcretes or silcretes in areas of inland drainage affected by evaporation and a fluctuating water table. Hardening of aluminium and iron oxide to form cuirasses, which armour the otherwise friable surface, is common in tropical terrains. As natural vegetation is cleared from such environments, essential nutrients are further leached and the top soil may be completely eroded.

Evidence exists that some deserts were formerly subjected to extreme leaching under tropical conditions during the Cretaceous and Tertiary (Butt & Zeegers 1992). Most desert soils are typically lithosols, the important constituents of which are resistate detrital quartz and accessory minerals such as zircon and secondary clay minerals. Organic material in desert soils is minimal and surface waters are highly oxidizing, with variable pH (frequently with high pH values). They generally have higher salinity than those in other environments, with chloride rather than bicarbonate as the main anionic species. Some generalized relationships between different surface environments are shown in Fig. 2.

Chemical elements tend to undergo more pronounced separation in tropical environments than in temperate regions, although this depends on their chemical mobility and on the nature of the local environment (Trescases 1992). Hence areas with a potential for deficiency or toxicity conditions are likely to be much more common in tropical countries than in temperate regions. The most deeply leached environments retain only the most inert elements such as Fe, Mn, Zr, Hf and the rare earths, either as secondary or primary oxides. On the other hand the most mobile elements, such as the alkalis, alkaline earth elements, the halogens and elements mobile in conditions of high pH, including anions and oxyanions, such as those of B, Se, Mo, V and U accumulate in arid environments generally, inland drainage systems and near the base of weathering profiles. The variable oxidation states of the first row transition elements Co, Cr, Cu, and Zn also favour removal from solution by ion exchange, precipitation and surface sorption. The lack of organic matter as a result of intense oxidation can mean that total levels of such elements in the regolith may be high, but they may be bound on Fe–Mn oxides and thus their bioavailable levels may be exceptionally low, resulting in potential deficiency for plants and animals, especially in deeply oxidized surface environments. The controls on the speciation of As, Cd and Pb are broadly similar to those of the first row transition metals, although the oxidation states of these elements, with the exception of As, exhibit less control on mobility (Ure & Davidson 1995).

Essential and potentially harmful chemical elements

Two main groups of chemical elements are of particular importance for health. Those identified as essential to animal life (according to Mills 1996) include the first row transition elements Fe, Mn, Ni, Cu, V, Zn, Co and Cr, together with Mo, Sn, Se, I and F. The disorders associated with deficiency of these elements are given in Mills (1996). Boron has not yet been shown to be necessary for animals, although it is essential for higher plants. By contrast potentially harmful elements (PHEs), known to have adverse physiological significance at relatively low levels, include As, Cd, Pb, Hg and some of the daughter products of U. Aluminium can also have adverse physiological effects in trace amounts in animals and particularly in fish and plants (Sposito 1989). All trace elements are toxic if ingested or inhaled at sufficiently high levels for long enough periods of time. Selenium, F and Mo are examples of elements which show a relatively narrow concentration range (of the order of a few $\mu g\,g^{-1}$) between essential and toxic levels.

The difficulties of diagnosing disease, and particularly subclinical conditions, in animals and man related to trace element deficiencies or excesses are discussed by Mills (1996). Except in some specific cases, for example Iodine Deficiency Disorders (IDD), symptoms may be non-specific and diagnosis, based on tissue or blood sampling, costly. Geochemical maps, however, can indicate areas where there is the potential for trace element deficiency or toxicity, enabling expensive veterinary or medical investigations to be better targeted.

Multi-media geochemical surveys (ideally incorporating soil, stream sediment and water data) can also be of considerable value in studies linking diet and health. Although many PHEs, such as Pb, are not phytotoxic, elevated concentrations in the regolith may be transmitted, through uptake by healthy plants, into the food chain and can thus lead to adverse health effects in animals and humans, the source of which can be difficult to diagnose (Mills 1996). The concentrations of certain essential trace elements (such as Se) in crops (rice, corn, soybean) has been shown to correlate with the concentrations in the soil in which they are grown (Levander 1986); and, regionally, levels in fine fraction stream sediments give a good indication of likely soil concentrations (Appleton 1992). Geochemical mapping can therefore be a cost-effective method of indirectly investigating the chemical composition of crops; and rural communities in developing countries offer a particularly valuable opportunity for examining the relationship between geochemistry, diet and health.

The need for baseline geochemical data in developing countries

In many developed countries geochemical mapping is now an integral component of strategic systematic geoscience surveys (Plant et al. 1988; Darnley et al. 1995; Plant & Hale 1995), with applications for a wide range of economic and environmental purposes (Webb 1964; Thornton 1983; Thornton & Howarth 1986). It is clear from these studies that geochemistry, particularly the surface distribution and concentration of trace elements, can be difficult to predict from geological maps (Darnley et al. 1995). Areas underlain by different granites, for example, may have comparable major element concentrations but the levels of PHEs, such as U, Mo, Be and Pb, can vary by factors of an order of magnitude or more (Plant et al. 1983). Levels of environmentally important trace elements are even more difficult to predict in areas underlain by sedimentary and metamorphic rocks. Moreover, the geochemistry of some elements, for example Se, is not fully understood, so that any predictions based on geological maps can prove misleading. In tropical countries the problem is frequently compounded by deep weathering whereby levels of chemical elements in the surface regolith may show little or no relationship to bedrock composition. Indeed, many elements, including Al, Fe, Mn, Co, U, P, Cr, Ni, Cu and Au, may be concentrated to ore grade in the lateritic mantle; and high concentrations of PHEs such as As and Sb may also occur (Smith et al. 1987). In Jamaica, for example, the distribution of high concentrations of radioactive and other PHEs has been shown to follow closely that of bauxite (Simpson et al. 1991).

Geochemical mapping has long been used in conjunction with geological mapping as a mineral exploration tool in developing countries. Regional geochemical maps, such as those of Zambia and Sierra Leone, were first published some 30 years ago (Webb et al. 1964; Nichol et al. 1966); and atlases, such as that of Uganda, more than 20 years ago (Reedman 1973). Many of such surveys, however, covered only the regions considered most prospective for metalliferous mineral deposits and were designed specifically to cover large areas at low density and cost, for example in Peru (Baldock 1977).

More recent programmes, for example in Zimbabwe (Dunkley 1987, 1988) and Sumatra (Stephenson et al. 1982; Coulson et al. 1988) have produced higher density multi-element data with some potential for application in environmental and animal and human health studies (Appleton & Ridgway 1993). Basic hydrogeochemical parameters and data for such environmentally significant elements as Se and I are generally lacking, however (Darnley et al. 1995).

There is now an urgent need in developing countries for high resolution geochemical data which are adequate for environmental and epidemiological studies, particularly in urban areas where conflicts between rapid development and environmental sustainability are severe and where surface water, groundwater and soils are increasingly becoming contaminated (WHO 1988).

The requirements and methodologies for preparing high resolution geochemical baseline data are discussed in detail in the final report of the IGCP Project 259 (Darnley et al. 1995) which includes internationally agreed standards for sampling, analysis and data processing in different terrains. Analytical methods aimed at providing comprehensive data for the most environmentally important elements, with limits of detection below estimated crustal abundance variance, are also recommended, together with a quality control system and recommendations for databasing and presenting data.

At present no international agency is responsible for mapping the distribution of chemical substances, other than radioactive elements for which the International Atomic Energy Agency continues to provide the scientific infrastructure. Hence geochemical mapping continues to be carried out using a range of different methods, often varying according to short-term goals and the practices established in the developed country or organization providing aid. Moreover, aid agencies continue to fund geochemical mapping only as a low cost adjunct to geological mapping for mineral exploration. This is unfortunate since geochemistry provides the most relevant geoscience data for environmental studies and land use planning and has a much lower cost per unit of area covered than other types of geoscience survey (Plant & Slater 1986).

The separate conduct of surveys of the chemistry of soils, water and stream sediments by different agencies – a legacy from the past when geochemistry developed as a component of geological, soil and hydrological survey organizations – also limits the effective application of geochemistry in addressing environmental problems. Ideally, comparable multi-element data for each sample medium should be collected as part of a single, holistic, multi-media geochemical campaign. This is particularly important since

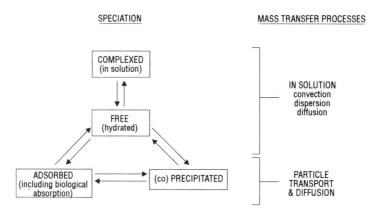

Fig. 3. Relationships between speciation of chemical elements and mass transfer processes (after Bourg 1988).

new data on speciation, combined with computer modelling and GIS, make possible the preparation of geochemical maps targeted at environmental problems.

Understanding the pathways of chemical elements from rocks and soils to man and animals

The total concentration of chemical elements indicated on geochemical maps can be of direct value in studies of relationships between disease and the levels of trace elements (e.g. Thornton & Plant, 1980; Plant & Stevenson 1985; Plant & Thornton 1986; Appleton 1992). However, the amount of each element which is bioavailable is more important for these studies than the total amount. For example Al, which is the commonest metal in the earth's surface layer, occurs in both inert and bioavailable forms, and its potential toxicity thus depends on its chemical form or speciation. It can occur in a large number of dissolved species, especially in conditions of low or high pH, or in colloids with organic carbon or silica which limit its toxicity. The toxicity of As and Sb also depends on their chemical form. They are most toxic in the M^{-3} gaseous state with decreasing toxicity in the sequence $M^{3+} > M^{5+} >$ methyl As/Sb (Abernathy 1993; Chen et al. 1994).

The speciation of chemical elements affects their distribution, mobility and toxicity (Fig. 3). This has been known for a long time in the case of the common anions (HCO_3^-, NO_3^-, SO_4^{2-}) and more recently for a wide range of chemical elements in the geosphere (Goldberg 1954) and surface environment (Stumm and Morgan 1981; Buffle 1988).

The importance of chemical speciation in relating geochemical data to health was first established by agricultural scientists (Underwood 1979; Lander 1986). Recently considerable developments have been made in the prediction of chemical speciation by modelling thermodynamic and kinetic equilibria (Basset & Melchior 1990; Stumm 1991) and by experimental determinations (Buffle 1988; Marabini et al. 1992). Biochemical processes, induced by microbial, plant and animal activity, are also important controls on both speciation and mobility, but were relatively poorly understood until recently (Ehrlich 1990; Deighton & Goodman 1995).

Some of the most important controls, particularly on trace element speciation and the mobility, include hydrogen ion activity (pH), redox potential (Eh), temperature, surface properties of solids, the abundance and speciation of potential ligands, major cations and anions, the presence or absence of dissolved and/or particulate organic matter, and biological activity. Two of the most important factors directly controlling mobility and solubility are Eh and pH (Fig. 4). The solution chemistry of an element is affected profoundly by changes in oxidation state, while dissolution reactions, including hydrolysis, inorganic complexation, complexation with smaller organic anions (such as oxalate) and sorption/desorption, are all pH controlled. Under conditions of high pH, anions and oxyanions (such as those of Te, Se, Mo, U, As, P and B) are more mobile and most cations (such as those of Cu, Pb, Hg and Cd) are less mobile, while at low pH the reverse is generally true. Where humic substances or biological byproducts are present, however, stable organo-

RELATIVE MOBILITIES	ENVIRONMENTAL CONDITIONS			
	oxidizing	acid	neutral-alkaline	reducing
VERY HIGH			Mo *U* Se	
HIGH	Mo *U* Se F *Ra* Zn	Mo *U* Se F *Ra* Zn Cu Co Ni *Hg*	F *Ra*	F *Ra*
MEDIUM	Cu Co Ni *Hg* *As Cd*	*As Cd*	*As Cd*	
LOW	*Pb Be Bi Sb Tl*	*Pb Be Bi Sb Tl* Fe Mn	*Pb Be Bi Sb Tl* Fe Mn	Fe Mn
VERY LOW TO IMMOBILE	Fe Mn Al Cr	Al Cr	Al Cr Zn Cu Co Ni *Hg*	Al Cr Mo *U* Se Zn Co Cu Ni *Hg* *As Cd* *Pb Be Bi Sb Tl*

Fig. 4. General relationships between Eh, pH and the mobility of some essential and potentially toxic elements (modified after Andrews-Jones 1968). Essential elements are shown in normal type and potentially hazardous elements in italics.

metallic complexes (which may behave as anions) are formed, increasing trace-element mobility. The kinetics of inter-species interactions also act as important controls on speciation, particularly where natural systems are disturbed by the influences of man.

Geochemical interaction between the geosphere/hydrosphere and the biosphere depends partly on sorption processes and partly on chemical speciation. Some elements, for example Al, Ti and Cr, are relatively poorly assimilated by plants, although others such as Cd, Se, Mo and Co, can readily cross the soil–plant barrier and enter the food cycle (Loehr 1987). In soils, sorption of elements by clay minerals and organic material is the predominant fixing mechanism, with soil pH controlling sorption processes and metal solubility/bioavailability. Many metals are more soluble in the low pH conditions induced by natural organic acids and root exudates (Fig. 5).

In higher animals, including man, assimilation occurs by ingestion of nutrients and contaminants, by absorption through the skin, and by respiration (WHO 1984). Speciation both in the natural environment and in the gastro-intestinal tract, exerts a major influence on the uptake and assimilation of trace nutrients and PHEs (WHO 1984, 1994).

Speciation studies are of particular importance in areas affected by land degradation, deforestation or pollution caused by mining, industrial activity or urbanization, where they can be used to optimize amelioration strategies and improve management practices. A knowledge of the factors controlling speciation in

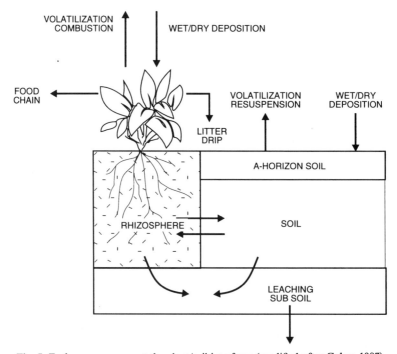

Fig. 5. Exchange processes at the plant/soil interfaces (modified after Cohen 1987).

different environments can be used to predict the potential for absorption of PHEs and as a guide to the need for trace element supplementation for crops and animals, particularly where baseline geochemical data are available. Ideally, environmental geochemical surveys, particularly in tropical terrains affected by intense chemical weathering, should be based on chemical speciation. Extraction techniques can be used to provide an indication of speciation (e.g. Tessier & Campbell 1988), but sensitivity to changes prior to analysis limits the applicability of such determinations where large numbers of samples are involved.

Prediction of speciation using a range of thermodynamic models such as WATEQ4F (Ball & Nortdstrom 1991), PHREEQEV (Crawford 1996) and EQ3/6 (Wolery 1992) has greater potential, provided that suitable thermodynamic and/or kinetic data are available (Bassett & Melchior 1990) and the effects of biochemical processes are quantified. An iterative approach using detailed experimental studies of selected elements to validate model predictions is thought to offer considerable potential for developing speciation maps from geochemical baseline data for large-scale environmental and epidemiological studies. Studies of Al and As speciation, for example, are being used to develop methods of processing regional geochemical data to indicate speciation (Simpson *et al.* 1996; Smith *et al.* 1996).

Selected case studies

The role of geochemistry in identifying, and in helping to ameliorate, environmental and health problems resulting from trace element imbalances or contamination in developing countries is illustrated by selected case studies.

Natural trace element imbalances and animal health

Although undernutrition is the major limiting factor to grazing livestock production in tropical areas, trace element deficiencies or imbalances in soils and forages are also responsible for low production and reproduction problems. For grazing livestock, deficiencies of Co, Cu, I, Fe, Mn, Se and Zn, together with excesses of Cu, F, Mn and Mo, may particularly lead to adverse effects; As, Pb, Cd, Hg and Al also cause toxicity (Appleton 1992).

The diagnosis and mapping of affected areas have generally been carried out using forage, animal tissue or fluid compositions, all of which are expensive and time consuming (Appleton 1992). Veterinary scientists and agronomists are generally familiar with the use of soil geochemistry, but the value of regional geochemical maps based on drainage samples is largely unknown, especially in developing countries (Appleton & Greally 1992). Studies in Bolivia and Zimbabwe have been undertaken in order to assess whether such data, either collected specially or previously (e.g. for exploration), can be used to predict problems of animal health, and whether low levels of trace elements in drainage sediments correlate with reported deficiencies in grazing livestock.

In northeastern Zimbabwe stream sediments, soils and forage all exhibit significant correlation and the same regional patterns for Zn (Fordyce *et al.* 1996). The lack of a significant correlation between Zn levels in those media and the levels in cattle blood serum is ascribed in part to a range of dietary and physiological factors. It is suggested that high concentrations of Fe and Mn in soil and forage inhibit the availability and therefore the uptake of trace elements, such as Cu or Zn. Consequently Zn levels in blood serum from cattle foraging in areas with moderate to high Zn levels may be equally low as those from cattle feeding in low Zn areas (Fig. 6). Likewise the availability of P to plants and animals may be significantly reduced by high Fe oxide levels in soils. Nevertheless stream sediment geochemical data are shown to be of value in helping to target areas for specific veterinary investigations, because they provide regional information indicating both where Zn and Cu are likely to be deficient and where high Fe and Mn may induce low Zn or Cu status in cattle, despite moderate to high levels of these trace elements in sediments, soils and forage.

This case study (Fordyce *et al.* 1996) therefore supports the review by McDowell *et al.* (1993) which identified a lack of direct correlation between trace element levels in soil and forage and those in animal blood serum. The findings do not fully corroborate the conclusions of other reviews (Thornton 1983; Aggett *et al.* 1988; Appleton 1992) which suggest that stream sediment geochemical maps can be used directly to indicate areas where ruminants may be subject to trace element deficiencies. However, the conclusion that high concentrations of Fe and Mn oxides in soil may inhibit the availability of P to plants and of Cu and Zn to cattle could have wide implications in developing countries because of the preponderance of ferralitic soils in tropical regions. It also underlines the need to understand speciation or phase partitioning in order to relate 'total' geochemical

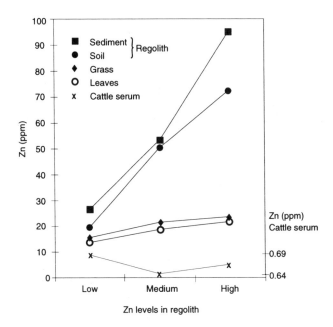

Fig. 6. Zinc concentrations in stream sediment, soil, grass leaves and cattle serum from northeast Zimbabwe (after Fordyce et al. 1996).

Environmental impact of coal-fired power stations

Many developing countries depend heavily on coal-burning power stations for energy generation and in some cases low grade or 'dirty' coal feedstocks are used, resulting in serious environmental degradation. Accordingly the fates and environmental impact of potentially harmful trace element emissions (PHEs) from coal-fired power stations have been studied, particularly in China.

The multi-element geochemistry of the coal feedstock has been compared with that of waste products and emissions at a power station in northeast China. Normalization of the output (slag, fly-ash) chemistry, to coal, characterizes element partitioning within the combustion products, which are generally enriched by up to an order of magnitude (Simpson et al. 1995), the slag containing particularly high levels of Be, As, Mo, Cd and W and the fly-ash high Li, As, Cd, U and Th. For example, the results of mass balance calculations (Fig. 7) show that of the 36 tonnes per annum of As consumed (6 ppm As in six million tonnes of coal feed) some 24 tonnes are output in solid wastes (slag 12 ppm, and fly-ash 15 ppm), but that almost 12 tonnes per annum of As are emitted from the stack. By contrast, most U and all Mo is shown to partition into the solid wastes. The high levels of certain trace elements from stack emissions may have a considerable impact on the geochemistry of surface soils in the surrounding area, particularly from pollution plumes in the prevailing wind directions. Such emissions could therefore have adverse effects on the surface environment and on plant, animal or human health, partly because the emitted, potentially toxic elements may well occur as readily available and ingestible chemical species; however, more studies are needed to confirm this conclusion.

A further environmental hazard may result from leaching of PHEs from (often unlined) lagoons, in which slag and fly-ash slurries are dumped. A study of lagoon leachates (Simpson et al. 1995), normalized to cooling-water chemistry, shows that there is strong enrichment in some elements that can pose environmental threats, such as Al, Mo, V and particularly F, which exceeds 25 mg/l (Fig. 7). Experimental sequential leaching again suggests that high levels of bioavailable and therefore easily ingested species of certain trace elements are present.

The surface environment around the power

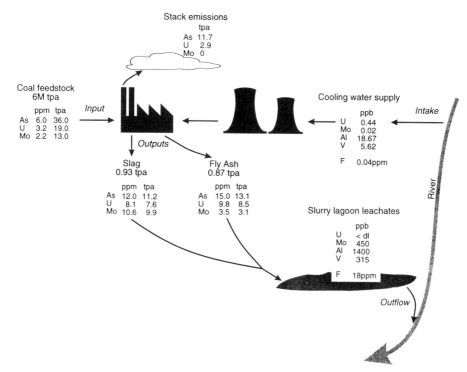

Fig. 7. Mass balance calculations for some PHEs in a power station in northeast China (after Simpson *et al.* 1995).

station has been geochemically characterized and compared with uncontaminated baseline regions of similar geology. The effects of emissions of such elements as As, F, Be, of gases such as SO_2, and of leachates from the slurry lagoon carrying elevated levels of available PHEs, are being assessed in terms of a critical loads analysis. Preliminary results confirm the importance of factors such as carbonate-rich rocks and well buffered soils in reducing both the potential for acidification and the mobility of PHEs (Simpson *et al.* 1992; Flight 1994).

Impact of mining

In many developing countries, the exploitation of mineral resources is of considerable importance for economic growth, employment and infrastructural development. It can also cause serious environmental problems, particularly in tropical regions characterized by high rates of weathering and biogeochemical cycling, because outputs of major constituents and impurities in the ore, or of chemicals used in processing, can accumulate to levels that may be harmful to plant, fish, animal and human health.

Studies of the environmental and health impacts of mining, undertaken at numerous representative mines in southeast Asia and southern Africa, have highlighted systematic relationships between mineral deposit geochemistry, mining methods and drainage quality (Williams 1993). The most important controls are water–rock interactions, sulphide oxidation rate at the fluid–rock interface and *in situ* buffering which is strongly influenced by gangue lithology/mineralogy and the weathering rate.

Mobilization of most heavy metals is controlled by local pH/Eh conditions and occurs principally at low pH under conditions of acid mine drainage (AMD). By contrast As and Sb are mobile across a much wider pH/Eh range (Fig. 4), a characteristic with important implications for remediation schemes, because conventional buffering, such as liming, may be ineffective for As contaminated mine-waters.

Studies have shown that the precipitation of Fe oxyhydroxides is, however, a very effective practical process for scavenging and immobilizing As. Significant As contamination in surface drainage, soil pore water and groundwater over considerable areas may particularly occur in conditions of low ambient Fe concentrations

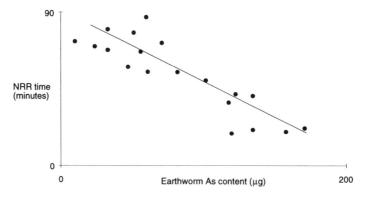

Fig. 8. Results of neutral red retention (NRR) pilot test: Wanderer Mine, Zimbabwe (after Williams & Breward 1995).

(e.g. Globe and Phoenix Mine, Zimbabwe). Some soils with low (< 5%) Fe content actually carry over 100 ppm water soluble As, resulting in marked uptake by crops (up to 10 ppm As in maize), whereas this is not observed in vegetation over sites with similarly high total As, but which also contain normal or high levels of Fe and in which the As is held as immobile ferric hydroxide/arsenate complexes (Williams 1994). The problem can be exacerbated by low soil P, as crops may assimilate As as a P substitute, due to the close chemical similarity of the phosphate (PO_4^{3-}) and arsenate (AsO_4^{3-}) anions (Williams & Breward 1995). In another example at Ron Phibun, southern Thailand (National Epidemiological Board of Thailand 1993), shallow groundwater exceeds $5 \, mg/l^{-1}$ As (compared with the WHO drinking water guideline of only $10 \, \mu g \, l^{-1}$) in areas where soils contain < 2% Fe. Where soil Fe increases to > 5%, As drops to < $100 \, \mu g \, l^{-1}$ (Fordyce & Williams 1994). Modelling of Fe oxide sorption properties suggests that precipitation of dissolved Fe as hydrous oxide would, in most instances, scavenge As particularly effectively, and consequently may provide a practical remediation strategy for As-rich mine waters, as the As would be immobilized in inert solid materials.

Mining or beneficiation technology may also be critical for metal mobility. In the presence of cyanide complexing agents, the mobility of As and heavy metals is maintained at neutral and even high pH. Drainage waters of pH 8, for example at Globe and Phoenix, can hold > $30 \, mg \, l^{-1}$ Cu, > $10 \, mg \, l^{-1}$ As and > $15 \, mg \, l^{-1}$ Sb in the presence of cyanides, even at low (< $10 \, mg \, l^{-1}$) concentrations (Williams 1994).

The effective assessment of the ecotoxicological impacts of mining contamination may be difficult and costly in developing countries. A practical pioneering method for assessing the toxicological impacts of exposures to As (and other PHEs) has recently been developed and tested successfully at Wanderer Mine, Zimbabwe, by the BGS working in collaboration with the Institute of Terrestrial Ecology (Williams & Breward 1995). Individual invertebrate cells are collected and spiked with neutral-red dye. Healthy cells retain the dye for lengthy periods (longer than one hour) but with increasing exposure to toxic conditions (As or heavy metals) the cells become stressed and release the dye increasingly rapidly (Fig. 8).

This study into the impacts of As or heavy metal contamination from mining again underlines the importance of understanding speciation/phase partitioning in characterizing the principal controls that determine sorption/precipitation or a high degree of mobility/availability, thus determining whether or not enhanced metal levels are likely to affect the health of plants, animals or humans. The study has also identified some potential methods for controlling As levels, as well as for assessing their ecotoxicological impact in practical, cost effective ways.

Conclusion and recommendations

Increases in our understanding of the chemistry of natural systems, coupled with improved computing power, analytical methods and information technology, now offer an opportunity for considerable advances in environmental geochemistry. For example, critical load analysis can be performed and regional geochemical data processed to provide an indication of speciation and hence bioavailability. Such new and sophisticated geochemical tools can be used as a basis for targeting resources into areas worst affected by environmental degradation, predicting potential hazards and identifying factors limiting agricultural productivity or those likely to cause health problems in man. In order to benefit from these new approaches, more multi-element, multi-parameter baseline geochemical data are needed, as well as specific research into the speciation of PHEs and essential nutrients in different geological and climatic terrains.

Many studies have tended to concentrate on potentially toxic heavy metals. More work is needed on those trace elements that are essential to human health but have harmful effects unless a critical, narrow range is maintained, especially Se and I. The deficiency of Se has been linked to endemic osteo-arthropathic and cardiovascular disorders in parts of China (Jianan 1990), and to heart disease in Ohio (Schlamberger 1980), while Iodine Deficiency Disorders (IDD) including goitre and cretinism, are estimated as likely to affect up to 800 million people in developing countries (WHO 1994). Inter-element effects also require more study. Until recently, high quality geochemical data on elements such as Se and I have rarely been available because of the difficulty or high cost of determining these elements at levels at and below the average abundances in natural materials. Analytical methods for Se and for several other environmentally important chemical elements are now being developed by the BGS and other organizations to enable their distribution to be mapped cost effectively. Methods for studying Al and As speciation in the environment have also been developed. Studies of links between the distribution and speciation of environmentally important elements and their epidemiological significance should now be pursued, ideally in countries with good epidemiological data such as China (China Map Press 1979).

Despite the fact that urbanization is a major factor in the degradation of the natural environment, there has been little or no systematic work on the sustainability of the geosphere and hydrosphere, or on their ability to absorb the effects of land degradation and pollution, in the very large cities developing in the third world. Geochemical studies of the urban environment (including soil, water, dust and air) are urgently needed if the environmental damage caused by the rapid urbanization of the developing world is to be contained, including the threats to the marine environment from major coastal cities.

Environmental geochemistry, particularly the preparation of high quality baseline data (Darnley et al. 1995), has a crucial role to play in aiding the understanding of land use problems and designing sustainable solutions appropriate for the economies of the developed and developing world.

The paper is published by permission of the Director, British Geological Survey (NERC). Dr J. D. Appleton and Dr H. W. Haslam are thanked for making helpful comments on the text.

References

ABERNATHY, C. O. 1993. *Draft drinking water criteria document on arsenic.* US–EPA Science Advisory Board report, contract 68-C8-0033.

AGGETT, P. J., MILLS, C. F., MORRISON, A., CALLAN, M., PLANT, J. A., et al. 1988. A study of environmental geochemistry and health in northeast Scotland. *In:* THORNTON, I. (ed.) *Geochemistry and Health. Proceedings of the Second International Symposium.* Science Reviews Limited, Northwood, 81–91.

ANDREWS-JONES, D. A. 1968. The application of geochemical techniques to mineral exploration. *Colorado School of Mines, Mining Industry Bulletin*, 11, No 6.

APPLETON, J. D. 1992. *Review of the use of regional geochemical maps for identifying areas where mineral deficiencies or excesses may affect cattle productivity in tropical countries.* British Geological Survey Technical Report WC/92/24 (unpublished).

—— & GREALLY, K. 1992. *Environmental Geochemistry R&D Project: A comparison of the trace element geochemistry of drainage sediments and soils in eastern Bolivia.* British Geological Survey Technical Report WC/92/67 (unpublished).

—— & RIDGWAY, J. 1993. Regional geochemical mapping in developing countries and its application to environmental studies. *Applied Geochemistry*, Supplement, **2**, 103–110.

BALDOCK, J. W. 1977. Low density geochemical reconnaissance in Peru to delineate individual mineral deposits. *Trans IMM*, **86**, B63–73.

BALL, J. W. & NORDSTROM, D. K. 1991. *Users manual for WATEQ4F with revised thermodynamic database and test cases for calculating speciation of major, trace and redox elements in natural waters.* US Geological Survey, Open-File Report, 91–183.

BASSETT, R. L. & MELCHIOR, D. C. 1990. Chemical modelling of aqueous systems: An overview. MELCHIOR, D. C. & BASSETT, R. L. (eds) *Chemical Modelling of Aqueous Systems II.* ACS Symposium Series, American Chemical Society, Washington DC, **416**, 1–14.

BOURG, A. C. M. 1988. Models in aquatic and terrestrial systems: sorption speciation and mobilisation. *In:* SALOMONS, W. & FORSTNER, U. (eds) *Chemistry and biology of solid water dredged materials and mine tailings.* Springer, New York.

BGS. 1978–95. *Regional geochemical atlas series.* British Geological Survey, Keyworth, Notting-

ham.
BUDEL, J. 1982. *Climatic geomorphology*. Princeton University Press.
BUFFLE, J. 1988. *Complexation Reaction in Aquatic Systems: An Analytical Approach*. Ellis Horwood, Chichester.
BUTT, C. R. M. 1989. Geomorphology and climatic history-keys to understanding geochemical dispersion in deeply weathered terrain. *In:* GARLAND, G. D. (ed.) *Proceedings of Exploration '87. Third Decennial International Conference on Geophysical and Geochemical Exploration for Minerals and Groundwater*. Special Volume 3, Ontario Geological Survey, Toronto, 323–334.
—— & ZEEGERS, H. 1992. Climate, geomorphological environment and geochemical dispersion models. *In:* BUTT, C. R. M. & ZEEGERS, H. (eds) *Regolith Exploration Geochemistry in Tropical and Subtropical Terrains. Handbook of Exploration Geochemistry, Volume 4*. Elsevier, Amsterdam, 3–24.
CHEN, S. L., DZENG, S. R. & YANG, M. H. 1994. Arsenic species in the Blackfoot disease area, Taiwan. *Environmental Science Technology*, **28**, 877–881.
CHINA MAP PRESS. 1979. *Atlas of Cancer Mortality in the People's Republic of China*. Shanghai.
COHEN, Y. 1987. Modelling pollutant transport and accumulation in a multimedia environment. *In:* DRAGGAN, S. *et al.* (eds) *Geochemical and hydrologic processes and their protection*. Praeger, New York.
COULSON, F. I. E., PEART, R. J. & JOHNSON C. C. 1988. *North Sumatra Geochemical and Mineral Exploration Project*. British Geological Survey, Report 25, 46.
CRAWFORD, M. B. 1996. PHREEQEV: the incorporation of Model V to model organic complexation in dilute solutions into the speciation code PHREEQEV. *Computers and Geosciences*, **22**, 2, 109–116.
DANIELS, J. L. 1975. Palaeogeographic development of Western Australia–Precambrian. *In: Geology of Western Australia*. Western Australia Geological Survey Memoir, **2**, 437–450.
DARNLEY, A. G., BJÖRKLUND, A., BØLVIKEN, B., GUSTAVSSON N., KOVAL, P. *et al.* 1995. *A Global Geochemical Database, Recommendations for International Geochemical Mapping*. Final Report of ICGP Project 259, UNESCO, Paris.
DEIGHTON, N. & GOODMAN, B. A. 1995. The speciation of metals in biological systems. *In:* URE, A. M. & DAVIDSON, C. M. (eds) *Chemical Speciation in the Environment*. Blackie, Glasgow, 307–334.
DUNKLEY, P. N. 1987. *Regional drainage geochemical exploration survey of the country between Rushinga and Nyamapanda, NE Zimbabwe*. British Geological Survey, Overseas Directorate, Report No MP/87/16.
—— 1988. *Regional drainage geochemical exploration survey of the Harare area, Zimbabwe*. British Geological Survey, Overseas Directorate, Report No MP/87/18.
EHRLICH, H. L. 1990. *Geo-microbiology*. Marcel Dekker, New York.
FLIGHT, D. M. A. 1994. *Report on visit to Chongqing, Sichuan Province, P. R. China*. British Geological Survey Technical Report WC/94/17R (unpublished).
FORDYCE, F. M. & WILLIAMS, T. M. 1994. *Impacts of mining and mineral processing with particular reference to gold and complex sulphide deposits*. British Geological Survey, Overseas Visit Report (unpublished).
——, MASARA, D. & APPLETON, J. D. 1996. Stream sediment, soil and forage chemistry as indicators of cattle mineral status in northeast Zimbabwe. *This volume*.
GOLDBERG, E. D. 1954. Marine Geochemistry 1: Chemical Scavengers of the Sea. *Journal of Geology*, **62**, 249.
JIANAN, T. 1990. The influence of selenium deficiency in the environment on human health in NE China. *In:* LAG, J. (ed.) *Excess and deficiency of trace elements in relation to human and animal health in Arctic and Sub-Arctic regions*. Norwegian Academy of Science & Letters, Oslo, 90–108.
KING, L. C. 1957. The geomorphology of Africa. *In: Erosion surfaces and their mode of origin*. Sci Prog, **45**, 672–681.
KÖPPEN, W. 1936. *Das geographische System der Klimate*. Handbuch der Klimatologie, **1**, C, Berlin.
LANDER, L. 1986. Speciation of Metals in Water, Sediment and Soil Systems. *In: Proceedings of an International Workshop (Sunne, October 15–16, 1986)*, Springer, Lecture Notes in Earth Sciences, **11**, 185.
LEVANDER, O. A. 1986. Selenium. *In:* MERTZ, W. (ed.) *Trace Elements in Human and Animal Nutrition*, Academic, Orlando, Fifth Edition, **2**, 209–279.
LOEHR, R. C. 1987. Advancing knowledge on protection of the land/soil resource: assimilative capacity for pollutants. *In:* DRAGGAN, S. *et al.* (eds) *Geochemical and hydrologic processes and their protection*. Praeger, New York.
MCDOWELL, L. R., CONRAD, J. H. & HEMBRY, F. G. 1993. *Minerals for Grazing Ruminants in Tropical Regions*. 2nd Edition, University of Florida.
MCMICHAEL, A. J. 1993. *Planetary Overload. Global environmental change and the health of the human species*. Cambridge University Press.
MARABINI, M. A., PASSARIELLO, B. & BARBARO, M. 1992. Inductively Coupled Plasma Mass Spectrometry: Capabilities and Applications. *Microchemistry Journal*, **46**, 302–312.
MARTYN, C. N., BARKER, D. J. P., OSMOND, C., HARRIS, E. C., EDWARDSON, J. A. & LACEY, R. F. 1989. Geographical relation between Alzheimer's disease and Al in drinking water. *Lancet*, **i**, 59–62.
MERTZ, W. 1986. *Trace Elements in Human and Animal Nutrition*. US Dept of Agriculture. Academic, Orlando.
MICHEL, P. 1973. *Les Bassins des Fleuves Sénégal et Gambie. Etude Géomorphologique*. ORSTOM, Paris, Mémoire 63.
MILLS, C. F. 1996. Geochemical aspects of the aetiology of trace element related diseases. *This*

volume.

MURDOCH, W. M. 1980. *The Poverty of Nations. The Political Economy of Hunger and Population.* Johns Hopkins University Press, Baltimore, 6–7.

NAHON, D. & TARDY, Y. 1992. The ferruginous laterites. *In:* BUTT, C. R. M. & ZEEGERS, H. (eds) *Regolith Exploration Geochemistry in Tropical and Subtropical Terrains.* Handbook of Exploration Geochemistry, Vol. 4, Elsevier, Amsterdam, 41–55

NATIONAL EPIDEMIOLOGICAL BOARD OF THAILAND. 1993. *Arsenic contamination in Southern Province of Thailand.* Information leaflet.

NICHOL, I., JAMES, L. D. & VIEWING, K. A. 1966. Regional geochemical reconnaissance in Sierra Leone. *Transactions of the Institute of Mining and Metallurgy* (Sect B: Applied Earth Science), **75**, May, B146–B161.

PEDRO, G. 1985. Grandes tendances des sols mondiaux. *Cultivar,* **184**, 78–81 [in French].

PLANT, J. A. & HALE, M. 1995. *Handbook of Exploration Geochemistry, Vol. 6, Drainage Geochemistry.* Elsevier, Amsterdam.

—— & SLATER, D. 1986. Regional geochemistry – potential developments. *Transactions of the Institute of Mining and Metallurgy,* Sect B, Applied Earth Science, **95**, 63–70.

—— & STEVENSON, A. G. 1985. Regional geochemistry and its role in epidemiological studies. *In:* MILLS, C. F., BREMNER, I. & CHESTERS, J. K. (eds) *Trace Element Metabolism in Man and Animals.* Rowett Research Institute, Aberdeen, 900–906.

—— & THORNTON, I. 1986. Geochemistry and health in the United Kingdom. *In:* THORNTON, I. (ed.) *Proceedings of the First International Symposium on Geochemistry and Health.* Science Reviews, Northwood, 5–15.

——, HALE, M. & RIDGWAY J. 1988. Developments in regional geochemistry for mineral exploration. *Transactions of the Institute of Mining and Metallurgy,* **97**, B116–B140.

——, SIMPSON, P. R., GREEN, P. M., WATSON, J. V. & FOWLER M. B. 1983. Metalliferous and mineralised Caledonian granites in relation to regional metamorphism and fracture systems in northern Scotland. *Transactions of the Institute of Mining and Metallurgy,* **92** B33–B42.

REEDMAN, A. J. 1973. *Geological Atlas of Uganda.* Uganda Geological Survey and Mines Department, Entebbe.

SCHLAMBERGER, R. J. 1980. Selenium in drinking water and cardiovascular disease. *Journal of Environmental Pathology and Toxicology.* **4–2, 3**, 305–308.

SIMPSON, P. R., FLIGHT, D. M. A. & BULL, K, in collaboration with ZENG RONGSHU, ZHANG YI-GANG & XU WEN-LONG. 1992. *Report on visit to Tangshan General Power Plant, Tangshan, Hebei Province, P. R. China.* British Geological Survey, Technical Report WC/92/49R (unpublished).

——, HURDLEY, J., LALOR, G. C., PLANT, J. A., ROBOTHAM, H. & THOMPSON, C. 1991. Orientation studies in Jamaica for multi-purpose geochemical mapping of Caribbean region. *Transactions of the Institute of Mining and Metallurgy,* **100**, B98–B110.

——, BREWARD, N., COOK, J. M., FLIGHT, D. M. A., HALL, G. E. M., LISTER, T. R. & SMITH, B. 1996. High resolution regional hydrochemical mapping of stream water for mineral exploration and environmental studies in Wales and the Welsh borders. *Applied Geochemistry.* **11**, 3, (in press).

—— & 14 others 1995. *Environmental impact of coal-burning power stations.* British Geological Survey Technical Report WC/95/44. (unpublished).

SMITH, B., BREWARD, N., CRAWFORD, M. B., GALIMAKA, D., MUSHIRI, S. M. & REEDER, S. 1996. The environmental geochemistry of aluminium in tropical terrains and its implications to health. *This volume.*

SMITH, R. E., BIRRELL, R. D. & BRIGDEN, J. 1987. The implication to exploration of chalcophile corridors in the Archaean Yilgarn Blocks, Western Australia, as revealed by laterite geochemistry. *In: 12th International Geochemical Exploration Symposium, Orleans, 1987.* Programme and Abstracts, BRGM, Orleans.

SPOSITO, G. 1989. *The Environmental Chemistry of Aluminium.* CRC, Boca Raton, Florida.

STEPHENSON, B., GHAZALI, S. A. & WIDJAJA, H. 1982. *Regional geochemical atlas of Northern Sumatra.* Institute of Geological Sciences.

STIGLIANI, W. & SALOMONS, W. 1993. Our fathers' toxic sins. *New Scientist,* December, 38–42.

STUMM, W. 1991. *Aquatic Chemical Kinetics.* Wiley Interscience, New York.

—— & MORGAN, W. 1981. *Aquatic Chemistry.* Wiley, New York.

TESSIER, A. & CAMPBELL, P. G. C. 1988. *Partitioning of Trace Elements in Sediments. Proceedings, Metal Speciation Theory, Analysis and Application.* Lewis, New York, 183–199.

THORNTON, I. 1983. *Applied Environmental Geochemistry.* Academic, London.

—— & HOWARTH, R. J. 1986. *Applied Geochemistry in the 1980s.* Graham and Trotman, London, 103–139.

—— & PLANT, J. A. 1980. Regional geochemical mapping and health in the United Kingdom. *Journal of the Geological Society, London,* **137**, 575–586.

TRESCASES, J. J. 1992. Chemical weathering. *In:* BUTT, C. R. M. & ZEEGERS, H. (eds) *Regolith Exploration Geochemistry in Tropical and Subtropical Terrains.* Handbook of Geochemistry, Elsevier, Amsterdam, 25–40.

UNDERWOOD, E. J. 1979. Trace elements and health: an overview. *Philosophical Transactions of the Royal Society of London,* **B 288**, 5–14.

UN. 1990. *World Urbanisation Prospects.* Department of International Economic and Social Affairs, UN, New York.

UN. 1992 *Long Range World Population Projections.* Department of International Economic and Social Affairs. UN, New York.

URE, A. M. & DAVIDSON, C. M. 1995. *Chemical Speciation in the Environment.* Blackie, Glasgow.

US GEOLOGICAL SURVEY. 1984. *Water Quality Issues in*

National Water Summary 1983, Hydrologic Events and Issues. Water Supply Paper 2250, Washington DC.

WEBB, J. S. 1964. Geochemistry and life. *New Scientist*, **23**, 504–507.

——, FORTESCUE, J. A. C., NICHOL, I. & TOOMS, J. S. 1964. *Regional geochemical maps of the Namwala Concession area, Zambia.* 1-X, Lusaka, Geological Survey.

WOLERY, T. J. 1992. *EQ3/6: Version 7. A software package for geochemical modelling of aqueous systems.* Lawrence Livermoor National Laboratory, Report UCRL:MA-110662, parts 1, 2 and 3.

WORLD BANK. 1992. *World Development Report: Development and the Environment.* Oxford University Press.

WHO. 1984. *Guidelines for Drinking Water Quality: Volume 2. Health Criteria and Other Supporting Information.* WHO, Geneva.

WHO. 1988. *Urbanisation and its implications for child health: potential for action.* WHO, Geneva.

WHO. 1994. *WHO Report Series: Environmental Health Criteria.* WHO, Geneva.

WILLIAMS, T. M. 1993. *Dispersal pathways of arsenic and mercury associated with gold mining and mineral processing. Field visit to Zimbabwe.* British Geological Survey, Technical Report, WP/93/7R. (unpublished).

—— 1994. *Impacts of mining and mineral processing with particular reference to gold and complex sulphide deposits and environmental geochemical mapping. Field visit to Zimbabwe.* British Geological Survey, Technical Report, WP/94/4 (unpublished).

—— & BREWARD, N. 1995. *Environmental impact of gold and complex sulphide mining (with particular reference to arsenic contamination).* British Geological Survey, Technical Report, WC/95/2 (unpublished).

Stream sediment, soil and forage chemistry as indicators of cattle mineral status in northeast Zimbabwe

F. M. FORDYCE,[1] D. MASARA[2] & J. D. APPLETON[1]

[1] *British Geological Survey, Keyworth, Nottingham NG1 5GG, UK*
[2] *Veterinary Research Laboratory, PO Box CY 551, Causeway, Harare, Zimbabwe*

Abstract: Results of previous studies investigating the use of soil and forage chemistry as indicators of cattle mineral status have been somewhat equivocal, possibly due to the limited range of trace element concentrations in the areas investigated. This paper describes an investigation of the relationship between trace element concentrations in stream sediments, soils, forage (grass and leaves) and cattle blood (serum) in northeast Zimbabwe in order to identify which, if any, of the sample media provide a reliable guide to cattle mineral status. Soil, forage and cattle serum were collected from an area characterized by a wide range of Zn in stream sediments. The area was subdivided into three regions of relatively low, medium and high Zn concentration on the basis of stream sediment data and variations in the chemistry of cattle serum, forage, soil and stream sediment samples were examined. Significant correlations exist between element concentrations in stream sediments, soils and forage but there are no significant correlations with cattle serum. Although this lack of direct correlation may, in part, be due to a range of biological factors, it is suggested that high concentrations of Fe in soil and forage inhibit (i) the availability of P to plants and (ii) the absorption of Cu and Zn in cattle. This may have wide ranging implications due to the predominance of ferrallitic soils in many countries in tropical regions.

The main causes of low production and reproduction rates amongst grazing livestock in many developing countries are probably linked to undernutrition. However, mineral deficiencies and imbalances in forages also have a negative effect. Trials in South America have demonstrated that mineral supplementation significantly improves calving rates and weight gains (McDowell *et al.* 1993). Although dietary mineral supplementation is commonly practised in developed countries, in the developing world grazing livestock are often totally dependent on indigenous forage for their mineral intake. As farmers are encouraged to seek higher levels of productivity from forage fed livestock, it will become increasingly important to identify those areas where trace element deficiencies are negatively affecting animal productivity.

The majority of trace element imbalances in animals do not result in diagnostic clinical symptoms; effects are sub-clinical and difficult to detect without thorough investigations. The assessment of areas with trace element deficiency or toxicity problems in grazing livestock has traditionally been executed by mapping spatial variations in soil, forage, animal tissue or fluid compositions. Soil and forage surveys generally employ high density, detailed sampling techniques in order to obtain representative results because soil chemistry can vary considerably on a local scale and because the trace element content of vegetation varies between species, ecotype and with plant maturity (McDowell *et al.* 1993). In addition, animal studies often require that samples are refrigerated and analysed soon after collection. In developing countries, where there is often a lack of biological and pedological information over large areas, these methods may prove too expensive and logistically impractical for reconnaissance assessment.

Stream sediment geochemical mapping may provide a more practical alternative to soil, forage and animal assessment methods. It is generally accepted that stream sediment sampling is more representative and cost effective than soil or vegetation sampling for rapid reconnaissance geochemical surveys (Levinson 1980). In addition, stream sediment data can be used for several purposes including mineral exploration and environmental studies. Previous studies in temperate regions have demonstrated that stream sediment geochemical mapping can be used to delineate areas where trace element deficiencies or excesses could prejudice animal health (Plant & Thornton 1986; Aggett *et al.* 1988). Regional stream sediment geochemical datasets collected principally for mineral exploration already exist in many developing countries (Plant *et al.* 1988). The application of these data for animal health studies in tropical regimes has been examined in general terms (Appleton 1992; CTVM 1992; Appleton & Ridgway 1993). However, quantitative correlations between stream sediment trace element

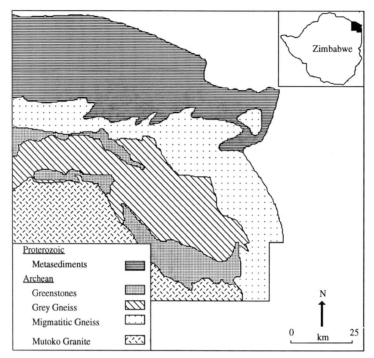

Fig. 1. Simplified geological map of northeast Zimbabwe (modified after Barton *et al.* 1991). Inset shows location of field area.

levels and livestock mineral levels have not been clearly established and no detailed investigations into these relationships in tropical environments have been carried out.

This paper describes an investigation of the relationship between trace element concentrations in stream sediments, soils, forage (grass and leaves) and cattle blood (serum) in northeast Zimbabwe in order to identify which, if any, of the sample media provide a reliable guide to cattle mineral status.

Methodology

Study area

An area of communal grazing land in northeast Zimbabwe, farmed on a subsistence basis by small family groups, was selected for the investigation because high quality stream sediment data already existed as a result of a previous ODA funded project (Dunkley 1987). The study area (Fig. 1) comprises 9000 km^2 of tropical, seasonally wet terrain in the districts of Mudzi, Mutoko, Murehwa and Rushinga. Annual rainfall of 600–800 mm occurs almost entirely in the months of November to March.

The study area includes five major rock types (Fig. 1; Barton *et al.* 1991). In the centre of the area, the Migmatitic Gneisses include biotite and hornblende rich migmatites, mafic to felsic granulites and tonalite gneisses. The Greenstones and Grey Gneisses form a volcano-plutonic complex separated from the Migmatitic Gneisses by a major Archean tectonic break. Greenstones range in composition from basaltic andesite to dacite whereas the Grey Gneisses comprise trondhjemitic and tonalitic granitoid intrusives. In the south of the area, the Greenstone–Grey Gneiss complex is intruded by the Mutoko Granite which itself is intruded in places by basic and ultrabasic rocks. Proterozoic metasedimentary rocks in the north of the area include leucomigmatite with horizons of mafic gneiss and garnet granulite.

Soils are mostly fersiallitic, with high Fe and Al contents (Thomson & Purves 1978). Greyish brown, coarse sands and sandy loams characterize areas underlain by granitic rocks whereas brown to reddish-brown sandy loams overlying sandy clays are more common over siliceous gneisses and schists. Reddish-brown granular clays occur over the greenstones and basic and ultrabasic intrusive rocks, such as those that

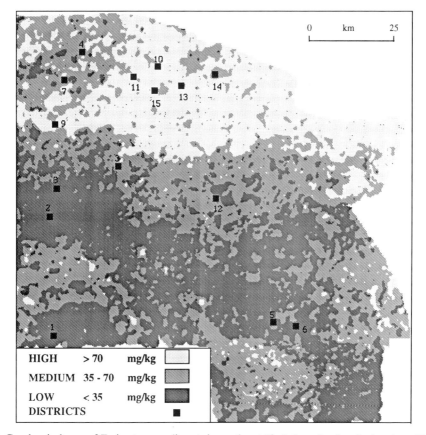

Fig. 2. Geochemical map of Zn in stream sediments in northeast Zimbabwe showing the location of blood, soil and forage sampling districts 1 to 15.

intrude the Mutoko Granite in the southwest. Lithosols characterize areas of rugged terrain. Soils are acid, with pH values of 4.4–6.4 (Nyamapfene 1991).

Much of the area is covered by medium to dense, mixed woodland savannah. Above 600 m the species *Julbernardia globiflora*, *Brachystegia bohemia* and *Brachystegia spiciformis* dominate whereas below 600 m, *Adansonia digitatas*, *Colophospermum mopane*, *Diopsyros*, *Terminalia*, *Combretum* and *Commiphore* (spp) are more common (Anderson 1986a,b; Brinn 1986).

Cattle form an important component of the local economy and family wealth is measured in terms of the number of cattle owned. In addition to the monetary revenue generated by the sale of animals, cattle provide a valuable source of milk for the family. Despite the importance of cattle for the local economy and family, no mineral supplementation is currently practised in the area.

Sampling and analysis

The study area in northeast Zimbabwe was selected to provide a wide range of Zn in stream sediments. The area was subdivided into three regions of relatively low (<35 mg/kg), medium (35–70 mg/kg) and high (>70 mg/kg) Zn, on the basis of existing stream sediment data (Fig. 2) and a sampling strategy was devised to test whether these geochemical differences were reflected in soil, forage and cattle serum. Cattle serum was selected for this study as it is easier to collect than bone and liver samples. Zn was the principal trace element investigated because Zn in cattle serum is generally accepted as a reliable indicator of cattle Zn status (McDowell *et al.* 1993).

Sampling was carried out from April to June 1993 over a period spanning the end of the wet season and the start of the dry season. Cattle were resampled for Zn analysis from April to

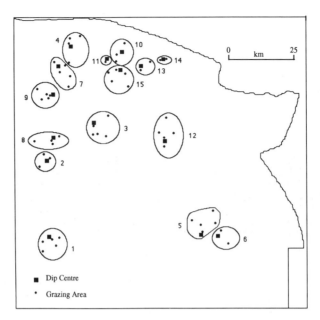

Fig. 3. Map of northeast Zimbabwe showing the relationship of grazing areas to dip centres in each district (1 to 15).

June 1994 because it was suspected the 1993 serum samples had been contaminated with Zn from the sample tube caps. This was confirmed by lower Zn concentrations in the 1994 serum samples.

Blood sampling was conducted at government-run dip centres when cattle from the surrounding villages were gathered together. Dip centres therefore form the focal points for sampling districts (Fig. 3). Fifteen sampling districts in northeast Zimbabwe were selected for the study, five in each of the three Zn regions. Although small-scale geochemical variations occur within districts, this is not considered to affect the overall classification of districts as low, medium and high Zn in stream sediments (Figs 2, 3). A total of 300 cattle, 20 from each district, were sampled in 1993. However, during the Zn resampling programme in 1994, it was only possible to collect 245 samples. None of the cattle had received mineral supplements. Ca, Cu, Mg and Zn were determined in serum by AAS and P in serum by colorimetry. Analyses for Ca, Cu, Mg and P in the 1993 samples were carried out by the Veterinary Research Laboratory in Zimbabwe whereas Zn analysis of the 1994 samples was conducted at the Ministry of Agriculture, Fisheries and Food Veterinary Investigation Centre, Sutton Bonnington, UK. Analytical precision for Zn was 11% (95% confidence level) (Fordyce et al. 1994).

Forage and soil samples were collected from the areas grazed by the cattle sampled in the study. Cattle often shared grazing areas therefore only up to five grazing areas were sampled in each district (Fig. 3). Separate grass and shrub leaf samples comprising a representative range of grazed and browsed species were collected. Unwashed forage samples were dried at 60°C, ground to < 1 mm and dry ashed at 550°C prior to digestion in hot concentrated hydrochloric acid (Fick et al. 1979) and analysis by ICP-AES for Ca, Cu, Fe, Mg, Mn, P and Zn. The effectiveness of the hydrochloric acid digestion method for forage was confirmed by the results for an international standard (Table 1). Mass balance calculations indicate that soil contamination is not a significant factor influencing major and trace element concentrations in forage samples.

Composite soil samples were collected from a depth of c. 10–15 cm, sieved to pass a 2 mm mesh, ground to < 120 mm, digested in hot concentrated hydrochloric acid and analysed by ICP-AES for Ca, Cu, Fe, Mg, Mn, P and Zn. A hydrochloric acid digestion was selected so that soil geochemical data would be directly comparable with data from the earlier reconnaissance stream sediment survey (Dunkley 1987).

Table 1. *Comparison of analytical results for international forage and soil standards with published data*

Standard	Mn	Fe	P	Mg	Ca	Zn	Cu	Co
				mg/kg				
B2/81	78.8	167	2407	1490	6644	31.5	8.1	na
B2/81*	81.6	164	na	na	na	31.5	9.6	na
GXR3	28817	216397	1167	8313	165896	224.7	16.6	50.4
GXR3 †	22308	190000	1100	8100	135800	207.0	15.0	43.0
GXR5	231	30327	220	6373	6655	40.8	333.5	21.7
GXR5 †	310	33900	310	11900	6400	49.0	354.0	30.0
GXR6	1241	63615	383	4744	1645	135.6	75.6	13.6
GXR6 †	1007	55800	350	6100	1800	118.0	66.0	13.8

B2/81, average of 5 analyses of International Rye Grass Standard B2/81; *, recommended values from Office of Reference Materials (1991); GXR3, average of 3 analyses of International Soil Standard GXR3; GXR5, average of 3 analyses of International Soil Standard GXR5; GXR6, average of 3 analyses of International Soil Standard GXR6; † recommended values from Potts *et al.* (1992); na, not available.

Table 2. *Limits of detection for soil and forage analysis by ICP-AES*

Element	Limit of detection (mg/kg)		
	Soil	Grass	Leaves
Mn	3.0	0.02	0.02
Fe	6.6	0.36	0.26
P	9.0	0.56	0.76
Mg	17.0	1.12	0.98
Ca	4.2	0.34	0.44
Zn	0.8	0.06	0.08
Cu	2.0	0.12	0.14
Co	3.2	nd	nd

nd, not determined.

Results suggest the digestion may be less effective for soils containing a significant proportion of primary minerals (GXR5, Table 1). However, since the soils in northeast Zimbabwe are fersiallitic, the HCl digestion is likely to extract elements such as Fe, Mn, Zn, Cu and Co which are contained, for the most part, in secondary Fe-oxides. Analytical precision, based in repeat analysis of samples, for soil and forage was 12% and 16% respectively (95% confidence level) whereas within sample site variation was generally ±20% (Fordyce *et al.* 1994). Detection limits for soil and forage analyses are listed in Table 2. Results for soil and forage samples are presented on a dry matter basis.

Stream sediments collected at an average density of one per km^2 were sieved to < 177 mm prior to digestion in hot concentrated hydrochloric acid and analysis by AAS for Co, Cu, Mn and Zn (Dunkley 1987). Data for streams draining the grazing areas were selected for comparison with soil, forage and cattle serum analytical data. Additional information on sampling and analytical methodologies is given in Fordyce *et al.* (1994).

Each district is represented by different numbers of samples for each sample medium. Relationships between media were therefore assessed by calculating district average values. Correlation coefficients were employed to identify significant inter-element and inter-media relationships. Spearman Rank non-parametric correlation coefficients were calculated as they are less sensitive to outlying values than product moment (Pearson) correlation coefficients.

Results and discussion

Rock, stream sediment and soil geochemistry

Comparison of the Zn in stream sediments geochemical map (Fig. 2) with the geological map (Fig. 1) and with maps for Co, Cu and Mn in stream sediments (Appleton 1992), shows that the Greenstones are characterized by elevated levels of Co, Cu, Mn and Zn. Copper, Mn and Zn concentrations are generally high over the metasedimentary rocks, with localized high Co values. The Migmatitic Gneisses and Grey Gneisses in the centre of the area are characterized by very low levels of Co, Cu, Mn and Zn reflecting the low levels of these trace elements in the parent rocks (Barton *et al.* 1991) and the sandy infertile soils derived from them. Similarly low levels of Co, Cu, Mn and Zn are associated with those parts of the Mutoko Granite in the

Table 3. *Comparison of average values for elements in rocks, sediments and soils in northeast Zimbabwe.*

Sample medium	Element	Average element composition in rock type (mg/kg)		
		Mutoko granite and grey gneiss*	Migmatitic gneiss	Metasedimentary
Rock	Ca	15700	32200	35100
	Fe	13200	40100	66300
	Mg	3860	17100	13800
	Mn	290	840	1300
	P	290	700	800
	Samples taken	22	16	37
Sediments	Co	6	10	12
	Cu	11	20	22
	Zn	28	34	86
	Mn	381	436	975
	Samples taken	34	50	101
Soils	Ca	1293	2189	3649
	Fe	8314	14718	36870
	Mg	1644	3356	4480
	Mn	249	321	800
	P	115	159	381
	Co	3	7	11
	Cu	6	14	20
	Zn	18	27	66
	Samples taken	15	14	27

Average rock compositions calculated from data in Barton *et al.* (1991). * No grazing areas were underlain by greenstones therefore average rock data do not include analyses of greenstones.

Table 4. *Spearman Rank correlation coefficients between elements in stream sediments and soils for the 15 districts*

	Co soil	Cu soil	Fe soil	Mg soil	Mn soil	P soil	Zn soil	Co sed	Cu sed	Mn sed	Zn sed
Ca soil	.821	.643	.671	.896	.750	.639	.814	.553	.715	.557	.725
Co soil		.793	.686	.929	.696	.521	.743	.713	.765	.564	.646
Cu soil			.786	.718	.711	.607	.689	.627	.722	.632	.675
Fe soil				.643	.936	.825	.896	.458	.543	.796	.786
Mg soil					.675	.507	.721	.647	.803	.489	.614
Mn soil						.825	.957	.415	.465	.821	.814
P soil							.861	.416	.468	.829	.757
Zn soil								.468	.511	.782	.796
Co sediment									.854	.517	.670
Cu sediment										.518	.706
Mn sediment											.921

r 95% = 0.457; r 99% = 0.612; r 99.95% = 0.780 (Koch & Link 1970).

southwest of the area that are not intruded by basic rocks. There is, therefore, a strong geological control on the levels of trace elements in stream sediments (Table 3). In addition, the chemistry of soils generally reflects bedrock chemistry, especially for Ca, Fe and Mn (Table 3).

The trace and major element chemistry of soils correlates strongly with stream sediment geochemistry thus confirming the dominant influence of bedrock composition on regional variation in both these sample media (Table 4). Sorption of trace elements by secondary Fe and Mn oxides also influences the trace element content of soils as indicated by the strong correlations between Fe and Mn with Co, Cu, P and Zn in soils (Table 4).

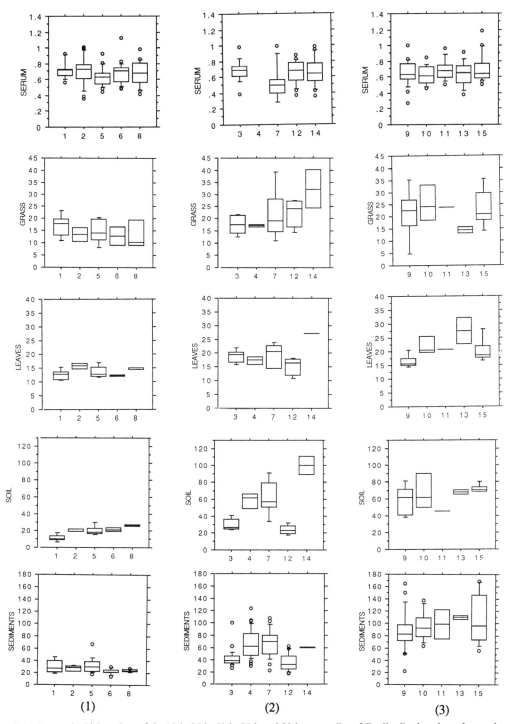

Fig. 4. Box and whisker plots of the 10th, 25th, 50th, 75th and 90th percentiles of Zn distributions in each sample type for each district (1 to 15) in the low (1), medium (2) and high (3) Zn regions. Circles indicate values < 10th and > 90th percentiles. Zn concentrations in mg/kg except serum (mg/l) (No cattle serum data are available for district number 4.)

Zinc

The distributions of Zn within each district and between districts vary considerably for all sample media (Fig. 4). Concentrations of Zn in sediment, soil, leaves and grass from districts in the low Zn region generally have lower Zn concentrations than samples from districts in the medium or high Zn regions. However, within each region there is no consistent relationship between district Zn concentrations in sediment, soil, leaves and grass. There is no apparent relationship between Zn in sediment, soil and forage and Zn in serum at the district level.

Table 5. *Results of ANOVA statistical probability test comparing populations for Zn in each sample media from the low, medium and high Zn regions*

ANOVA	F-value	P-value
Zn sediment v. Zn region		
224 samples	119.37	< 0.001
15 district mean values	45.939	< 0.001
Zn soil v. Zn region		
56 samples	29.662	< 0.001
15 district mean values	6.754	0.0108
Zn grass v. Zn region		
56 samples	5.429	0.0072
15 district mean values	5.343	0.0219
Zn leaves v. Zn region		
56 samples	13.254	< 0.001
15 district mean values	6.612	0.0116
Zn serum v. Zn region		
245 samples	1.424	0.2428
15 district mean values	1.691	0.2289

At the regional level, there are statistically significant differences between the regions for Zn in sediment, soil, grass and leaves but not in cattle serum (Table 5). Zn concentrations in soil and forage reflect the geochemical trend from low to high Zn in stream sediments (Fig. 5). In grass and leaves, more overlap occurs between the concentration ranges for the low, medium and high Zn regions. In contrast to the other sample media, the ranges for Zn in cattle serum are approximately the same in the three regions. These observations are reiterated in Table 6 which shows there are significant (95% confidence level) correlations between district average values for Zn in grass, leaves, soils and sediments but no significant correlations between Zn in serum and the other sample media. Therefore, at both the district and regional levels, cattle serum Zn does not directly reflect environmental Zn concentration.

Many biological factors such as the species type and state of maturity of plants, and the age and gender of animals, exert significant controls on the uptake of major and trace elements by living organisms (Mertz 1987). In addition, infection and vaccination are known to enhance Cu and deplete Zn in cattle serum. A large number of serum samples were collected in this study to minimize the effects of biological factors on district means. Despite this precaution, these biological factors may largely explain why the levels of Zn in cattle serum do not directly reflect environmental Zn in northeast Zimbabwe. Another explanation for the lack of correlation between cattle serum Zn and forage Zn may be the uncertain contribution of individual grass and browse species and the proportion of grass and browse material in the dietary intake of cattle in this study.

Table 6. *Spearman Rank correlation coefficients between Zn in serum, grass, soil and leaves for the 15 districts and other elements in various media (based on district mean concentrations).*

	Zn serum	Zn grass	Zn leaves	Zn soil
Zn grass	−.200			
Zn leaves	−.293	.575		
Zn soil	.392	.732	.886	
Zn sediment	−.359	.600	.836	.796
Cu grass	.150	.461	.300	.271
Cu leaves	−.117	.521	.793	.654
Cu soil	−.112	.586	.725	.689
Fe grass	−.299	.511	.843	.775
Fe leaves	−.112	.357	.764	.618
Fe soil	−.392	.621	.857	.896
Mn grass	−.035	.214	.411	.214
Mn leaves	− 273	.368	.735	.175
Mn soil	−.366	.632	.893	.957
Ca leaves	−.350	−.461	−.909	−.186
Ca soil	−.442	.582	.625	.814
P soil	−.317	.807	.825	.861

r 95% = 0.457; r 99% = 0.612; r 99.95% = 0.780 (Koch & Link 1970).

Although statistical correlations do not prove a causal relationship, the strong spatial and statistical correlations between sediment, soil and forage Zn suggest that increased levels of Zn in soils result in increased uptake of Zn by plants. However, ingestion of plants containing higher levels of Zn does not result in an increase in the Zn levels found in cattle serum. In fact, Zn in serum appears to decrease slightly as Zn in

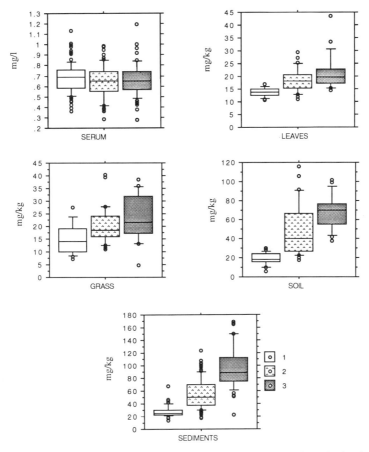

Fig. 5. Box and whisker plots of the 10th, 25th, 50th, 75th and 90th percentile of Zn distributions in each sample type for the low (1), medium (2) and high (3) Zn regions. Circles indicate values < 10th and > 90th percentiles.

forage increases (Table 6). One possible explanation may be the antagonistic relationships present between elements as they are absorbed during digestion by the cattle. Fe and Zn are known to have an antagonistic relationship during absorption in humans (Sandstrom et al. 1985; Mertz 1987) and in rats (Quarterman 1985). In addition, antagonistic relationships between Mn, Fe and Zn have been reported in humans (Christophersen 1994). Lebdosoekojo et al. (1980) suggest that high levels of Fe and Mn may interfere with the metabolism of other trace elements in cattle. Several clinical studies in animals and humans have identified a mutually antagonistic relationship between Cu and Zn (Mertz 1987). In northeast Zimbabwe, the area of high Zn in soils and forage coincides with high Cu, Fe and Mn in soils and forage (Tables 4, 6), therefore it is possible that uptake of these elements is inhibiting the absorption of Zn in cattle.

Zn concentrations in serum tend to decrease slightly as the Ca content of leaves and soil and the P content of soil increase (Table 6). High Ca and P ingestion have been shown to reduce Zn absorption in humans, pigs and poultry but these relationships are less clear in cattle (Mertz 1987).

Copper

There are significant correlations between district average values for Cu in sediment and soil (Table 4) and between Cu in soil and leaves (Table 7). However, Cu in serum exhibits no correlation with Cu in soil, grass or leaves (Table 7) or average forage (Fig. 6). Therefore although there is some evidence to suggest that higher levels of Cu in the environment are taken up by vegetation, these higher levels are not reflected in the Cu content of serum. Similar results are reported by McDowell (1976) who found that the Cu content of soils and forage did not correlate with the Cu status of cattle. Evidence from the present study suggests that, as with Zn, this lack of correlation may partly reflect antagonistic relationships between elements during uptake.

Table 7. *Spearman Rank correlation coefficients between Cu in serum, grass and leaves for the 15 districts and other elements in various media (based on district mean concentrations)*

	Cu serum	Cu grass	Cu leaves
Cu grass	−.195		
Cu leaves	.032	.446	
Cu soil	.009	.371	.696
Cu sediment	−.019	.452	.374
Zn grass	−.496	.461	.521
Zn leaves	−.356	.300	.793
Zn soil	−.347	.271	.645
Zn sediment	−.489	−.398	.500
Mn grass	−.593	.204	.354
Mn leaves	−.675	.464	.139
Mn soil	−.340	.204	.614
Mn sediment	−.640	.504	.543
Fe grass	−.381	.329	.657
Fe leaves	−.349	.189	.486
Fe soil	−.277	.336	.629

r 95% = 0.457; r 99% = 0.612; r 99.95% = 0.780 (Koch & Link 1970).

The most significant negative correlations are between serum Cu and (i) Mn in stream sediment, grass and leaves and (ii) Zn in stream sediment and grass (Table 7). In addition the correlations between Cu in serum and Fe in soil and forage are slightly negative (Fig. 6). These negative relationships suggest that Zn, Mn and possibly Fe ingested in forage and soil may be inhibiting the absorption of Cu in cattle. Clinical trials have demonstrated that high levels of dietary Zn reduce Cu absorption in rats, pigs and sheep although the relationship in cattle is less clear (Mertz 1987). The inhibitory effects of dietary Fe on Cu absorption are well documented (Mertz 1987). For example, Humphries et al. (1985) found dietary intakes of $350\,\mathrm{mg\,kg^{-1}}$ of Fe in forage were sufficient to significantly reduce the Cu content of the liver in young calves. Russell et al. (1985) demonstrated that Fe ingested from soil has a negative effect on Cu absorption in sheep. Up to 25% of the Fe content of soils can be extracted during simulated digestion in sheep (Brebner et al. 1985). Since the level of Fe in soils is 100 times the content of forage in northeast Zimbabwe (Table 8), ingestion of high Fe soils may exert a greater inhibitory effect on Cu uptake than Fe in forage.

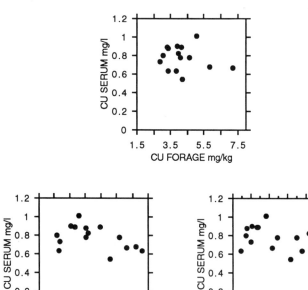

Fig. 6. Plots of district average values of Cu in serum versus Cu in forage, Fe in forage and Fe in soil. (Average forage values represent the average of grass and leaf compositions for each district.)

Table 8. *Average element values in serum, forage and soils from the present study compared to studies from other areas of mineral deficiency in grazing ruminants*

Media		Country	Ca	Mg	P	Fe	Mn	Co	Cu	Zn
Serum (mg/l)	a	Zimbabwe	80	23	45	nd	nd	nd	0.8	0.7
	b	Swaziland	na	na	39	na	na	na	na	na
	h	Malawi	76	na	56	na	na	na	0.3	na
	d	Indonesia	na	na	na	na	na	na	0.6	0.8
	e	Bolivia	112	18	70	na	na	na	0.9	1.0
	f	Bolivia	80	21	67	na	na	na	0.6	1.1
Forage (mg/kg)	a	Zimbabwe*	4409	1713	1444	234	109	nd	1.9	19.0
		†	16189	3575	1750	165	179	nd	6.1	17.0
	b	Swaziland	4050	1500	5500	224	86	na	5.6	9.0
	c	Guatemala	3100	1850	2500	520	92	0.4	14.0	32.5
	d	Indonesia	na	na	na	742	88	0.3	9.3	5.0
	e	Bolivia	4400	1850	1900	170	390	0.5	5.2	24.0
	f	Bolivia	2100	1600	1500	134	133	0.2	5.9	30.0
	g	Colombia	1300	1550	1300	563	143	0.1	1.8	13.5
	g	Brazil	5050	2550	5000	238	145	0.1	4.0	26.5
Soil ‡ (mg/kg)	a	Zimbabwe	2807	3653	251	23537	517	8.1	15.4	43.1
	b	Swaziland	524	191	12	na	4	na	1.0	1.1
	c	Guatemala	1860	387	10	52	59	na	2.4	5.8
	d	Indonesia	na	na	na	98	24	na	1.6	4.6
	e	Bolivia	648	204	4	115	17	na	0.5	3.0
	f	Bolivia	192	34	1.2	24	0.3	na	0.3	1.3

a, present study; b, Ogwang (1988); c, Tejada *et al.* (1985); d, Prabowo *et al.* (1991); e, McDowell *et al.* (1982); f, Peducassé *et al.* (1983); g, Miles *et al.* (1989); h, Mtimuni *et al.* (1983). nd, not determined; na, not available. * Mean concentrations for grass samples; † mean concentrations for leaf samples; ‡ soils were digested with hot concentrated HCl in the present study whereas published results from Guatemala, Indonesia, Bolivia and the USA are for partial extraction with $0.05\,N\,HCl + 0.025\,N\,H_2SO_4$. Soil data for Swaziland are for partial extraction with ammonium acetate.

Table 9. *Spearman Rank correlation coefficients between Ca in serum, grass and leaves and Mg in serum for the 15 districts and other elements in various media (based on district mean concentrations)*

	Ca grass	Ca leaves	Ca soil	P grass	P leaves	Mg grass	Mg leaves
Ca serum	−.279	−.039	−.093	−.461	−.007	−.254	.321
Ca grass		−.207	.136	.257	−.218	.446	.089
Ca leaves			−.209	−.096	.543	−.286	−.479
Mg serum	−.157	.389	.043	.318	.625	−.296	−.282

r 95% = 0.457; r 99% = 0.612; r 99.95% = 0.780 (Koch & Link 1970).

Table 10. *Spearman Rank correlation coefficients between P in serum for the 15 districts and other elements in various media (based on district mean concentrations)*

	P soil	P grass	P leaves	Fe soil	Fe grass	Fe leaves
P serum	−.079	−.025	.243	−.311	−.343	−.321

r 95% = 0.457; r 99% = 0.612; r 99.95% = 0.780 (Koch & Link 1970).

Calcium and magnesium

Ca and Mg in soils correlate closely with all the other elements in soils as a result of the strong influence of bedrock geochemical variations (Table 4). Ca and Mg in soils do not correlate significantly with Ca and Mg in forage or serum samples (Table 9). Increased uptake of these elements by plants and animals as a result of higher levels in soils is not evident.

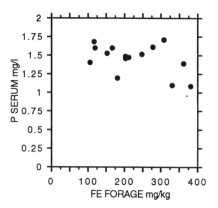

Fig. 7. Plots of district average values of P in serum versus Fe in forage. (Average forage values represent the average of grass and leaf compositions for each district.)

Phosphorus

P in serum does not correlate significantly with P in any of the other sample media (Table 10). Although there is a strong positive correlation between Fe and P in soil, strong sorption of P by Fe oxides in soils normally reduces the availability of P to plants. This may explain why P in forage and cattle serum does not increase with P in soils; indeed P in serum tends to decrease slightly as Fe in soil and forage increases (Table 10; Fig. 7). This suggests that elevated levels of Fe in soil and forage may cause P deficiency in cattle. Additional significant relationships between P and other elements are mentioned in previous sections.

Iron and manganese

In addition to the relationships discussed above, significant correlations exist between Fe in soil with grass and leaves (Table 11) and Mn in sediment with soil (Table 4), grass and leaves (Table 11). This demonstrates that Fe and Mn in sediments and soils can be used as indicators of levels of these elements in forage. No indicators of cattle mineral status were obtained for Fe and Mn.

Deficiency levels

Table 8 shows the average element concentrations in cattle serum, forage and soil from the present study compared with published data from other parts of the world where deficiencies in cattle serum, forage and soils have been recorded. Cattle mineral status studies commonly classify animal, forage and soil samples with respect to deficient, marginal and toxic element concentrations (McDowell et al. 1993). Samples below the critical level are termed deficient. The use of marginal bands acknowledges the uncertainty of predicting the precise level at which deficiency is induced. Applying the critical levels and marginal bands used in the published studies (Table 12) to the data from northeast Zimbabwe, it is clear that a high proportion of cattle serum and forage samples are marginal in Zn (Table 13).

Discrepancies between the percentage of samples below the critical levels and marginal bands in serum, forage and soil have been observed in previous studies (references from Table 8). This is confirmed by the present study

Table 11. Spearman Rank correlation coefficients between Fe in grass, leaves and soil for the 15 districts and Mn in various media (based on district mean concentrations).

	Fe leaves	Fe soil	Mn leaves	Mn soil	Mn sediment
Fe grass	.428	.743	.346	.828	.825
Fe leaves		.671	.089	.628	.632
Fe soil			.314	.936	.796
Mn grass			.578	.246	.486
Mn leaves				.225	.493

r 95% = 0.457; r 99% = 0.612; r 99.95% = 0.780 (Koch & Link 1970).

in which only 26% of the cattle are Cu deficient whereas nearly 100% of forage samples are below the critical concentration (Table 13).

Table 12. *Deficiency critical levels and marginal bands for elements in cattle serum and forage*

Media	Element	Critical value deficient	Marginal band
Serum (mg/l)	Ca*	< 80	
	Mg		< 10–20
	P*	< 45	
	Cu	< 0.65	
	Zn		< 0.6–0.8
Forage (mg/kg)	Ca	< 3000	
	Mg	< 2000	
	P	< 2500	
	Co	< 0.1	
	Cu	< 10	
	Fe	< 30	
	Mn		< 30–40
	Zn	< 30	

All values taken from McDowell et al. (1993) except * from Peducassé et al. (1983)

Summary and conclusions

1. Stream sediment geochemical maps for Zn and Mn provide a reliable indication of the relative distribution of these elements in soil and forage (grass and leaves). Cu in stream sediments can be used to predict the levels of these elements in soil and grass, but less reliably in leaves. Fe in soils provides a reliable indication of levels in grass and leaves. Although hot hydrochloric acid extractable Ca, P and Mg in soil reflect variations in the chemical composition of the underlying rocks, this information cannot be used to predict relative concentrations of these elements in forage.
2. Ca, Cu, Mg, P and Zn in cattle serum do not correlate positively with these elements in forage, soil and sediment samples. Therefore, it appears that it is not possible to use the concentrations of these elements in forage, soil or stream sediments to predict the mineral status of cattle.
3. The lack of direct relationship between cattle mineral status and forage status probably reflects the influence of biological factors including plant species and maturity and cattle gender and age on element uptake. The relationship may also be influenced by the uncertain contribution of grass and leaf species in the dietary intake of the cattle.
4. The results of this study suggest that interactions and antagonistic effects between elements as they are absorbed by plants and animals may influence the mineral status of cattle in northeast Zimbabwe. High concentrations of Fe and Mn in soil and forage appear to inhibit the availability of P to plants and the absorption of Cu and Zn in cattle. These findings may have wide ranging implications due to the preponderance of ferrallitic soils in many countries in tropical regions.
5. Although stream sediment, soil and forage geochemical maps cannot be used to predict the mineral status of grazing ruminants directly, they can serve to indicate those areas where Zn, Cu and P are likely to be low in soils and forage and those areas where higher levels of Fe (and/or Mn) may induce low Zn, Cu and P status in cattle despite higher levels of these elements in soils and forage.
6. Discrepancies between the percentage of deficient serum and percentage of deficient forage samples found in previous studies are also apparent in northeast Zimbabwe. This suggests that the critical levels used to determine deficiency may require further investigation.
7. The limitations inherent in the use of univariate rank statistical tests to assess what are almost certainly multivariate relationships are acknowledged. A multivariate

Table 13. *Percentages of northeast Zimbabwe serum and forage samples with elements below critical levels and marginal bands (Table 12)*

	% Deficient						% Marginal		
	Ca	Mg	P	Cu	Fe	Zn	Mg	Mn	Zn
Serum	47	–	55	26	nd	–	17	nd	86
Grass	10	77	94	100	0	92	–	6	–
Leaves	0	2	88	98	0	100	–	0	-

Marginal, < 20 mg/l Mg serum; < 0.8 mg/l Zn serum and < 40 mg/kg Mn forage. nd = no data.

statistical approach to data interpretation is currently under investigation and will include consideration of Fe in cattle serum.
8. It is recommended that additional studies should be carried out over a variety of geographical and geological conditions in order to confirm the findings of this study.

This study forms part of the British Government Overseas Development Administration (ODA) Technology Development and Research Programme 'Environmental Geochemical Mapping' Project (R5547, 91/6).

The advice and cooperation of the following people is very gratefully acknowledged: U. Ushewokunse-Obatolu, P. Gamble, M. Simpson, C. Mahanda, E. Anderson and regional office staff (Veterinary Services Dept, Harare); N. Ndiweni (Preclinical Veterinary Science Dept, University of Zimbabwe); J. Topps (Animal Science Dept, University of Zimbabwe); R. Mbofana, P. Pitfield, P. Fey and S. Ncube (Geological Survey Dept, Harare); A. Hunt (MAFF Veterinary Inspection Centre, Sutton Bonnington); G. Freeland (ODA Senior Animal Health Adviser); C. Livesey (Head of Biochemistry, Central Veterinary Laboratory); C. Mills (Rowett Research Institute, Aberdeen); L. McDowell (Animal Science Dept, University of Florida); K. Ryder (ODA Biometrics Unit, Rothamsted) and N. Suttle (Moredun Institute, Edinburgh). J. Baldock, R. Bowell, N. Breward, L. McDowell, N. Suttle and M. Williams are thanked for their written comments and suggestions.

This paper is published by permission of the Director of the British Geological Survey (NERC) and the Director of Veterinary Services, Harare.

References

AGGETT, P. J., MILLS, C. F., PLANT, J. A., SIMPSON, P. R., STEVENSON, A., DINGWALL-FORDYCE, I. & HALLIDAY, C. F. 1988. A study of environmental geochemistry and health in northeast Scotland. In: THORNTON, I. (ed.) Geochemistry and Health. Proceedings of the Second International Symposium. Science Reviews, Northwood, 81–91

ANDERSON, I. 1986a. Communal Land Physical Resource Inventory Mudzi District. Zimbabwe Chemistry and Soil Research Institute, Soils Report, **A519**.

——— 1986b. Communal Land Physical Resource Inventory Darwin District. Zimbabwe Chemistry and Soil Research Institute, Soils Report, **A529**.

APPLETON, J. D. 1992. Review of the Use of Regional Geochemical Maps for Identifying Areas Where Mineral Deficiencies or Excesses May Affect Cattle Productivity in Tropical Countries. British Geological Survey, Overseas Geology Series, Technical Report, **WC/92/24**.

——— & RIDGWAY, J. 1993. Regional geochemical mapping in developing countries and its application to environmental studies. Applied Geochemistry, Supplementary Issue, **2**, 103–110

BARTON, C. M., CAIRNEY, J. N., CROW, M. J.,
DUNKLEY, P. N. & SIMANGO, S. 1991. The Geology of the Country Around Rushinga and Nyampanda. Zimbabwe Geological Survey Bulletin, **92**.

BREBNER, J., THORNTON, I., MCDONALD, P. & SUTTLE, N. F. 1985. The release of trace elements from soil under conditions of simulated rumenal and abomasal digestion. In: MILLS, C. F., BREMNER, I. & CHESTERS, J. K. (eds) Trace Elements in Man and Animals – TEMA 5. Commonwealth Agricultural Bureaux, Slough, 850–852.

BRINN, P. J. 1986. Communal Land Physical Resource Inventory Rushinga District. Zimbabwe Chemistry and Soil Research Institute, Soils Report, **A518**.

CTVM (Centre for Tropical Veterinary Medicine, Edinburgh University). 1992. Report on the Mineral Status of Animals in Some Tropical Countries and Their Relationship to Drainage Geochemical Maps of Minerals in Those Countries. British Geological Survey, Overseas Geology Series, Technical Report, **WC/92/60**.

CHRISTOPHERSEN, O. A. 1994. Some aspects of the biogeochemical behaviour of Mn as compared to other elements. In: LAG, J. (ed.) Geomedical Problems related to Al, Fe and Mn. The Norwegian Academy of Sciences and Letters, Oslo, 85–99.

DUNKLEY, P. N. 1987. A Regional Geochemical Survey of the Country Between Rushinga and Nyampanda, North-east Zimbabwe. British Geological Survey, Overseas Directorate, Open File Report, **MP/87/16**.

FICK, K. R., MCDOWELL, L. R., MILES, P. H., WILKINSON, N. S., FUNK, J. D. & CONRAD, J. H. 1979. Methods of Mineral Analysis for Plant and Animal Tissues. University of Florida.

FORDYCE, F. M., APPLETON, J. D. & MASARA, D. 1994. Final Report on Stream Sediment, Soil and Forage Chemistry as Indicators of Cattle Mineral Status in Northeast Zimbabwe. British Geological Survey, Overseas Geology Series, Technical Report, **WC/94/3**

HUMPHRIES, W. R., BREMNER, I. & PHILLIPPO, M. 1985. The influence of dietary iron on copper metabolism in the calf. In: MILLS, C. F., BREMNER, I. & CHESTERS, J. K. (eds) Trace Elements in Man and Animals – TEMA 5. Commonwealth Agricultural Bureaux, Slough, 371–373.

KOCH, G. S. & LINK, R. F. 1970. Statistical Analysis of Geological Data. Wiley, New York.

LEBDOSOEKOJO, S., AMMERMAN, C. B., RAUN, N. S., GOMEZ, J. & LITELL, R. C. 1980. Mineral nutrition of beef cattle grazing native pastures on the eastern plains of Colombia. Journal of Animal Science, **51**, 1249–1260.

LEVINSON, A. A. 1980. Introduction to Exploration Geochemistry, Applied Publishing, Wilmette.

MCDOWELL, L. R. 1976. Mineral Deficiencies and Toxicities and Their Effects on Beef Production in Developing Countries. In: SMITH, A. J. (ed.) Beef Cattle Production in Developing Countries. University of Edinburgh Press, 216.

———, CONRAD, J. H. & HEMBRY, F. G. 1993. Minerals for Grazing Ruminants in Tropical Regions. Second Edition, University of Florida.

——, BAUER, B., GAIDO, E., KOGER, M., LOOSLI, J. K. & CONRAD, J. H. 1982. Mineral supplementation of beef cattle in the Bolivian tropics. *Journal of Animal Science*, **55**, 964–970.

MERTZ, W. 1987. *Trace Elements in Human and Animal Nutrition*. Academic, London.

MILES, W. H., MCDOWELL, L. R. & RAUN, N. S. 1989. More complete mineral supplements key technology for tropical America. *In:* POPE, L. S. & MOORE, P. (eds) *World Animal Review*, *1*. Agriservices Foundation, Clovis, California.

MTIMUNI, J. P., CONRAD, J. H. & MCDOWELL, L. R. 1983. Effect of season on mineral concentrations on beef cattle in Malawi. *South African Journal of Animal Science*, **13**, 1–2.

NYAMAPFENE, K. 1991. *Soils of Zimbabwe*. Nehanda, Harare.

OFFICE OF REFERENCE MATERIALS. 1991. *Certified Reference Materials*. Issue 2. Laboratory of the Government Chemist, London.

OGWANG, B. H. 1988. The mineral status of soil, forage and cattle tissues in the middleveld of Swaziland. *Exploratory Agriculture*, **24**, 177–182.

PEDUCASSÉ, A. C., MCDOWELL, L. R., PARRA, L., WILKINS, J. V., MARTIN, F. G., LOOSLI, J. K. & CONRAD, J. H. 1983. Mineral status of grazing beef cattle in the tropics of Bolivia. *Tropical Animal Production*, **8**, 118–130.

PLANT, J. A. & THORNTON, I. 1986. Geochemistry and Health in the United Kingdom. *In:* THORNTON, I. (ed.) *Proceedings of the First Symposium on Geochemistry and Health*. Science Reviews, Northwood, 5–16.

——, HALE, M. & RIDGWAY, J. 1988. Developments in regional geochemistry for mineral exploration. *Transactions of the Institute of Mining and Metallurgy*, **97**, B116–B140.

POTTS, P. J., TINDLE, A. G. & WEBB, P. C. 1992. *Geochemical Reference Material Compositions*. CRC, London.

PRABOWO, A., MCDOWELL, L. R., WILKINSON, N. S., WILCOX, C. J. & CONRAD, J. H. 1991. Mineral status of grazing cattle in South Sulawesi, Indonesia: 2: Microminerals. *Asian Australian Journal of Animal Science*, **4**, 121–130.

QUARTERMAN, J. 1985. The role of intestinal mucus on metal absorption. *In:* MILLS, C. F., BREMNER, I. & CHESTERS, J. K. (eds) *Trace Elements in Man and Animals – TEMA 5*. Commonwealth Agricultural Bureaux, Slough, 400–401.

RUSSELL, K., BREBNER, J., THORNTON, I. & SUTTLE, N. F. 1985. The influence of soil ingestion on the intake of potentially toxic metals and absorption of essential trace elements by grazing livestock. *In:* MILLS, C. F., BREMNER, I. & CHESTERS, J. K. (eds) *Trace Elements in Man and Animals - TEMA 5*. Commonwealth Agricultural Bureaux, Slough, 847–849.

SANDSTROM, B., DAVIDSSON, L., CEDERBLAD, A. & LONNERDAL, B. 1985. Effects of inorganic iron on the absorption of zinc from a test solution and a composite meal. *In:* MILLS, C. F., BREMNER, I. & CHESTERS, J. K. (eds) *Trace Elements in Man and Animals – TEMA 5*. Commonwealth Agricultural Bureaux, Slough, 414–415.

TEJADA, R., MCDOWELL, L. R., MARTIN, F. G. & CONRAD, J. H. 1985. Mineral element analyses of various tropical forages in Guatemala and their relationship to soil concentrations. *Nutrition Reports International*, **32**, 313–323.

THOMSON, J. G. & PURVES, W. D. 1978. *A Guide to the Soils of Rhodesia*. Rhodesian Journal of Agriculture, Technical Handbook **3**.

Effects of botanical composition, soil origin and composition on mineral concentrations in dry season pastures in western Kenya

I. O. JUMBA,[1] N. F. SUTTLE,[2] E. A. HUNTER[3] & S. O. WANDIGA[1]

[1] Dept of Chemistry, University of Nairobi, PO Box 31097, Kenya
[2] Moredun Research Institute, Edinburgh EH17 7JH, UK
[3] Biomathematics and Statistics Scotland, Edinburgh EH9 3JZ, UK

Abstract: The influence of botanical (pasture species), geographical (altitude) and pedological (bedrock type, soil pH and extractable mineral concentration) factors on mineral concentrations in dry season pasture was studied in samples of topsoil and herbage from 135 sites on 84 farms in the Mt Elgon region of western Kenya. Of the four major elements measured in herbage, only S was affected by geology, low values being found above Tertiary volcanic (TV) and metamorphic gneiss (MG) bedrocks. By contrast, only P was not affected by species, Setaria being low and Kikuyu grass usually high in macro minerals. Of the seven trace elements analysed in herbage, geology influenced only one: Cu; low values were again found above TV and MG but Cu availability to grazing ruminants would be relatively high because of the associated low S values. By contrast, only Se was unaffected by species, Kikuyu grass being high in all but Mn. Soil bedrock had a greater influence on soil composition but correlations between extractable soil and herbage mineral concentrations were poor for all elements, even within botanical species after correction for soil pH and soil contamination. Concentrations of Ca, P, Mg, S, Cu and Zn were often less than tabulated requirements for grazing livestock. However, risks of deficiency could not be predicted from the pedological factors measured.

The prediction of mineral deficiencies in grazing livestock requires good correlations between convenient markers of mineral status and animal health or productivity. Correlations are likely to become weaker in moving from animal to pasture to soil in pursuit of a predictor because of the many factors which influence mineral uptake at each interface. However, soils are the easiest to characterize and correlations might be improved by removing the effects of known sources of variation. The influence of botanical (pasture species), geographical (altitude) and pedological (bedrock type, soil pH and extractable mineral concentration) factors on mineral concentrations in dry season pasture was, therefore, assessed in a region of Kenya where the forages are commonly deficient in several macro- and micro-elements (Jumba et al. 1996a,b) The effects of botanical composition on the mineral content of dry season forage were substantial (Jumba et al. 1995a,b) and in this paper we contrast those effects with the minor influence of soil bedrock and extractable mineral concentrations in the soil.

Materials and methods

At least three samples of topsoil and herbage were gathered from each of 135 sites on 84 farms in the Mt Elgon region of western Kenya in the dry season (wet seasons surveys present logistical problems) between January and March 1987 (Jumba 1989). The underlying parent bedrock was determined from 1:125 000 geological survey maps and altitude from topographical maps. The botanical composition of the pasture sample was recorded. Distribution of principal botanical species and all bedrock types amongst the sample sites are indicated in Table 1. Most of the sites (51%) lie between 2000 and 2333 m above sea level (a.s.l.) with 28% between 1666 and 2000 m, 16% between 1333 and 1666 m and the remainder (5%) between 2333 and 2777 m. Soil pH and total Se or extractable concentrations of Ca, P, Co, Cu, Fe, Mn, Mo and Zn were determined in duplicate for the pooled samples from each site by standard methods as were total mineral concentrations in unwashed herbage (Jumba et al. 1995a,b, 1996a,b). Methods for extractable mineral concentrations in soils were chosen after comparisons for each element based on optimal recovery of the added element from soil solutions (Jumba 1989). Table 2 gives the extraction method selected and those rejected. Soil Mg and S were not measured.

Statistical analysis used a residual maximum likelihood (REML) approach (Patterson & Thompson 1971) to fit a mixed model to obtain fitted means for the unbalanced datasets, Wald statistics (distributed as a Chi square) being used to assess significance (Genstat 5.3 program,

Table 1. *The associations between herbage species and soil bedrock at 132** of the sample sites*

Herbage species (abbreviation)	Common name	Bedrock*					
		TV	IG	MS	MG	SSG	AD
Pennisetum purpureum (Pp)	Napier	4	8	5	15	1	6
Setaria sphacelata (Ss)	Setaria	2	0	0	6	0	1
Chloris guyana (Cg)	Rhodes	7	4	0	23	5	9
Pennisetum clandestinum (Pc)	Kikuyu	4	1	0	2	1	1
Ss/Cg mixtures		2	1	1	8	1	0
Mixed natural grasses (mng)		1	3	2	5	1	2

** Single samples of three sown species are excluded from this table and Tables 3–5.
* TV, Tertiary volcanic; IG, igneous granite; MS, metamorphosed sedimentary; MG, metamorphic gneiss; SGG, sedimentary sandstone and grits; AD, alluvial deposits.

Table 2. *Outcome of preliminary screening of extraction methods for assessing the available mineral concentrations in soils*

Method	Evaluated	Preferred for	Preferred to
A	1 M Ammonium acetate pH 7	Ca, Mn, Zn, Mo	E, F
B	1 M Ammonium acetate pH .8	Fe	C, F
C	1 M Ammonium acetate in 0.05 M HCl; pH 3	None	–
D	0.5 M $NaHCO_3$, pH 8.5	None	–
E	0.5 M Acetic acid, pH 2.5	Co	None
F	0.1 M HCl in 0.0125 M H_2SO_4	P, Cu	A, E, F(P); A, G(Cu)
G	0.05 M EDTA: pH 7	None	–
H	1 M Ammonium oxalate pH 3	None	–

copyright 1994; Lawes Agricultural Trust). The model allowed the mean pH (Tables 3 and 4) or extractable mineral concentration in soils associated with a particular species or soil bedrock to be calculated, since any such associative effects would be confounded with the main effects of species and bedrock. Where appropriate, the model was augmented by including the continuous variables, altitude, soil pH and soil extractable mineral concentration as covariates. In the augmented model factors were fitted in the order altitude, extractable concentration in soil, soil pH, soil bedrock and finally herbage species. For each covariate, Wald statistics were calculated ignoring terms fitted later in the model but standard errors for each term were generated on a 'best fit' basis allowing their significance to be accurately assessed.

Results

Pasture minerals and cattle requirements

The overall mean and range of mineral concentrations in the pastures are presented with estimates of the dietary requirements of beef cattle for those minerals in Table 3. Pasture concentrations of Ca and Cu were generally below the requirement of ruminants, Mg, P, S and Zn were marginal, Co and Se occasionally low while Fe and Mn were adequate. The mineral composition of the herbage in relation to soil bedrock and pasture species are shown in Tables 4 and 5 respectively.

Table 3. *Means and ranges of mineral concentrations in western Kenya pastures compared with dietary requirements published by ARC (1980) for beef cattle*

Element	Mean	Range	Requirement
		($g\,kg^{-1}$ DM)	
Ca	1.5	0.3–3.2	1.7–6.7
P	1.4	0.04–3.5	1.1–3.4
Mg	1.6	0.6–2.6	1.0–2.0
S	1.5	0.5–3.1	1.1–1.6
		($mg\,kg^{-1}$ DM)	
Co	0.2	0.03–0.96	0.08–0.10
Cu	4.0	1.8–9.4	8–15
Fe	300	30–270	30–100
Mn	220	50–590	20–25
Mo	1.1	0.1–4.2	?
Se	0.1	0.009–0.62	0.03–0.05
Zn	24	5–97	12–35

Table 4. *Effects of soil bedrock on the macro- ($g\,kg^{-1}\,DM$) and micro- ($mg\,kg^{-1}\,DM$) mineral concentrations in dry season herbage*

	TV	IG	MS	MG	SSG	AD	s.e.d.m.*		
	n=20	17	8	59	9	19	min	max	W
				($mg\,kg^{-1}$ DM)					
Mo	1.1	1.4	0.9	1.3	1.4	1.0	0.22	0.40	4.0
Fe	355	270	280	349	341	338	42.9	80.4	9.2
Mn	210	240	213	236	231	228	22.0	41.4	5.7
Zn	22.7	25.1	24.7	27.6	27.6	26.1	3.02	5.68	1.1
Se†	67	52	70	75	95	61	0.105	0.217	3.0
Cu	4.0	4.6	5.5	3.9	4.5	4.3	0.36	0.65	19.1
Co	0.29	0.25	0.24	0.25	0.25	0.24	0.037	0.068	4.0
				($g\,kg^{-1}$ DM)					
Ca	1.2	1.5	1.6	1.3	1.4	1.4	0.13	0.24	2.9
P	1.8	1.4	1.7	1.3	1.8	1.3	0.14	0.30	9.3
Mg	1.7	1.7	1.5	1.6	1.9	1.5	0.13	0.23	6.6
S	1.3	1.7	1.8	1.2	1.9	1.5	0.13	0.23	28.6
Soil pH	5.7	5.2	4.9	5.2	5.0	5.2	0.16	0.30	13.2

For explanation of soil bedrock abbreviations, see Table 1.
*Standard errors of the differences between means (s.e.d.m.) are given for the smallest and largest classes along with Wald (W) statistics to indicate the significance of class differences; with $W > 11.07$, p is < 0.05; $W > 15.09$, $p < 0.01$; $W > 20.52$, $p < 0.001$.
† For Se units are $\mu g\,kg^{-1}$ DM and derived means are given with s.e.d.m. for $\log_{10}$ transformed data.

Table 5. *Effects of botanical composition on the mineral concentrations in dry season herbage (mean soil pH at sites occupied by the species also given)*

	Pp	Ss	Cg	Pc	mng	s.e.d.m.		
	n=39	9	48	9	14	min	max	W*
			($mg\,kg^{-1}$ DM)					
Mo	1.1	0.9	0.9	1.7	1.4	0.16	0.34	15.1
Fe	179	264	311	416	406	34.3	73.0	37.8
Mn	181	304	239	152	205	17.8	38.0	32.6
Zn	22.8	17.7	19.5	30.1	27.7	2.5	5.3	29.5
Se†	65	47	74	74	80	0.058	0.124	7.0
Cu	4.2	3.9	3.7	6.0	4.9	0.24	0.50	39.4
Co	0.19	0.23	0.21	0.35	0.29	0.025	0.055	19.2
			($g\,kg^{-1}$ DM)					
Ca	1.3	1.1	1.7	1.4	1.4	0.10	0.21	26.2
P	1.5	1.3	1.4	1.9	1.7	0.14	0.30	5.8
Mg	1.7	1.4	1.5	1.9	1.7	0.10	0.21	11.4
S	1.1	1.4	2.0	1.7	1.5	0.10	0.21	82.6
Soil pH	5.2	5.1	5.1	5.4	5.2	0.07	0.16	9.5

See Table 1 for explanation of herbage species abbreviations.
* Levels of significance are the same as for Table 4.
† Units for Se in $\mu g\,kg^{-1}$ DM with s.e.d.m. for $\log_{10}$ transformed data.

Variation in pasture mineral concentrations

Soil bedrock had little influence on herbage composition. Of the four macro-elements, only S was affected by geology, low values being found above TV and MG bedrock (Table 4). Of the seven trace elements analysed, geology influenced only one (Cu) (Table 4); low values were again found above TV and MG but Cu availability to grazing ruminants would be relatively high because of the associated low S values. By contrast, P was the only macro-mineral not

Table 6. *Effects of soil bedrock on the extractable trace mineral composition of soils (mg kg^{-1} DM) and soil pH at 135 sites in western Kenya*

	TV	IG	MS	MG	SSG	AD	s.e.d.m.		W	p
	n=20	17	10	60	9	19	min (60)	max (9)		
pH	5.67	5.24	4.95	5.23	4.96	5.17	0.163	0.303	13.2	*
Mo	2.26	1.14	1.26	1.74	1.64	1.63	0.250	0.464	15.6	**
Fe	289	616	554	559	341	509	97.5	181.5	11.3	*
Mn	1408	629	770	1014	1194	1043	129	239	27.1	***
Zn	4.23	1.75	1.81	2.70	3.29	2.09	0.058‡	0.158‡	18.6	**
Se†	298	269	333	305	297	301	49.8	92.5	0.9	ns
Cu	4.65	0.68	1.17	1.56	5.61	1.33	0.120‡	0.222‡	45.6	***
Co	1.47	0.70	0.79	1.02	1.86	1.17	0.189	0.349	22.5	***

See Table 1 for explanation of soil bedrock abbreviations.
W, Wald statistic significance levels were: $p < 0.05$, *; $p < 0.01$, **; $p < 0.001$, ***; $p > 0.05$, ns.
†Total Se in µg kg^{-1} DM.
‡Log$_{10}$ transformation needed: back transformed means given but standard errors are for the transformed values.

affected by species, Setaria (Ss) being low and Kikuyu (Pc) usually high in macro minerals (Table 5). Similarly, only Se of the micro elements was unaffected by species, Kikuyu (Pc) being rich in all but Mn (Table 5).

Variation in soil mineral composition

Soil bedrock influenced soil composition particularly extractable Cu, Mn and Co (Table 6) but simple correlations between soil and herbage usually accounted for less than 10% of the variation in pasture composition (maximum r values 0.48 for P; 0.35 for Co). When corrections for species, altitude and pH effects were made in the augmented model, the correlation coefficients for P, Cu and Co showed little or no increase (for example, the corrected r value for P was 0.53). The influence of species on herbage Co decreased if herbage Fe was used as a covariate, the W value decreasing from 19.2 to 10.5. Since soil is richer in Co and Fe than the herbage, this suggests that contamination by soil Co varied between species of different growth habit (at one extreme, Napier is a tall, erect forage species while at the other, Kikuyu has a short, prostrate habit ideal for grazing). Soil pH was generally low and its use as a covariate did little to improve soil/plant relationships.

Herbage Cu increased and Se decreased in curvilinear relationships with altitude (Table 7) but the maximal effect of altitude over the range encountered remained small.

Table 7. *Regression coefficients for the effects of altitude (A) on herbage Cu and Se ($mg\,kg^{-1}$ DM) (where A = metres above sea level and A^2 is the variate)*

	Coefficient	s.e.	W*
Herbage Cu ($\times 10^{-6}$)	1.56	0.600	18.5
$\log_{10}$ herbage Se	−0.375	0.1689	8.5

s.e., standard error.
* W, Wald statistic for the continuous variate altitude; when W is > 4.1, p is < 0.05.
Effects of species, bedrock and soil pH were taken into account as covariates.

Discussion

The results indicate that there are prospects of multiple mineral deficiencies in the pastures of western Kenya but geochemical variables had surprisingly little influence on the macro- or micro-mineral composition of the dry season forage. To some extent the results may be peculiar to the region which is dominated by one particular bedrock (MG) and consists of heavily-weathered, acid soils in which the influence of geological origin may have been dissipated. However, there were marked differences in extractable mineral concentrations between soils of different geological origin and their influences were either masked by the overriding influence of plant uptake or were artefacts of the extraction methods. Removal of the larger species effects on herbage composition did little to improve soil/plant relationships. Although evidence of species variation in soil contamination of the herbage was obtained, the correlation between herbage and soil Co remained weak after using both species and herbage Fe as covariates for Co. Three particular aspects of the prediction of mineral deficiencies in livestock from geochemical or plant data merit further discussion.

Deficiencies of soil extraction methods

The double-acid method chosen for soil Cu (and P) is similar to that used by McDowell *et al.* (1982) and Fig. 1 illustrates its inability to predict marginally deficient Cu values in herbage. Since good correlations were found between each of the chosen extraction methods and those rejected for a particular element (Table 2; Jumba 1989), predictions would not have been greatly improved by using different methods. McLaren *et al.* (1984) showed that another extractant, EDTA, was increasingly imprecise when it came to predicting the Cu, Mn and Zn concentrations of single species (lucerne) pastures in New Zealand, the respective correlation coefficients being 0.53, 0.31 and 0.26. For elements such as Cu for which soil organic matter (OM) has a strong influence on availability to plants, the predictive value of the EDTA extraction test can be improved by correcting for soil OM content (Silanpaa 1982). It would, however, be far more sensible to use either a weaker extractant (e.g. dilute acids) which, like the plant, has difficulty in removing organic matter-bound Cu from the soil or a strong acid mixture to give total extraction, avoiding unwarranted assumptions about availability.

The poor predictive value of soil extraction methods for minerals other than Cu in soils from western Kenya agrees with the findings of a global study involving maize and wheat (Silanpaa 1982). If soil tests are of as little use in predicting mineral concentrations in wet season

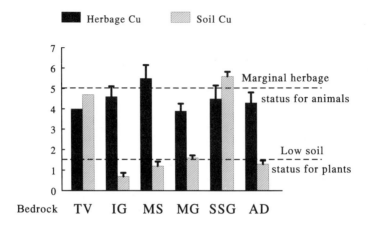

Fig. 1. A comparison of extractable copper concentrations in soils, total copper concentrations in herbage and respective standards of adequacy or plants (COSAC 1982) and livestock (ARC 1980; low molybdenum diets). See footnote to Table 1 for explanation of bedrock abbreviations: values for TV are derived means.

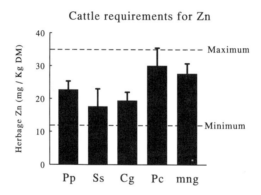

Fig. 2. A comparison of species differences in herbage Zn set against the range of tabulated Zn requirements given by ARC (1980). The interpretation of species effects depends greatly on the class of livestock which vary widely in their estimated requirement for Zn. See Table 1 for explanation of herbage species abbreviations.

forage as they are in the dry season, then herbage analyses must be relied upon for assessing soil mineral availability to herbage species in the region.

Species differences and geochemical reconnaissance

The usefulness of geochemical reconnaissance data for predicting the mineral status of grazing livestock has yet to be established (Appleton 1992). In view of the large effects of species on the herbage concentrations of most of the elements studied in western Kenya, geochemical reconnaissance data should be evaluated either by sampling the commonest herbage species in a region or by using the statistical approaches presented here to correct for species bias where it is impractical to assess mineral status in livestock directly. The robustness of the REML technique was confirmed by omitting poorly represented subsets from the statistical analysis and showing that there were minimal changes in predicted mean values or W statistics for any element in any species.

Species differences and animal requirements

Predictions of mineral deficiences in animals are complicated by variations in the seasonal contribution different species make to total dietary intake and by seasonal fluctuations in animal requirement (Table 3). The nutritive value of dry season forage is so low that it would not be expected to sustain high levels of production without supplementation. Caution must, therefore, be exercized when selecting the standard of nutritional requirement. Figure 2 illustrates the species variation in herbage Zn concentration in relation to the range of Zn requirements quoted by one authority (ARC 1980). The lower limit of the requirement range is for maintenance, which is all that could be expected of dry season Setaria or Rhodes grass: on average, the Zn provided by these species would probably meet the maintenance need. White et al. (1991) showed that where pasture quality and quantity are poor, Zn requirements for growth and wool production in ewes and lambs can be met on pastures containing only 10–20 mg Zn/kg DM. Similarly with P, cattle are able to tolerate diets with concentrations well below all published nutritional standards for long periods (e.g. Call et al. 1986) and they are less tolerant than sheep of P deficiency (Read et al. 1986). The presence of Zn and P concentrations well below standards of nutritional requirement does not guarantee that production would benefit from Zn or P supplementation and the same is true for other minerals.

Conclusions

Macro- and trace mineral deficiencies may occur in grazing livestock during the dry season in western Kenya since pasture mineral concentrations fall well below published standards of requirement for several elements. This conclusion needs to be verified by gathering data on wet season pastures and on the mineral status of animals grazing in the area. Any risk of deficiency is likely to be influenced by botanical composition of the pastures and to a much lesser extent by the topographical and pedological factors considered in this study.

References

AGRICULTURAL RESEARCH COUNCIL. 1980. *The nutrient requirements of ruminant livestock.* Commonwealth Agricultural Bureaux, Farnham Royal, UK.

APPLETON, J. D. 1992. *Review of the use of regional geochemical maps for identifying areas where trace element deficiencies or excesses may affect cattle productivity in tropical countries.* British Geological Survey, Technical Report, **WC/92/24/R**.

CALL, J. W., BUTCHER, J. E., SHUPE, J. L., BLAKE, J. T. & OLSON, A. E. 1986. Dietary phosphorus for beef cows. *American Journal of Veterinary Research*, **47**, 475–481.

COSAC (Council of Scottish Agricultural Colleges). 1982. Trace element deficiency in ruminants. Report of a Scottish Agricultural Colleges and Research Institutes Study Group, Edinburgh.

JUMBA, I. 1989. *Tropical soil-plant interactions in relation to mineral imbalances in grazing livestock.* PhD Thesis, University of Nairobi.

———, SUTTLE, N. F. & WANDIGA, S. O. 1996a. The mineral composition of forages in the Mount Elgon region of Kenya: 1. Macro minerals. *Tropical Agriculture*, in press.

———, ——— & ——— 1996b. The mineral composition of forages in the Mount Elgon region of Kenya: 2. Trace minerals. *Tropical Agriculture*, in press.

———, ———, HUNTER, E. A. & WANDIGA, S. O. 1995a. Effects of soil origin, soil mineral composition and herbage species on the mineral composition of forages in the Mt Elgon region of W. Kenya. 1. Macro elements. *Tropical Grasslands*, **29**, 40–46.

———, ———, ——— & ——— 1995b. Effects of soil origin, soil mineral composition and herbage species on the mineral composition of forages in the Mt Elgon region of W. Kenya. 2. Trace elements. *Tropical Grasslands*, **29**, 47–52.

McDOWELL, L. R., BAUER, B., GALDO, E., ROGER, M., LOOSLI, J. K. & CONRAD, J. H. 1982. Mineral supplementation of beef cattle in the Bolivian tropics. *Journal of Animal Science*, **55**, 964–970.

McLAREN, R. G., SWIFT, R. S. & QUIN, B. F. 1984. EDTA-extractable copper, zinc and manganese in the soils of the Canterbury Plains, New Zealand. *New Zealand Journal of Agricultural Research*, **27**, 207–217.

PATTERSON, H. D. & THOMPSON, R. 1971. Recovery of inter-block information when block sizes are unequal. *Biometrika*, **58**, 545–554.

SILANPAA, M. 1982. *Micronutrients and the nutrient status of soils: a global study.* F.A.O., Rome.

READ, M. P., ENGELS, E. A. N. & SMITH, W. A. 1986. Phosphorus and the grazing ruminant. I. The effect of supplementary P on sheep at Armoedsvlakte. *South African Journal of Animal Science*, **16**, 1–6.

WHITE, C. L., CHANDLER, C. L. & PETER, D. W. 1991. Zinc supplementation of lactating ewes and weaned lambs grazing improved mediterranean pasture. *Australian Journal of Experimental Husbandry*, **31**, 183–189.

The distribution of trace and major elements in Kenyan soil profiles and implications for wildlife nutrition

JOHN MASKALL* & IAIN THORNTON

Environmental Geochemistry Research Group, Centre for Environmental Technology, Royal School of Mines, Imperial College of Science, Technology and Medicine, London SW7 2BP, UK

Abstract: Concentrations of trace and major elements are examined in several soil profiles from national parks and wildlife reserves in Kenya. Broad variations in soil trace element concentrations between locations are largely attributable to differences in parent material and variations in soil pH are related to sodium and calcium concentrations. However, element concentrations and distributions are also influenced by soil forming processes. The process of sodication in alkaline solonetz soils in Lake Nakuru National Park appears to have lowered the concentrations of copper, cobalt and nickel in the surface horizon. At Amboseli National Park, a marked accumulation of molybdenum, sodium, potassium, calcium and magnesium in the surface horizon of an alkaline solonchak is probably due to salinization processes. Apparent mobilization of copper, cobalt and nickel down the profile of a humic nitisol in the Aberdares Salient is associated with eluviation and leaching processes. In two andosols and an ando-humic nitisol, copper, cobalt and nickel tend to accumulate in the surface horizon in association with organic matter. In vertisols from Amboseli National Park and Lewa Downs Wildlife Reserve, the relatively constant trace element concentrations in the A and B horizons are linked to the self-swallowing processes that characterize this soil type. The elevated pH in the solonetz and solonchak soils at Lake Nakuru and Amboseli National Parks results in enhanced uptake of molybdenum in the grass species *Sporobulus spicatus*. At Lake Nakuru National Park, high molybdenum concentrations in this and other plant species are associated with low copper status of impala. The implications of soil geochemistry for the trace element nutrition of wild animals in small conservation areas are discussed.

The trace element status of animals depends strongly on that of their diet (Underwood 1977). For wildlife, studies have demonstrated that the distribution and seasonal movements of some species are related to concentrations of major and trace elements in grasses (McNaughton 1988, 1990; Ben-Shahar & Coe 1992). The enclosure of wildlife within relatively small national parks has focused attention on the capacity of soils to release adequate trace elements into the food chain. The uptake of a trace element into a plant depends on the plant species and on a number of soil properties which determine the bioavailability of the element in the soil. These soil properties include the total concentration of the element, soil pH, moisture content, organic matter content, clay content and redox conditions and are ultimately determined by both the parent material and the soil forming processes.

Previous geochemical studies in national parks and wildlife reserves in Kenya have shown that broad variations of the trace element content of soils were largely attributable to differences in soil parent material (Maskall & Thornton 1991). In a study of Mole National Park in Ghana, Bowell & Ansah (1993) found that bedrock geology was the major control on concentrations of cobalt, copper, manganese and selenium in soils. However, within particular locations, local pedogenic and hydrological factors appeared to influence the total concentration of trace elements in surface soils and their bioavailability to plants (Maskall & Thornton 1991). To investigate the influence of these factors in greater detail, the geochemistry and pedology of a number of soil profiles from wildlife conservation areas in Kenya was examined, and the implications for the trace element nutrition of wild animals evaluated.

Methodology

Soil profiles were examined at naturally occurring or man-made soil exposures or by digging a pit to a maximum depth of 2 m. Individual horizons were characterized for colour using Munsell soil colour charts and for texture and

---

*Present address: School of the Environment, Benedict Building, St George's Way, University of Sunderland, Sunderland SR2 7BW, UK

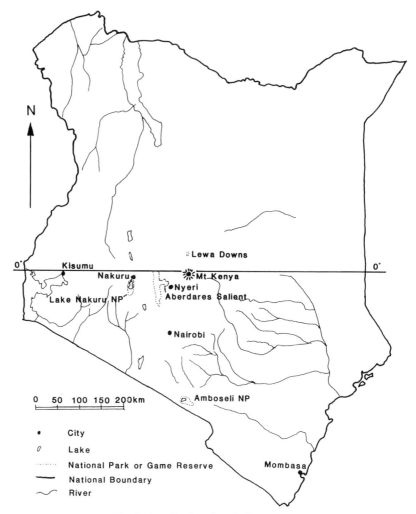

Fig. 1. Sampling locations in Kenya.

wetness by hand (Faniran & Areola 1978). One soil sample was taken from each horizon or sub-horizon which varied in thickness according to the profile characteristics. Each sample comprised between 800 and 900 g of material which was removed from a freshly exposed face of the profile using a trowel. All soil samples were sun-dried, disaggregated, homogenized and passed through a 2 mm sieve in Kenya. For the soil samples, c. 400 g of material was transported to Imperial College, London. On arrival, samples were redried, rehomogenized and c. 35 g of material ground to less than 180 mm. 'Total' concentrations of trace and major elements were determined by digesting a 0.25 g sample of the $<180\,\mu m$ soil fraction with a concentrated nitric/perchloric acid mixture and analysing by Inductively Coupled Plasma Atomic Emission Spectrometry (ICP-AES) (Thompson & Walsh 1983). The total soil molybdenum concentration was determined by ICP-AES after extraction into heptan-2-one giving a detection limit of $0.06\,\mu g\,g^{-1}$ (Thompson & Zao 1985). Soils were analysed for pH by the method of Allen et al. (1974). Organic matter in non-sodic, non-calcic soils was determined using a soil ignition method. The loss in weight of a 5 g soil sample was measured after ignition at a temperature of 450°C (Allen et al. 1974). At Amboseli National Park, surface soils (0–15 cm) and grass samples were also taken as part of the reconnaissance survey of the area. Surface soils were collected using a 2.5 cm diameter soil auger and each sample comprised a composite of nine subsam-

Table 1. *Climate and altitude of sampling sites*

Location		Mean temp. (°C)	Annual rainfall (mm)	Altitude (m)
Lake Nakuru N.P.		18	876	1800
Aberdares Salient	East	18	700	1900
	West	13	1300	2300
Amboseli N.P.		23	350–400	1250
Lewa Downs W.R.		18–20	450–900	1800

Sources: Sombroek *et al.* (1982); Vareschi (1982); Western (1969).

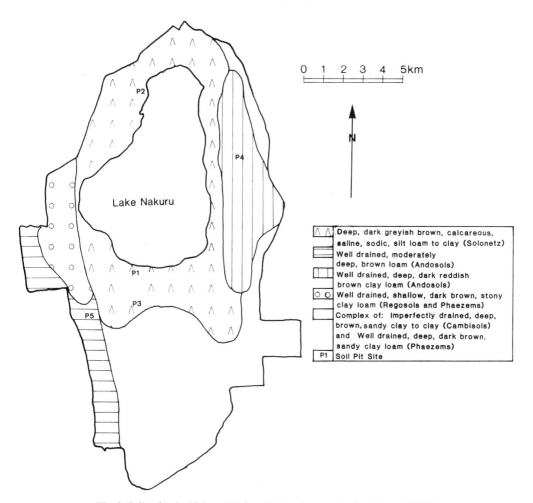

Fig. 2. Soils of Lake Nakuru National Park. Source: Sombroek *et al.* (1982).

ples taken on a 4 × 4 m square. Soils were then treated as described above. Grass samples were collected using scissors within the 4 × 4 m square, sun-dried in Kenya and transported to Imperial College. Plant samples were redried, milled and analysed for trace element content by ICP-AES after digestion with nitric and perchloric acids (Thompson & Walsh 1983). Data were assessed for accuracy and precision using a quality control system integral to the analytical procedure (Ramsey *et al.* 1987). All data achieved the precision and accuracy targets of 10%.

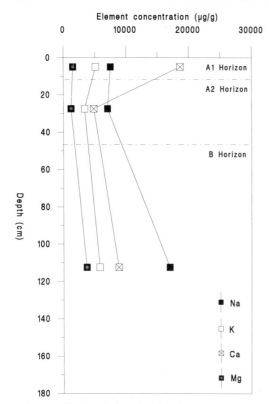

Fig. 3. Profile 1: variation of major element concentrations with depth.

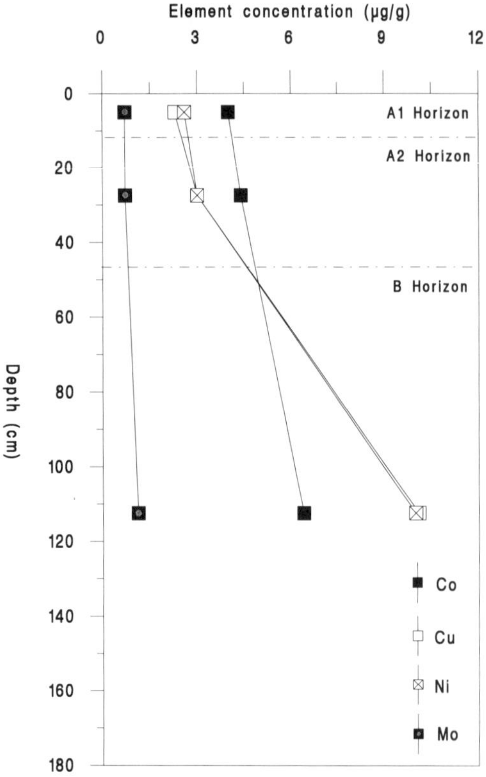

Fig. 4. Profile 1: variation of trace element concentrations with depth.

Site descriptions and results

Four conservation areas were studied, their locations in Kenya are shown in Fig. 1 and climatic data are given in Table 1. Ten soil profiles were examined and the full datasets have been deposited with the Geological Society Library and the British Library Document Supply Centre, Boston Spa, Wetherby LS23 7BQ, UK, as Supplementary Publication No. SUP18110 (13 pp). The Supplementary Publication contains data on pH and the following elements in the ten profiles: sodium, potassium, calcium, magnesium, aluminium, manganese, iron, cobalt, copper, nickel, zinc and molybdenum. In addition, the organic matter content is given for Profiles 5, 6, 9 and 10. In this paper, data for eight elements and soil pH are presented for Profiles 1, 5, 6, 8 and 10 only.

Lake Nakuru National Park

The park covers an area of 180 km^2 in the Rift Valley of which 40 km^2 are occupied by the lake. The floor of the Rift Valley around Nakuru belongs to a Tertiary–Quaternary volcanic suite with associated alkaline sediments (McCall 1967). Lake Nakuru is the low point in a catchment basin of c. 1800 km^2 bounded to the west by the fault scarp of the Mau. To the east of the lake lies Lion Hill, originally a subsidiary volcano of the lower Menengai succession. Lake Nakuru is a hypereutrophic, alkaline–saline closed basin system which deposits trona at times of low water (Vareschi 1982). Soils on the Rift Valley floor are derived from sediments of volcanic and lacustrine origin, giving rise to dark, poorly drained clays being more calcareous, saline and sodic near to the lake (Fig. 2). The Mau escarpment and Lion Hill are mantled by andosols developed over phonolites and phonolitic trachytes (Sombroek et al. 1982).

Five profiles were studied:
Profile 1: solonetz developed on lake sediments located 200 m south of Lake Nakuru on flats with a patchy covering of the grass *Sporobulus spicatus.*
Profile 2: solonetz developed on lake sediments located 500 m west of Lake Nakuru in dense *Acacia xanthophloea* forest.

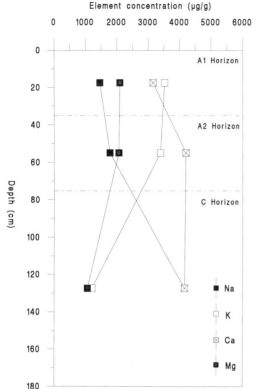

Fig. 5. Profile 5: variation of major element concentrations with depth.

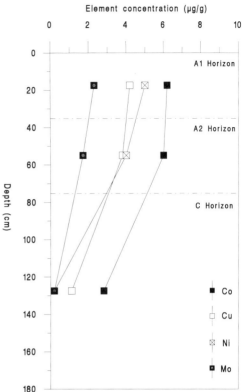

Fig. 6. Profile 5: variation of trace element concentrations with depth.

Profile 3: solonetz developed on lake sediments located 2000 m south of Lake Nakuru in open *Cynodon dactylon/Chloris gayana* grassland.

Profile 4: humic andosol developed on a phonolitic trachyte located on Lion Hill in *Tarchonanthus camphoratus* scrub.

Profile 5: vitric andosol developed on a phonolite located on the Mau Escarpment in mixed *Acacia xanthophloea/Tarchonanthus camphoratus* woodland.

The solonetz soils feature a relatively narrow A horizon with a sandy loam or sandy silt loam texture and a thicker B horizon with a clay-rich texture. The profiles have a high pH (6.6–11.1) and elevated concentrations of sodium (4190–19660 $\mu g\,g^{-1}$) and calcium (1705–51800 $\mu g\,g^{-1}$). Trace element concentrations in the surface horizon of the solonetz soils are low; mean concentrations of cobalt, copper, nickel and molybdenum are 3.9, 3.5, 2.9, and 0.9 $\mu g\,g^{-1}$ respectively. In general, elemental concentrations tend to reach a maximum in the B horizon and this is illustrated for Profile 1 in Figs 3 and 4. Soil pH also increases with depth reaching a maximum in the B horizon (Fig. 13).

The andosols both comprise a dark reddish–brown humic A horizon with a silty clay loam texture which directly overlies the C horizon. The geochemistry of both profiles is similar and the elemental data for Profile 5 are presented here. Sodium and calcium concentrations increase down the profile whereas potassium and magnesium decrease with depth (Fig. 5). The pH of the A_1 horizon is 6.7 which increases to 8.7 in the C horizon (Fig. 13). Trace element concentrations in the surface horizon are 6.2 $\mu g\,g^{-1}$ cobalt, 4.2 $\mu g\,g^{-1}$ copper, 5.0 $\mu g\,g^{-1}$ nickel and 2.3 $\mu g\,g^{-1}$ molybdenum. The trace elements are accumulated in the A horizon relative to the C horizon (Fig. 6) and may be associated with organic matter.

Amboseli National Park

Amboseli is located in the south of Kenya a few kilometres from the Tanzanian border (Fig. 1). The area has a lower rainfall, lower altitude and higher temperatures compared with the other

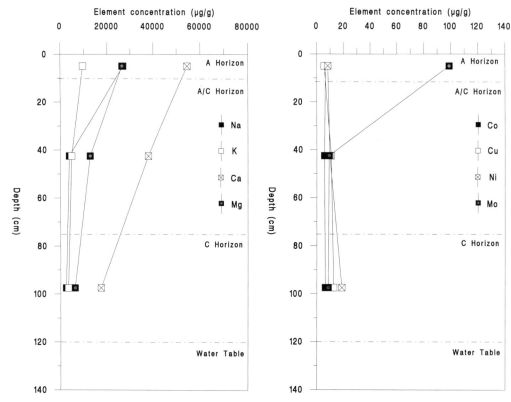

Fig. 7. Profile 6: variation of major element concentrations with depth.

Fig. 8. Profile 6: variation of trace element concentrations with depth.

sites (Table 1). A small part of the park of about 30 km² in area was studied, located in the Ol Tukai/Longinye area. The emergence of the Kilimanjaro volcanics in the late Pliocene produced basaltic lavas which flowed into the south of the study area. Subsequent infilling of the Amboseli basin occurred by deposition of gravels, silts and sands from rivers to the north and west and deposition of ashes, clays and conglomerates from Kilimanjaro (Williams 1972).

Two soil profiles were studied.

Profile 6: orthic solonchak located on the Ol Tukai salt pan developed on a parent material of clays, calcareous silts and silty clays. The pan supports a patchy covering of the grass *Sporobulus spicatus.*

Profile 7: chromic vertisol located to the south of the Engong Narok swamp developed on infill from basaltic rocks. No vegetation cover.

The solonchak comprises a light grey crust underlain by a narrow, sandy, pale brown A horizon, an intermediate A/C horizon and a dark greyish-brown C horizon with a silty sand loam texture. The water table was present at a depth of 1.2 m. The profile has a high pH (9.2–10.7) and high concentrations of sodium (2372–26980 $\mu g\,g^{-1}$) and calcium (17240–54500 $\mu g\,g^{-1}$). Sodium, potassium, calcium and magnesium are accumulated in the A horizon (Fig. 7) which also has the highest pH (Fig. 13). The molybdenum concentration in the A horizon is nearly 100 $\mu g\,g^{-1}$, over ten times that in the C horizon (Fig. 8). In contrast, cobalt, copper and nickel concentrations are low and tend to increase with depth.

Aberdares Salient

The Salient is a wedge-shaped area on the slopes of the Aberdare mountain range composed of the footridges of old basaltic volcanoes. It covers an area of about 70 km², the predominant vegetation being montane forest and associated scrub and clearings. The area receives a relatively high level of rainfall particularly on its

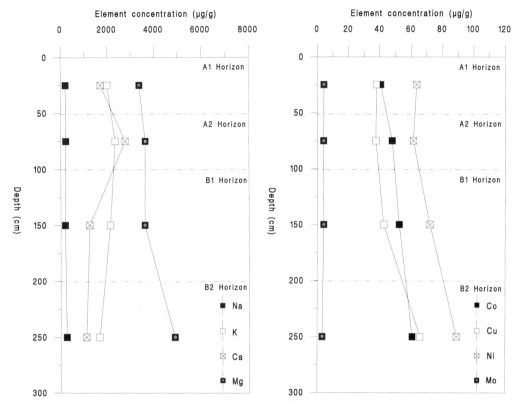

Fig. 9. Profile 8: variation of major element concentrations with depth.

Fig. 10. Profile 8: variation of trace element concentrations with depth.

upper slopes to the west (Table 1). The Salient is entirely underlain by lavas of the Laikipian and Sattima series (Shackleton 1945). In the east of the Salient, soils are classed as humic nitisols and in the west as a complex of ando-humic nitisols and humic andosols (Sombroek et al. 1982). This classification reflects the increase in the proportion of volcanic ash in the soil parent material in the west.

Two profiles were studied.

Profile 8: humic nitisol developed on basalt from the extreme east of the area at an altitude of 1920 m. The exposure is located in a clearing in dry intermediate forest.

Profile 9: ando-humic nitisol developed on basalt/volcanic ash in the northwest of the salient at an altitude of 2280 m. Vegetation is dominated by dense montane forest.

The nitisols are both relatively deep, dark reddish-brown, silty clay loams and the humic nitisol has a thick argillic B horizon. Soil organic matter is generally in the range 10–12% but is markedly higher (21.5%) in the A_1 horizon of the ando-humic nitisol. Base cation concentrations are relatively low (Fig. 9), particularly for sodium (89–261 $\mu g\,g^{-1}$) and calcium (163–2761 $\mu g\,g^{-1}$) and the soils have a relatively low pH (4.2–6.4). The nitisols are relatively rich in trace elements; the A_1 horizon of the humic nitisol contains 40.6, 37.9, 63.4 and 4.2 $\mu g\,g^{-1}$ of cobalt, copper, nickel and molybdenum respectively. In the humic nitisol (Profile 8) cobalt, copper and nickel concentrations increase down the soil profile (Fig. 10) but in the ando-humic nitisol (Profile 9), trace element concentrations are higher in the A horizon.

Lewa Downs Wildlife Reserve

Lewa Downs is a small wildlife reserve located on ranchland in the central highlands at an altitude of about 1800 m (Table 1). The 14 km² area is predominantly underlain by the Osirua basalts which vary from olivine–augite basalts to mugearites and basanites (Hackman 1988). Data are presented here for the predominant soil type:

Profile 10: chromic vertisol developed on basalt located in open grassland.

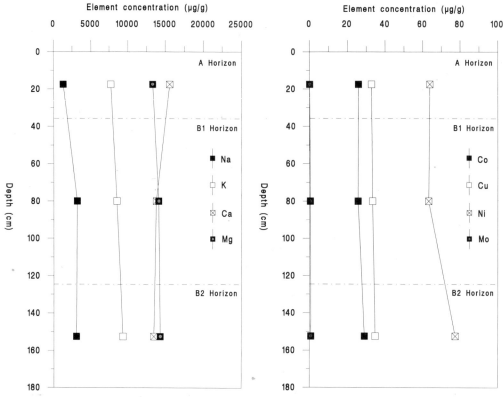

Fig. 11. Profile 10: variation of major element concentrations with depth.

Fig. 12. Profile 10: variation of trace element concentrations with depth.

The soil profile features a dark grey A horizon with a clay texture and significant cracking and crumbly consistency. The B_1 and B_2 horizons are relatively deep and show relatively little differentiation although there is some cracking in the B_1. Soil pH remains at a constant 8.5 (Fig. 13). The soil contains high concentrations of calcium (13430–15560 $\mu g\,g^{-1}$) and magnesium (13280–14290 $\mu g\,g^{-1}$) throughout the profile (Fig. 11). Concentrations of copper, cobalt, zinc and nickel show little significant variation with depth (Fig. 12). However, molybdenum is depleted in the A horizon (0.1 $\mu g\,g^{-1}$) relative to the B_1 horizon (0.4 $\mu g\,g^{-1}$) and the B_2 horizon (0.7 $\mu g\,g^{-1}$).

Discussion

The influence of parent material on element concentrations in soils

Throughout Africa, relatively high concentrations of copper and cobalt have been reported in a variety of soils derived from basic rocks; from basalts and amphibolites in Nigeria (Cottenie et al. 1981); from amphibolites in the Central African Republic (Boulvert 1966); from dolerite and basalt in Chad (Pias 1968); and from basic rocks in Ghana (Burridge & Ahn 1965) and Angola (Fragoso 1959). Low concentrations of copper and cobalt have been reported in soils developed on volcanic ash and lake sediments in Kenya (Chamberlain 1959; Nyandat & Ochieng 1976; Maskall & Thornton 1991). The broad variations in soil trace element concentrations between the study locations are largely attributable to differences in parent material. Thus at Lake Nakuru National Park, the solonetz soils which are developed on old lake sediments and the andosols which are developed on phonolites and phonolitic trachytes contain relatively low concentrations of copper, cobalt and nickel. Conversely, soils developed on basalts at the Aberdares Salient, Amboseli and Lewa Downs have relatively high copper, cobalt and nickel concentrations. It is interesting to note that concentrations of iron correlate significantly with those of copper ($r = 0.59$), cobalt (0.85) and nickel (0.53) for all sites. In some profiles,

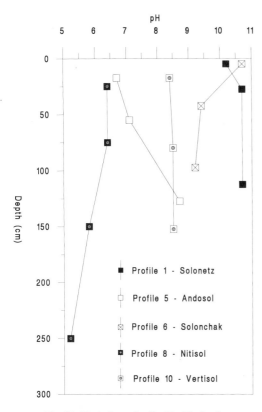

Fig. 13. Variation of soil pH with depth.

the concentrations of major elements in soil are influenced by the chemical composition of the parent material as indicated by the geochemistry of the C horizon. The high calcium concentrations in the solonchak at Amboseli appear to derive from the calcareous nature of the sedimentary parent material. Similarly, the high concentrations of calcium and magnesium in the vertisol at Lewa Downs probably originate from the underlying basaltic rocks. The relatively high sodium concentrations in the andosols at Lake Nakuru National Park reflect the composition of the underlying phonolitic trachytes and phonolites which are both relatively rich in sodium (McCall & Hornung 1972).

The influence of pedogenic processes on element distribution in soil profiles

Previous work has indicated that variations in soil trace element concentrations were influenced by the action of pedogenic processes, particularly in ferrisols and ferralsols (Aubert & Pinta 1977). In the Central African Republic ferrisols derived from charnockite, gneiss or migmatite had either very low (traces–1 $\mu g\,g^{-1}$), average (10–30 $\mu g\,g^{-1}$) or very high (100–200 $\mu g\,g^{-1}$) total copper concentrations, depending on the degree of rock weathering and soil leaching (Boulvert 1966). For ferralsols derived from basalts in Polynesia, the cobalt concentration varied from 5–25 $\mu g\,g^{-1}$ depending on the state of soil degradation (Tercenier 1963). Nalovic (1969) reported that in Madagascar, the range of soil cobalt concentrations expected from soil parent materials appeared reversed and attributed this to soil degradation factors; ferralsolic soils developed on basalts had lower concentrations of cobalt (trace–3 $\mu g\,g^{-1}$) than those developed on granite or limestone (15–20 $\mu g\,g^{-1}$). In this study, there is evidence that pedogenic factors can affect the trace element distribution in a range of soil types.

Solonetz soils develop in the presence of sodium cations and carbonate and bicarbonate anions, high concentrations of which have been reported in Lake Nakuru (McCall 1967). In the lake water, sodium accounts for 96% of the total cations whilst carbonate and bicarbonate account for 84% of the total anions (Vareschi 1982). According to Buringh (1979) the presence of highly soluble salts such as $NaHCO_3$ and Na_2CO_3 results in some calcium ions being replaced by sodium ions on the soil exchange complex. This process, termed sodication, allows clay particles and humus to be easily eluviated from the A horizon to the B horizon given sufficient rainfall. Thus, trace elements bonded to clay particles and humus would also be eluviated, or possibly leached, from the A horizon to the B horizon. This process creates a typical solonetz profile with an A horizon of a mineral, coarse-grained nature and a natric B horizon, rich in clay with higher concentrations of trace elements. Profile 1, 200 m from Lake Nakuru, fits this description almost exactly, suggesting that sodication is active in this area. At Profiles 2 and 3, the textural and chemical gradients associated with the solonetz profile are also present but are not as marked as in Profile 1.

In semi-arid areas, rainfall can be insufficient to remove soluble salts from the soil (Bridges 1978). Instead, salts in solution are drawn upwards through the soil profile by capillary action and are deposited at the surface as the water is evaporated. This process of salinization produces solonchak soils such as that examined at Amboseli National Park. In this case, the upward movement of water containing high

Table 2. *Element concentrations ($\mu g\,g^{-1}$) in surface soils (0–15 cm): orthic solonchak, Amboseli National Park*

	Arith. mean	Range	SD	n
Sodium	13897	6010–26120	6535	6
Potassium	13038	7080–22740	5005	6
Calcium	53375	23870–85300	19230	6
Magnesium	27322	14910–38500	7744	6
Molybdenum	25.2	3.0–41.0	13.6	6

concentrations of sodium, potassium, calcium and magnesium has resulted in the accumulation of these elements in the A horizon. In addition, the same process appears to have led to the accumulation in the A horizon of molybdenum, a phenomenon which has not been previously reported. At the prevailing pH of 9.2–10.7 in this soil, the most stable form of molybdenum over a wide range of Eh is the highly soluble molybdate anion (Brookins 1988). The accumulation of sodium, potassium, calcium, magnesium and molybdenum in surface soils may be occurring throughout the Ol Tukai Salt Pan. In the reconnaissance survey of the area, where samples were taken on a 2 km grid, elevated concentrations of these elements were found in the top 0–15 cm of the solonchak soil (Table 2). Relatively high molybdenum concentrations of between 11 and 16 $\mu g\,g^{-1}$ have also been reported in the top 30 cm of a solonchak in Mole National Park, Ghana (Bowell & Ansah 1993).

Nitisols develop in a relatively intense weathering regime and in the Aberdares Salient this is reflected in the soils by the low concentrations of base cations. Similar geochemistry has been recorded for ferralsols (Bowell 1993) although in general these soils develop in hotter, more humid environments than nitisols. In the eastern Salient, the characteristic argillic B horizon in the humic nitisol results from the eluviation of clay down the profile (Sombroek et al. 1982). Trace elements associated with clay particles appear to have been eluviated from the A horizon to the B horizon although leaching may also be responsible. Variations in trace element concentration have been associated with clay content in other tropical soils (Nalovic 1969; Bleeker & Austin, 1970). In the andosols in Lake Nakuru National Park and the andohumic nitisol in the western Aberdares Salient, some trace elements are accumulated in the A horizon. Combining the data for these three profiles, concentrations of cobalt, copper and nickel show a significant relationship to organic matter content ($p < 0.05$). Similar trace element enrichment by organic matter has also been reported in ferralsols in Ghana (Bowell 1993).

Vertisols develop on parent materials rich in calcium and magnesium in hot climates with pronounced seasonal contrasts (Duchaufour 1982). A characteristic homogeneous profile is produced by the 'self-swallowing' process resulting from the alternate shrinking and swelling of montmorillonite clays. In the dry season, surface soil falls down the deep cracks formed by clay shrinkage whilst in the wet season, clay swelling causes the cracks to close and squeezes material back up to the surface. The vertisols at both Amboseli and Lewa Downs have an A horizon with a loose, powdery consistency and have cracks extending to a depth of at least 50 cm. In addition, both profiles have the high calcium and magnesium concentrations associated with montmorillonite clays. Concentrations of cobalt, copper and nickel are relatively constant in the A and B horizons at both sites. This may be related to the homogenization of soils due to the self-swallowing processes active in the profiles.

Soil–plant uptake of trace elements

The transfer of a trace element from soil to plant depends not only on the total amount of the element present but on soil factors which affect its bioavailability and plant factors which determine its rate of uptake. The soil parent material has been found to have some influence on trace element bioavailability such as in the Kenyan Rift Valley where soils derived from volcanic ash, pumice and lake sediments have been found to be low in acetic acid extractable cobalt (Chamberlain 1959) and in EDTA extractable copper (Nyandat & Ochieng 1976; Maskall & Thornton 1992). Soil pH has a strong influence on metal availability and in alkali soils the elevated pH generally results in increased bioavailability of molybdenum and selenium whilst the bio-availability of copper, cobalt and nickel decrease (Adriano 1986). Copper uptake in wheat was found to decrease with increasing pH in several Kenyan soils (Nyandat & Ochieng 1976). An increase in the trace element content of plants in the wet season has been observed for pastures in the Kenya highlands (Howard et al. 1962), for grass species adjacent to Lake Nakuru

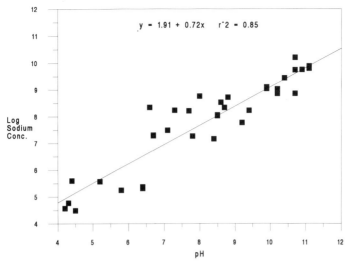

Fig. 14. Variation of soil pH with log sodium concentration.

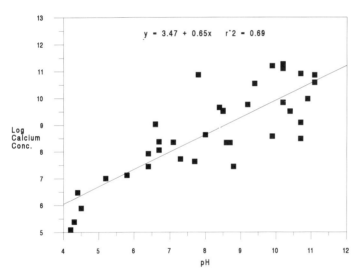

Fig. 15. Variation of soil pH with log calcium concentration.

in the Rift Valley (Maskall 1991) and for several grass and browse species in Mole National Park in Ghana (Bowell & Ansah 1993). Trace element concentrations in tropical pastures can fall as the plant matures (Gomide et al. 1969) and during periods of rapid growth (Fleming 1973). Significant differences in trace element content can occur between plant species in tropical areas, particularly between grasses and browse plants (Tartour 1966; Reid et al. 1979; Ben-Shahar & Coe 1992).

In the soils under study the bioavailability of trace elements appears to be strongly influenced by the soil pH which varies considerably and is related to the concentrations of sodium and calcium (Figs 14, 15). At Lake Nakuru National Park, the elevated pH of soils has been linked with the high molybdenum content of several plant species (Maskall & Thornton 1992). Particularly high concentrations of molybdenum up to a maximum of 69 $\mu g\,g^{-1}$ were found in the alkali tolerant grass species *Sporobulus spicatus* which grows on the solonetz soils adjacent to the lake (Table 3). Grasses sampled in the park tended to contain higher concentrations of copper and cobalt and lower concentrations of

Table 3. Trace element concentrations ($\mu g\, g^{-1}$ D.M.) in plants, Lake Nakuru National Park

		Mean[1]	n[1]	Mean[2]	n[2]
Cynodon dactylon	Cu	11.1	89	3.5	33
(Star Grass)	Co	0.4	89	0.2	33
	Mo	2.2	89	3.4	33
	Se	0.2	80	–	–
Themeda triandra	Cu	18.5	6	7.6	8
(Red Oat Grass)	Co	0.5	6	0.2	8
	Mo	2.4	6	2.6	8
	Se	0.1	6	–	–
Sporobolus spicatus	Cu	11.7	4	5.4	18
(Soda Grass)	Co	0.7	4	0.2	18
	Mo	8.9	4	18.1	18
	Se	0.3	4	–	–
Acacia xanthophloea	Cu	2.6	50	1.9	9
(Yellowthorn Tree)	Co	0.1	50	0.2	9
	Mo	2.5	50	1.7	9
	Se	0.6	40	–	–
Solanum incanum	Cu	5.7	24	5.1	23
(Sodom Apple Bush)	Co	0.3	24	0.3	23
	Mo	3.2	24	2.3	23
	Se	0.4	24	–	–

[1] Reconnaissance survey, Maskall & Thornton (1991).
[2] Biogeochemical zones survey, Maskall & Thornton (1992).

Table 4. Trace element concentrations ($\mu g\, g^{-1}$ D.M.) in *Sporobulus spicatus*, Amboseli National Park

	Arith Mean	Range	SD	n
Cobalt	0.4	0.3–0.8	0.2	5
Copper	3.6	1.7–5.4	1.7	5
Molybdenum	86.2	19.4–152.0	50.9	5

selenium than did the browse plants (Table 3). High molybdenum levels averaging over 80 $\mu g\, g^{-1}$ were also found in *Sporobolus spicatus* on the alkaline solonchak soils at Amboseli National Park (Table 4). This was attributed to a combination of enhanced molybdenum uptake into the grass and the presence of windblown soil particles on the leaf surfaces of the plant.

Implications for wildlife nutrition

McNaughton & Georgiadis (1986) proposed that many African herbivores were existing in a vague, qualitative state of undernutrition and that this had a strong influence on their foraging behaviour and dynamics. A subsequent study showed that the spatial distribution of animals in the Serengeti National Park in Tanzania was related to the mineral content of forages and that magnesium, sodium and phosphorus appeared particularly important (McNaughton 1988). A further study found that the seasonal movements of migratory grazers in the Serengeti were also related to grass mineral content (McNaughton 1990). In this case, the important elements identified were calcium, copper, nitrogen, sodium, zinc and also, for lactating females and growing young, magnesium and phosphorus. In South Africa, Ben-Shahar & Coe (1992) showed that the movements of migratory grazers were related to monthly variations in the nitrogen and phosphorus content of grasses. Thus the enclosure of wildlife within relatively small national parks may restrict their opportunity through migration to acquire adequate major and trace elements.

Soils in the area of the Kenyan Rift Valley around Nakuru have long been associated with mineral problems including copper deficiency in wheat (Pinkerton 1967) and copper and cobalt deficiencies in cattle (Hudson 1944; Howard 1970). Lake Nakuru National Park has been

Table 5. *Copper concentrations in animals ($\mu mol\, L^{-1}$ blood plasma)*

		Mean	Range	n
Impala				
Lake Nakuru N.P.	1986	11.0	4.0–18.9	25 (D)
	1987	14.2	4.8–29.7	7 (D)
All Kenya		19.8		81 (S)
		20.3		21 (D)
Waterbuck				
Lake Nakuru N.P.	1987	13.4	11.2–17.0	3 (D)
Black Rhino				
Solio W.R.	1987	16.9	3.8–24.6	14 (D)
Zimbabwe		25.6	19.2–36.4	20 (D)
White Rhino				81 (S)
Solio W.R.	1987	20.8	18.6–23.1	21 (D)

D, drug immobilized; S, shot.
From Maskall & Thornton (1991).

associated with suspected copper deficiencies in waterbuck (*Kobus defassa*) and impala (*Aepyceros melampus*) since the early 1980s. A low copper status of impala in the Park was reported by Maskall & Thornton (1991) based on concentrations of copper in blood plasma (Table 5). This was attributed to a high dietary intake of molybdenum which can interfere with the utilization of copper in ruminant animals (Underwood 1977). The mean molybdenum contents of plants in Lake Nakuru National Park (Table 3) are generally in excess of the $2\,\mu g\,g^{-1}$ level considered by Thornton (1977) to be sufficient to induce copper deficiency in domestic ruminants. In summary, the low copper status of impala at Lake Nakuru National Park is related to the low copper status of soils combined with an elevated soil pH, particularly in the solonetz soil, which results in enhanced molybdenum uptake into plants.

Molybdenum-induced copper deficiency has been reported in other wild animals including Grant's Gazelle (*Gazelle granti*) from the Kenyan Rift Valley; in this case the plant molybdenum content ranged from 0.5–5.6 $\mu g\,g^{-1}$ (Hedger et al. 1964). Plant molybdenum concentrations of less than $2\,\mu g\,g^{-1}$ were considered contributory to copper deficiency in moose (*Alces alces gigas*) in Alaska (Kubota 1974; Flynn et al. 1977a,b). In San Diego Wild Animal Park, USA, hypocuprosis in several exotic species was associated with alfalfa with a molybdenum content of 11–16 $\mu g\,g^{-1}$ (Nelson 1981). On the basis of these data, molybdenum concentrations in *Sporobulus spicatus* at Amboseli National Park, which have an average of 86 $\mu g\,g^{-1}$ (Table 4), are probably capable of inducing copper deficiency in certain wild ruminant species.

The high molybdenum content of solonchak soils at Amboseli could also affect animals which directly ingest earth at salt licks. In some cases, wildlife species preferentially ingest sodium-rich soils perhaps as a response to sodium deficiency (Fraser et al. 1980). Mule deer in Colorado were estimated to be ingesting between 8 and 30 g of soil per day from deliberate consumption of earth (Arthur & Alldredge 1979). In domestic ruminants soil ingestion has been shown to increase cobalt and selenium status (Healy 1973; MacPherson et al. 1978) although copper absorption and utilization can be inhibited (Suttle et al. 1984). However, little is known at present of the effects of earth eating behaviour in wild animals in terms of nutritional benefits or otherwise.

Conclusions

Broad variations in the concentrations of some trace and major elements in soils between the wildlife conservation areas studied are attributable to differences in soil parent material. However, pedogenic processes influence the distribution in soil profiles of trace and major elements, of which the latter can affect soil pH and trace element uptake into plants and animals. At Lake Nakuru National Park, the process of sodication at the lake margins has produced an alkaline solonetz soil which has particularly low concentrations of several nutritionally essential trace elements in the surface horizon. Salinization processes in an alkaline solonchak at Amboseli National Park have contributed to a marked accumulation of molybdenum, sodium, potassium, calcium and magnesium in the surface horizon. At the Aberdares Salient, apparent mobilization of trace elements in a humic nitisol is associated with eluviation and leaching processes. Trace elements are accumulated in the A horizon in two andosols and an ando-humic nitisol possibly in association with organic matter. In vertisols at Lewa Downs Wildlife Reserve and Amboseli National Park, the relatively constant concentrations of trace elements throughout the A and B horizons may be linked to homogenization of the profile by self-swallowing processes.

The conservation of wild animals and the associated tourism industry represent a significant source of income for many countries in the developing world and particularly in Africa. There is increasing evidence to suggest that the seasonal movements of grazing wildlife in Africa are related to changes in mineral status of forage species. The enclosure of wild animals within relatively small, enclosed National Parks may restrict their opportunities through migration to acquire an adequate intake of major and trace nutrients. In such cases, the health of the animal population, which may include rare or endangered species, depends on the ability of the conservation area to supply sufficient minerals. This in turn is influenced by the concentrations and bioavailabilities of major and trace elements in soils and the local vegetation and climate. The pressure on land resources for agriculture and settlement in developing countries increases the likelihood that conservation areas will be located on land of marginal quality. This study has shown that the presence of sodic and saline soils of low agricultural quality in conservation areas may have deleterious effects on the health of particular species. At Lake Nakuru National Park, the presence of solonetz soils of low trace element status and high pH lead to an elevated molybdenum content of plants and a low copper status in impala due to excess dietary molybdenum. At Amboseli National Park, the high pH and molybdenum content of a solonchak have resulted in elevated molybdenum concentrations in plants which are capable of inducing copper deficiency in grazing ruminants. The incidence of molybdenum induced copper deficiency in livestock can be alleviated through supplementation with mineral mixes. Additional major and trace elements could be provided to wild animals in areas of low mineral status by addition of mineral supplements to salt-lick soils.

The authors are grateful to the People's Trust for Endangered Species for funding the work. Transport in Kenya was provided by the Rhino Rescue Trust, the African Wildlife Foundation and Toyota Kenya. Many thanks go to the Kenya Wildlife Service for their co-operation and support. We are grateful to the staff of the Geology Department at Imperial College for help with elemental analysis. Thanks are also due to Dr Don Appleton of the British Geological Survey for editorial comments.

References

ADRIANO, D. C. 1986. *Trace Elements in the Terrestrial Environment.* Springer, New York.

ALLEN, S. E., GRIMSHAW, H. M., PARKINSON, J. A. & QUARMBY, C. 1974. *In:* ALLEN, S. E. (ed.) *Chemical Analysis of Ecological Materials.* Blackwell, Oxford.

ARTHUR, W. J. & ALLDREDGE, A. W. 1979. Soil ingestion by mule deer in Northcentral Colorado. *Journal of Range Management,* **32,** 67–71.

AUBERT, H. & PINTA, M. 1977. *Trace Elements in Soils.* Developments in Soil Science, 7, Elsevier, Amsterdam.

BEN-SHAHAR, R. & COE, M. J. 1992. The relationships between soil factors, grass nutrients and the foraging behaviour of wildebeest and zebra. *Oecologica,* **90,** 422–428.

BLEEKER, P. & AUSTIN, M. P. 1970. Relationship between trace element contents and other soil variables in some Papua-New Guinea soils as shown by regression analysis. *Australian Journal of Soil Research,* **8,** 133–143.

BOULVERT, Y. 1966. *Sols de Republique Centrafricaine.* Unpublished ORSTOM Report. Cited in Aubert & Pinta (1977).

BOWELL, R. J. 1993. Mineralogy and geochemistry of tropical rain forest soils: Ashanti, Ghana. *Chemical Geology,* **106,** 345–358.

—— & ANSAH, R. K. 1993. Trace element budget in an African savannah ecosystem. *Biogeochemistry,* **20,** 103–126.

BRIDGES, E. M. 1978. *World Soils.* 2nd edn. Cambridge University Press.

BROOKINS D. G. 1988. *Eh-pH Diagrams for Geochemistry.* Springer, New York.

BURINGH, P. 1979. *Introduction to the Study of Soils in Tropical and Subtropical Regions.* 3rd edn. Centre for Agricultural Publishing and Documentation, Wageningen, The Netherlands.

BURRIDGE, J. C. & AHN P. M. 1965. A spectrographic survey of representative Ghanan forest soils. *Journal of Soil Science,* **16,** 296–309.

CHAMBERLAIN, G. T. 1959. Trace elements in some East African soils and plants. 1. Cobalt, beryllium, lead, nickel and zinc. *East African Agricultural Journal,* **25,** 121–125.

COTTENIE, A., KANG, B. T., KIEKENS, L. & SAJJANPONGSE, A. 1981. Micronutrient status. *In:* GREENLAND, D. J. (ed.) *Characterisation of soils in relation to their classification and management.* Oxford University Press, 149–163.

Duchaufour, P. 1982. *Pedology*. Allen and Unwin, London.

Faniran, A. & Areola, O. 1978. *Essentials of Soil Study*. Heineman, London.

Fleming, G. A. 1973. Mineral composition of herbage. *In:* Butler, G. W. & Bailey, R. W. (eds) *Chemistry and Biochemistry of Herbage Vol. 1*. Academic, London, 529–566.

Flynn, A., Franzman, A. W. & Arenson, P. D. 1977a. Molybdenum–sulphur interactions in the utilisation of marginal dietary copper in Alaskan Moose. *In:* Chappell, W. R. & Petersen, K. K. (eds) *Molybdenum in the Environment*, Vol. 1, Marcel Dekker, New York, 116–124.

——, ——, —— & Oldemeyer, J. L. 1977b. Indications of copper deficiency in a sub-population of Alaskan moose. *Journal of Nutrition*, **107**, 1182–1189.

Fragoso, M. A. C. 1959. Microelements em solos de Angola. *Mem. Junta Invest. Ultram. Lisbon*, Second Series, No. 11.

Fraser, D., Reardon, E., Dieken, F. & Loescher, B. 1980. Sampling problems and interpretation of mineral springs used by wildlife. *Journal of Wildlife Management*, **44**, 623–631.

Gomide, J. A., Noller, C. H., Mott, G. O., Conrad, J. H. & Hill, D. L. 1969. Mineral composition of six tropical grasses as influenced by plant age and nitrogen fertilisation. *Agronomy Journal*, **61**, 120–123.

Hackman, B. D. 1988. *Geology of the Baringo–Laikipia Area*. Report No. **104**. Ministry of Environment and Natural Resources, Republic of Kenya.

Healy, W. B. 1973. Nutritional aspects of soil ingestion by grazing animals. *In:* Butler, G. W. & Bailey, R. W. (eds) *Chemistry and Biochemistry of Herbage Vol. 1*. Academic, London, 567–588.

Hedger, R. S., Howard, D. A. & Burdin, M. L. 1964. The occurrence in goats and sheep of a disease closely similar to swayback. *Veterinary Record*, **76**, 493–497.

Howard, D. A. 1970. The effects of copper and cobalt treatment on the weight gains and blood constituents of cattle in Kenya. *Veterinary Record*, **87**, 771–774.

——, Burdin, M. L. & Lampkin, G. H. 1962. Variation in the mineral and crude protein content of pastures at Muguga in the Kenya highlands. *Journal of Agricultural Science*, **59**, 251–256.

Hudson, J. R. 1944. Notes on animal diseases. 13. Deficiency diseases. *East African Agricultural Journal*, **10**, 51–55.

Kubota, J. 1974. Mineral composition of browse plants for moose. *Naturaliste Canadien*, **101**, 291–305.

McCall, G. J. H. 1967. *Geology of the Nakuru–Thompson's Falls–Lake Hannington area*. Report **78**, Geological Survey of Kenya. Government Printer, Nairobi.

—— & Hornung, G. 1972. A geochemical study of Silali volcano, Kenya, with special reference to the origin of the intermediate-acid eruptives of the Central Rift Valley. *In:* Girdler, R. W. (ed.) *East African Rifts*. Technophysics, **15**, 97–113.

McNaughton, S. J. 1988. Mineral nutrition and spatial concentrations of African ungulates. *Nature*, **334**, 343–345.

—— 1990. Mineral nutrition and seasonal movements of African migratory ungulates. *Nature*, **345**, 613–615.

—— & Georgiadis, N. J. 1986. Ecology of African browsing mammals. *Annual Review of Ecological Systems*, **17**, 39–65.

MacPherson, A., Voss, R. C. & Dixon, J. 1978. The response of young grazing calves to supplementation with copper, cobalt and vitamin B_{12}. *In:* Kirchgessner, M. (ed.) *Trace Element Metabolism in Man and Animals 3*. Arbeitskreis fur Tiercernahrungsforschung, Weihenstephan, 490–493.

Maskall, J. E. 1991. *The influence of geochemistry on trace elements in soils and plants in wildlife conservation areas of Kenya*. PhD Thesis, University of London.

—— & Thornton, I. 1991. Trace element geochemistry of soils and plants in Kenyan conservation areas and implications for wildlife nutrition. *Environmental Geochemistry and Health*, **13**, 93–107.

—— & —— 1992. Geochemistry and wildlife nutrition: factors affecting trace element uptake from soils to plants. *Trace Substances in Environmental Health*. **XXV**, 217–232.

Nalovic, L. 1969. Etude spectrographique des elements traces et leur distribution dans quelques types de sols de Madagascar. *Cab ORSTOM Sec. Pedol.*, **VII** (2), 133–181.

Nelson, L. S. 1981. *Secondary hypocuprosis in an exotic animal park*. Internal Report, San Diego Wild Animal Park, California, USA.

Nyandat, N. N. & Ochieng, P. N. 1976. Copper content and availability of arable and range areas of Kenya. *East African Agricultural Journal*, **42**, 1–7.

Pias, J. 1968. *Contribution à l'étude des formations sedimentaires tertaires et quaternaires de la cuvette tchadienne et des sols qui en derivent*. Thesis, Paris. Cited in Aubert & Pinta (1977) above.

Pinkerton, A. 1967. Copper deficiency of wheat in the Rift Valley, Kenya. *Journal of Soil Science*, **18**, 18–26.

Ramsey, M. H., Thompson, M. & Banerjee, E. K. 1987. Realistic assessment of analytical data quality from Inductively Coupled Plasma Atomic Emission Spectrometry. *Analytical Proceedings*, **24**, 260–265.

Reid, R. L., Post, A. J. & Olson, F. J. 1979. *Chemical Composition and Quality of Tropical Forages*. West Virginia University, Bulletin **669T**.

Shackleton, R. M. 1945. *Geology of the Nyeri Area*. Report 12, Geological Survey of Kenya, Government Printer, Nairobi.

Sombroek, W. G., Brown, H. M. H. & van der Pouw, B. J. A. 1982. *Exploratory Soil Map of Kenya*. Exploratory Soil Survey, Report E1, Kenya Soil Survey, Nairobi.

SUTTLE, N. F., ABRAHAMS, P. & THORNTON, I. 1984. The role of a soil × dietary sulphur interaction in the impairment of copper absorption by ingested soil in sheep. *Journal of Agricultural Science*, **103**, 81–86.

TARTOUR, G. 1966. *Study of the role of certain trace elements in relation to the health of livestock in the Sudan.* PhD Thesis, University of London.

TERCENIER, G. 1963. *Sols de Polynesie – Ile Moorea.* Unpublished ORSTOM report. Cited in Aubert & Pinta (1977) above.

THOMPSON, M. & WALSH, J. N. 1983. *A Handbook of Inductively Coupled Plasma Atomic Emission Spectrometry.* Blackie, Glasgow.

—— & ZAO, L. 1985. Rapid determination of molybdenum in soils, sediments and rocks by solvent extraction with Inductively Coupled Plasma Atomic Emission Spectrometry. *Analyst*, **110**, 229–235.

THORNTON, I. 1977. Biogeochemical studies on molybdenum in the United Kingdom. *In:* CHAPPELL, W. R. & PETERSEN, K. K. (eds) *Molybdenum in the Environment, Vol. 2*, Marcel Dekker, New York, 341–369.

UNDERWOOD, E. J. 1977. *Trace Elements in Human and Animal Nutrition* (4th ed.) Academic, New York.

VARESCHI, E. 1982. The ecology of Lake Nakuru (Kenya) 3: Abiotic factors and primary production. *Oecologica*, **55**, 81–101.

WESTERN, D. 1969. *The Structure, Dynamics and Changes of the Amboseli Ecosystem.* PhD Thesis, University of Nairobi, Kenya.

WILLIAMS, L. A. J. 1972. *Geology of the Amboseli Area.* Report **90**, Geological Survey of Kenya, Government Printer, Nairobi.

Formation of cave salts and utilization by elephants in the Mount Elgon region, Kenya

R. J. BOWELL,[1] A. WARREN[2] & I. REDMOND[3]

[1] *Steffan Robertson and Kirsten (UK) Ltd, Cardiff CF1 3BX, UK*
[2] *Department of Zoology, The Natural History Museum, London SW7 5BD, UK*
[3] *The African Ele-fund, PO Box 308, Bristol BS99 7LQ, UK*

Abstract: Herbivores such as the African elephant receive most of their nutrient uptake through digested vegetation or water. When these nutrient sources do not fulfil dietary requirements, eating and digestion of soil and rock can be a common phenomenon. In the Mount Elgon National Park on the Kenya–Uganda border, elephants have taken this activity one step further. Deposits of calcium–sodium-rich alkaline rocks show evidence of quarrying by elephants on the surface, but most of the activity takes place underground in caves. The Na–Ca–Mg-rich rocks are leached by groundwater which reacts with animal excreta in the humid environment of the cave floor to form a series of secondary carbonate, sulphate, halide, nitrate and phosphate minerals by evaporation. Additionally some salts are precipitated on the cave walls by direct evaporation of the cave waters. Inside the caves, elephants tusk and ingest the salt-enriched rock fragments. In an area like Mount Elgon, where supergene processes leach chemical elements essential for dietary requirements from the surface ecosystem, the secondary salts in the caves are potentially an important mineral supplement for wildlife nutrition. Although elephants are the principal exploiters, other wildlife species and humans (for livestock) also utilize the cave salts and surface diggings.

Wildlife species, such as the African elephant (*Loxodonta africana* (Blumenbach 1797)), spend most of their time foraging for food and vital nutrients (Wyatt & Eltringham 1974) with most of their nutritional requirements satisfied through vegetation and water. Little is known, however, of the effects of nutritional imbalances on wildlife or of their ability to respond to such deficiencies (Bell 1982; Maskall & Thornton 1989, 1991; McNaughton 1990; Ben-Sharhar & Coe 1992; Bowell & Ansah 1993, 1994). In a series of studies on the nutritional needs of the African elephant, Weir (1969, 1972, 1973) identified the importance of water soluble sodium in its daily diet and the importance of salt licks. All animals require a balanced intake of mineral salts and if there is a deficit they can go to great lengths to correct it (Denton 1972; Weeks & Kirkpatrick 1976; Dethier 1977). The dietary requirement for Ca, I, Mg, K and Na may be met by forage, water soluble forms or by soil ingestion (FAO 1961; Dougall & Sheldrick 1964; McCullagh 1969a,b; Weir 1972, 1973). The use of mineral salts tends to be seasonal with the most frequent utilization during the dry season when water and forage, the principal sources of macronutrients for elephants, are in short supply (Weir & Davidson 1965). The utilization of mineral salts by wildlife has been documented for other herbivores in Africa (French 1945; Weir 1969; Henshaw & Ayeni 1971; Ansah 1990; McNaughton 1990; Schulkin 1991) and elsewhere (Cowan *et al.* 1949; Dalke *et al.* 1965). As African wildlife is increasingly confined to well fenced reserves, both to protect the animals from poaching and to protect farms and crops, the utilization of salt-licks is more apparent. Consequently animals can no longer range over a wide area and become increasingly reliant on their immediate environment to satisfy their nutritional requirements (Maskall & Thornton 1989, 1991; Bowell & Ansah 1993, 1994).

In the Mount Elgon National Park on the Kenya–Uganda border (Fig. 1) elephants have taken this activity a stage further, exploiting mineral salts deposited in the caves (Redmond 1982, 1991, 1992; Redmond & Shoshani 1987). To reach and exploit this nutrient source, elephants have worn a narrow path to several caves, the largest of which is the Kitum cave. From a 40 m wide, letterbox-shaped entrance in a cliff at the head of a small valley, Kitum cave extends more or less horizontally into the mountain for 160 m (Ollier & Harrop 1958; Sutcliffe 1973; Redmond 1982). Inside the cave widens to more than 100 m, a cul-de-sac with interior walls scalloped by years of mining activity (Fig. 2). As well as tusking directly from the cave walls, elephants have been seen tusking

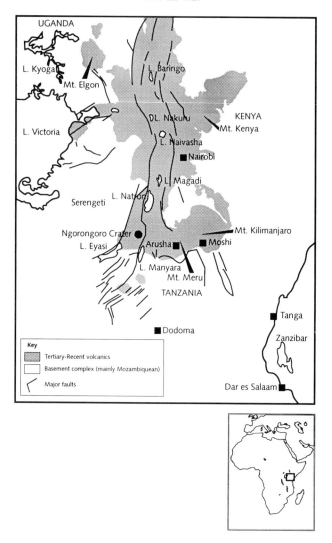

Fig. 1. Simplified topographic and geology map of Kenya and North West Tanzania showing location of the study area

at huge sections of fallen roof and picking up smaller, broken rock fragments from the floor to eat. They have also been observed pulling large fragments of rock from the cave roof, which then lie on the cave floor. Salt crusts form on the fallen rock fragments. Rock types quarried by the elephants are characterized in appearance by the presence of calcite–zeolite veins and vugs. Where no zeolites are present there is little or no evidence of elephant exploitation above or below ground. These rock types are softer than the other lithologies and so are relatively easy to mine.

The temperature of the cave interior remains a constant 13.5°C, so extended visits by the elephants during the night could be to keep warm, given that at this altitude (2 600 m) night temperatures can fall to 8°C. The caves represent a warm safe haven for the elephants with available water as well as salts. Also, the other main cave occupants are bats (chiefly fruit bats, *Rousettus* spp.) which are most active at night so the elephants tend to have the caves to themselves. Other cave visitors such as local tribesmen, bushbuck (*Tragelphus scriptus* (Pallas)), baboons (*Papio cynocephalus* (Linnaeus 1758)) and leopard (*Panthera pardus* (Linnaeus 1758)) only make use of the caves during the day. The knowledge of the caves and their mineral content is passed down from elephant to elephant through the generations, from mother to calf. The extensive poaching of the 1980s reduced elephant numbers from a 1970 estimate of 1200 to as few as 100 today (Redmond & Shoshani 1987; Redmond 1992). Poachers sometimes ambushed elephants in the cave mouth, which led to a reduced use of the cave salts and greater use of the surface soil licks. Intriguingly this period coincided with poorer tusk development in many elephants possibly due to a deficiency in mineral salts in their diet (P. Malisa, pers. comm.). Greater control on poaching in the area has led elephants to return once again to the caves. Use of the cave salts imprints its characteristic mark on salt-digging elephants who show worn down tusks (Fig. 3). Although elephants are the principal exploiters of the salts, other wildlife species, particularly bushbuck, are reported to visit the caves and, in addition, make use of the elephants diggings on the surface (for both salt and water). Local people also use the salts as dietary supplements for livestock.

The purpose of this contribution is to describe the geochemistry of the cave salts and surface salt-licks, to present possible models for their formation and to discuss the possible benefit of utilizing the caves' salts as a source of nutrients by the elephants and other herbivores.

Methodology

Field sampling and analysis

Samples of rocks, soils and waters were collected from the Mount Elgon National Park and from a government borehole 5 km to the east of the park. Soil and salt lick samples in the park were collected by use of a soil auger with at least 1 kg of material collected at each site. Water samples were collected from salt licks and areas away from salt licks by means of a soil auger and extraction using the procedure of Patterson et al. (1978). Groundwater was sampled by means of a hand pump and deep drilling by the Kenya Water for Health Organization, just outside the Mount Elgon National Park. Three 1 l water samples were collected at each site: unfiltered; unfiltered and acidified with 10% (v/v) HNO_3 acid; filtered through a 0.45 μm Durapore membrane filter and acidified with 10% (v/v) HNO_3 acid. Measurements of Eh, pH, electrical conductivity (E.C.), temperature and dissolved oxygen were made at each sample site using Oakton field instruments.

Faecal coliforms were enumerated in the field using a DelAgua water testing kit (DelAgua Ltd., University of Surrey, Guildford, UK). Immediately upon collection, water samples (100 ml, 50 ml or 10 ml) were passed through a 0.45 μm sterile filter membrane. Each membrane was placed onto a pad containing Membrane Lauryl Sulphate Broth and incubated for 14–18 hours at 44°C. Colony forming units of faecal coliform bacteria were then counted.

Mineralogy and lithogeochemistry

Mineral identification was carried out by optical and electron microscopy (Hitachi 2500S) and confirmed by X-ray diffraction (XRD, Philips 1820 with Co-filtered Cu–Kα radiation) using the programme PCIDENTIFY and single grain determinations. Fourier transform infra-red spectroscopy (Perkin Elmer 1720) was also carried out on some of the cave salts. Chemical analysis of the rocks and soils was carried out by inductively coupled plasma atomic emission spectrometry (ICPAES, Fisons ARL3410 Minitorch). Trace elements were extracted from 5 g of material by digestion with 15 ml of nitric acid (70%), 15 ml of hydrofluoric acid (40% v/v) and 15 ml of perchloric acid (70%). Precision of the technique was checked against known standards and found to be 2%.

Hydrogeochemistry

Waters were analyzed by ion-chromatography (Dionex-300) using for anion analysis, an AS4A-AMMS column with Na_2CO_3 (1.8 mM) eluent at a flow rate of 2.5 ml/min and for group I/II cation analyses, a CS 12 column with methane sulphonic acid eluent (20 mM) at a flow rate of 2 ml/min and detection by a pulsed electrochemical detector in conductivity mode. Transition metal analysis was accomplished with a Dionex CS 5 column with pyridine-di-carboxylic acid eluent and 4 (2-pyridylazo) resorcinol post column derivitization and measurement by a variable wavelength detector in the range 520–530 nm. All water samples were also analyzed by ICPAES.

Results

Mineralogy and geochemistry of the Kitum cave volcanics

The Kitum caves occur within the Elgon volcanics which are largely alkaline–mafic lavas and tuffs of foidite, phonolite, tephriphonolite and phonotephrite compositions (Davies 1952;

(a)

(b)

(c)

(d)

Fig. 2. Photographs of elephant and bushbuck utilization of cave salts in the Kitum cave. Photographed by IR.

Fig. 3. Photograph of elephant with one worn tusk, probably due to extensive rock and soil digging (photographed by IR).

Searle 1952; Le Bas 1977). These Miocene to Pliocene volcanics are underlain by metamorphosed Proterozoic granitoids and amphibolites.

In the Kitum Cave, the main rock type is a melilitite lava overlain by melilite melanephelinite lavas and a phonolitic nephelinite tuff (Table 1). The phonolite has the appearance of a friable clay–calcareous agglomerate which is highly porous. Following emplacement of the volcanic sequence, hydrothermal activity led to the precipitation of zeolites and other minerals. The zeolite assemblage in the alkaline rocks comprises natrolite, analcime, laumonite, apophyllite and gmelinite. Calcite is also a common vesicle mineral phase. Natrolite is the most common zeolite with minor harmotome, analcime, thomsonite, laumontite, stilbite, apophyllite and gmelinite. Calcite is the major component of the breccia matrix which also includes apatite, biotite, fluorite, ilmenite, magnetite, pectolite and rutile. Apatite, ilmenite, magnetite and rutile are probably relict magmatic phases while biotite and hornblende are alteration products of magmatic ferromagnesian silicates. The general order of infill in the veins are: Na-zeolite and biotite then Na,Ca-zeolites, then calcite and fluorite. All the rocks in Kitum cave have been subjected to a lesser or greater degree to interaction with groundwater.

The dominant mineralogy of the nephelinite rocks of Mount Elgon is given by Davies (1952) as olivine, clinopyroxene (augite and diopside), nepheline, melilite, ilmenite, magnetite and rutile. For the phonolite rocks the dominant minerals are melilite, nepheline, feldspar, augite, apatite, magnetite and rutile along with silicate glass. These minerals are present in the Kitum rocks along with the products of hydrothermal alteration (zeolite minerals, calcite, biotite, hornblende and quartz) in parts of the phonolite tuff, which has also been variously described as a 'dyke' and 'breccia' (Davies 1952; Sutcliffe 1973; Redmond 1982). In the Kitum rocks other minerals, presumably formed by supergene processes, are also present and have been observed partially to replace the primary minerals. These secondary minerals include the clay minerals kaolinite and smectite (others have also been reported by Davies (1952)), quartz, goethite, allophane, leucoxene, gypsum and

Table 1. *Geochemistry of silicate rocks from the Kitum cave, Mount Elgon*

Component	Phonolitic nephelinite	Melilite melanephelinite	Melilitite
wt% oxide			
SiO_2	43.0	40.3	37.5
TiO_2	2.33	2.59	2.51
Al_2O_3	12.7	10.2	9.23
Fe_2O_3	13.1	14.7	13.1
MnO	0.23	0.28	0.67
MgO	5.02	6.72	10.8
CaO	11.4	14.6	19.0
Na_2O	5.52	3.67	2.16
K_2O	2.52	2.18	1.62
P_2O_5	0.41	0.58	0.51
LOI	3.77	4.23	2.92
ppm element			
S	509	219	244
Sc	29	26	21
V	259	198	232
Cr	61	189	219
Co	80	62	79
Ni	29	79	111
Cu	13	36	44
Zn	60	113	88
Ga	12	16	16
Rb	49	33	38
Sr	650	556	529
Y	33	29	21
Zr	214	226	136
Ba	1120	971	965
Pb	1.4	3.2	2.0
La	89	98	82
Ce	153	165	118

possibly secondary calcite. The major element changes which result from the supergene alteration are loss of Ca and Mg and consequent relative enrichment of Si, Al and Fe.

On the assumption that the elements leached from the rocks are then precipitated from solution at the rock surface in the chamber, the encrustations that the elephants find palatable must be rich in salts of Ca, Na and Mg. In order to study the nature of the salts, samples were collected from a fallen roof block on the floor of the cave and precipitates on the wall of the cave where tusking was regularly observed.

Mineralogy and geochemistry of the cave salts

A greater number of mineral species (Table 2) and a greater volume of salts are formed on the fallen rocks than are precipitated on the cave wall. Only those salts which were observed by more than one technique are shown; several other phases were also observed by bulk X-ray diffraction but await confirmation.

Anhydrite, calcite, epsomite, hexahydrite, mirabilite, natron, polyhalite, sylvite and syngenite are present on the cave walls. Although no reliable bulk sample could be collected which was not contaminated by wallrock phases the major element chemistry, based on the mineralogy, is essentially Ca–Mg–K–Na–Cl–S. These salts are precipitated on the surface of the wallrock minerals.

The salt crust developed on the rock debris on the floor of the cave differs in being considerably thicker (30 cm as opposed to < 1 cm on the wall) and is mineralogically more complex. The major differences are: gypsum is present in the cave floor assemblage but anhydrite is absent; the presence of Fe–Al secondary salts (halotrichite and a phase of the tschermigite–lonecreekite series) and of N-salts (stercorite, mohrlite, nitromagnesite and tschermigite-lonecreekite); also trona is present along with natron. Other possible phases include nitrates, phosphates, other sulphates and a calcium iodide (seen in an infra-red pattern). NH_4 was observed in the

Table 2. *Mineralogy of salt crusts*

Mineral	Wall salt crust	Floor-rock boulder salt crust
Calcite ($CaCO_3$)*	xx	xx
Halite (NaCl)	xx	xx
Gypsum ($CaSO_4.2H_2O$)		xx
Anhydrite ($CaSO_4$)	xx	
Arcanite (K_2SO_4)		xx
Polyhalite ($K_2MgCa_2(SO_4).2H_2O$)	xx	xx
Epsomite ($MgSO_4.7H_2O$)	x	xx
Natron (Na_2CO_3)	xx	xx
Aphthitalite $(K, Na)_3Na(SO_4)_2$		xx
Sal-ammoniac (NH_4Cl)		xx
Magnesite ($MgCO_3$)		x
Mirabilite ($Na_2SO_4.10H_2O$)	xx	xx
Hexahydrite ($MgSO_4.6H_2O$)	x	xx
Halotrichite ($Fe^{2+}Al_2(SO_4)_4.22H_2O$)		x
Sylvite (KCl)	x	x
Mohrlite (($NH_4)_2SO_4.3H_2O$)		x
Stercorite($NaNH_4HPO_4 4H_2O$)		x
Blodite ($Na_2Mg(SO_4)_2.4H_2O$)		x
Syngenite ($K_2Ca(SO_4)_2.H_2O$)	x	x
Nitromagnesite ($Mg(NO_3)_2.6H_2O$)		x
Trona ($Na_3(CO_3)(HCO_3).2H_2O$)		x
Tschermigite–Lonecreekite (($NH_4)(Fe^{3+},Al)(SO_4)_2.12H_2O$)	.	x

* Calcite may be a residual hydrothermal phase or precipitated by evaporation/supergene mineral–water reaction.
xx, common mineral; x, uncommon/trace mineral.

infra-red spectra of some natrolite grains but further quantitative work has so far failed to confirm this. The salt crust can be divided into a 'mixed layer or zone' in which the secondary salts (listed in Table 2) are intermixed with primary minerals and supergene products (such as the zeolites and smectite; Fig. 4 a, b) and an 'outer crust of hydromorphic salts' with very little (< 20%) wallrock material.

Using this division, a number of chemical changes can be observed in the composition of the overall salt crust. Cl^-, I^-, SO_4^{2-} and P, increase in concentration going from unreacted rock to the outer crust of salts (Fig. 5). An increase in Na, K, Ca and Mg is also apparent going from rock to salt crust with a large increase in the mixed layer and, as predicted from the mineralogy, further increases in the salt layer. Conversely, the concentrations of SiO_2, Al, Fe, all decrease substantially as do the trace elements Cr, V, Zr, Cu and Co (Fig. 5).

Mineralogy and geochemistry of surface salt licks

During the late Pliocene to Recent, active chemical weathering in a tropical regime has taken place in the area, leading to extensive leaching of alkalis and other mobile elements from the rocks and formation of thick residual lateritic and volcanic soils (up to 4 m thick). The surface soils consist largely of kaolinite, smectite, mica, goethite, quartz, magnetite, leucoxene and chlorite. The soil geochemistry is essentially SiO_2–Al_2O_3–Fe_2O_3 with minor CaO and TiO_2 (Table 3). At a few surface locations, such as at Saito dam, a surface salt pan is formed with halite, gypsum and natron present as a surface crust. These crusts are comparable in bulk geochemistry with the floor-rock salt crusts in the Kitum cave. The higher P concentration (Table 3) of the salt-licks compared to the cave floor-rock salt crust is probably a reflection of the higher volume of apatite in surface soils compared to the precursor rock and possibly a greater volume of animal excreta as the diversity and biomass of species visiting the surface salt-licks are that much greater than in the Kitum cave. The surface salt-lick sites also have higher concentrations of SiO_2, TiO_2, V, W, Ni and Zr (Table 3) reflecting the presence of soil minerals in the salt pans. It should be noted that the sites of the salt pans are often in more open terrain so utilization of the salts by animals may also be influenced by the location, with a lower chance of ambush in these more open sites.

Table 3. *Geochemistry of surface salt-licks, soils and Kitum cave salt crust, Mount Elgon National Park*

Element	Salt-lick 1		Saito Dam		Soil† near Kitum		Cave Salt‡
	surface	120 cm	surface	100 cm	surface	100 cm	
wt%							
SiO_2	10.6	48.9	22.4	40.8	50.9	49.7	16.9
TiO_2	0.64	2.52	0.96	2.73	1.77	1.80	0.38
Al_2O_3	10.6	18.3	11.8	21.8	21.3	19.8	6.23
Fe_2O_3	5.46	7.92	4.98	8.33	14.5	14.0	4.78
MnO	0.11	0.21	0.13	0.20	0.12	0.19	0.30
MgO	10.3	2.34	6.43	2.23	0.29	0.89	10.2
CaO	11.3	4.54	10.6	5.42	2.71	5.22	7.10
Na_2O	14.9	6.76	12.1	7.94	2.31	3.22	18.3
K_2O	10.6	2.11	9.89	2.14	0.86	0.99	8.23
P_2O_5	4.98	2.34	8.16	2.77	0.31	0.29	1.78
LOI*	20.6	4.03	12.6	5.65	4.98	3.90	25.8
ppm							
SO_4^{2-}	779	428	622	312	408	327	10000
V	258	264	270	279	261	282	78
Cr	67	80	29	52	139	149	60
W	29	33	34	51	25	33	19
Co	49	50	60	72	71	85	48
Ni	105	118	83	91	96	131	65
Cu	44	67	62	72	18	31	59
Zn	35	44	49	64	79	105	37
Rb	48	56	39	49	26	41	39
Sr	960	670	689	669	312	397	689
Zr	124	124	120	123	112	120	98
Ba	989	959	668	690	528	608	722

* LOI, Loss on ignition (measure of volatile content). † Soils collected 150 m west of Kitum cave. ‡ Cave salt from surface of fallen rock in Kitum cave.

Hydrogeochemistry

The surface and groundwater of the Mount Elgon area are essentially alkaline Na–Ca–HCO_3 waters with minor Mg, K, SO_4^{2-}, Cl^- and variable F^- content. In comparison with waters at major salt occurrences elsewhere in East Africa (such as Lake Magadi; Eugster 1970; Jones *et al.* 1977) all the waters, except the Saito dam, are dilute.

The soil porewaters from the National Park tend to be more acidic and less saline than other waters, except at the salt-lick sites (Figs 6, 7). However, the soil porewaters extracted from the borehole at Saboti village 5 km to the east of Mount Elgon are much more saline and more acidic than those in the park and also have much higher total Al and Fe concentrations (Table 4). The salt-lick soil porewaters from Saito dam are more saline and alkaline than any other waters analyzed from the park and will be greatly influenced by the saline chemistry of Saito dam (total dissolved solid, (TDS) content of 3690 mg l^{-1}, Table 4).

Non-saline soil porewaters have a lower Na content than groundwater but similar ranges of Ca, K, Mg, Al and Fe concentrations and only slightly lower SiO_2 content (Table 4). This would suggest that the groundwaters are largely meteoric with little or no geothermal contribution as has been proposed in earlier studies (Davies 1952). The highest Al and Fe concentrations are associated with low pH and high salinity in porewaters extracted from soils of the Saboti borehole. These are total metal concentrations and will include particulate material as well. Both Al and Fe were below the 0.1 mg l^{-1} detection limit in filtered porewater and groundwater extracts. In the salt-lick porewater, Fe and Al concentrations are in the range 3–5 mg l^{-1}, which is higher than non-saline acidic soils from elsewhere in the National Park (Table 4). The high mobility of Fe and Al may be influenced by the high salinity of the waters.

Nitrate and phosphate levels are highest in surface soils and decrease with depth; for example in the park soil porewaters concentrations in the top 20 cm are 1.4 mg l^{-1} for nitrate

(a)

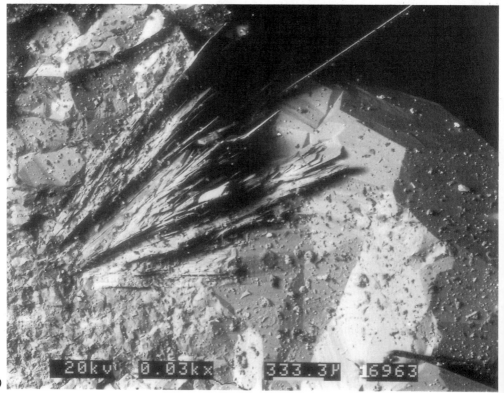

(b)

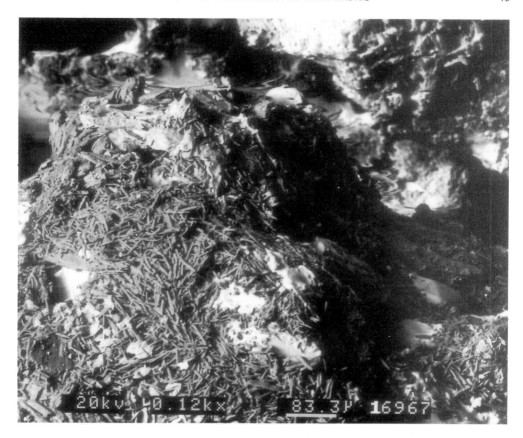

(c)

Fig. 4. Electron micrographs of salt crusts from Kitum caves: (a) analcime, natrolite and mesolite in vug; (b) smectite, gypsum and polyhalite on natrolite and calcite; (c) salt crust surface.

and $0.12\,mg\,l^{-1}$ for phosphate. At a depth of 40–60 cm these have dropped to $0.34\,mg\,l^{-1}$ nitrate and $0.06\,mg\,l^{-1}$ phosphate (Table 4). Despite the high soil P level at the salt-licks (8.16 wt% P_2O_5, Table 3), the porewater concentration is low ($0.29\,mg\,l^{-1}$ PO_4^{3-}) suggesting that much of the P is in a non-labile form, such as apatite. By comparison the cave waters show a much higher level of phosphate in solution (0.37 and $0.99\,mg\,l^{-1}$, Table 4) despite a lower content of P in the cave salts (1.78 wt% P_2O_5, Table 3). This P is largely held in guano and precipitated phosphate evaporites and as such is more readily leached.

The Kitum cave waters are similar to groundwater collected from the Saboti borehole with similar pH and bicarbonate ranges but higher chloride contents (Figs 6, 7). The cave waters also contain higher levels of nitrate, phosphate, sulphate, iodide, Na, K, Ca and Mg, with a similar concentration range for TDS, Al, Fe and SiO_2 but much lower fluoride (Table 4).

Waters were tested for the presence of faecal coliform bacteria, which are indicators of faecal contamination. The presence of high numbers of faecal coliforms (> 250 per 100 ml) in the stream and soil porewater samples is not surprising given that a variety of animals regularly utilize these resources. However, their presence in deep groundwater, albeit at low concentrations, was not expected. These findings suggest that nutrient levels in the groundwater will influence, and to some extent be influenced by, the biota (protists and bacteria) present in the subsurface. However, the very high faecal coliform count of 65 per 100 ml at 45 m depth in the borehole also coincides with increased concentration of chloride, iodide, nitrate, K, Al and Fe (Table 4). These results should be treated with caution as they may be due to contamination of the sample by faecal matter or soil porewaters during collection, rather than to high microbial activity

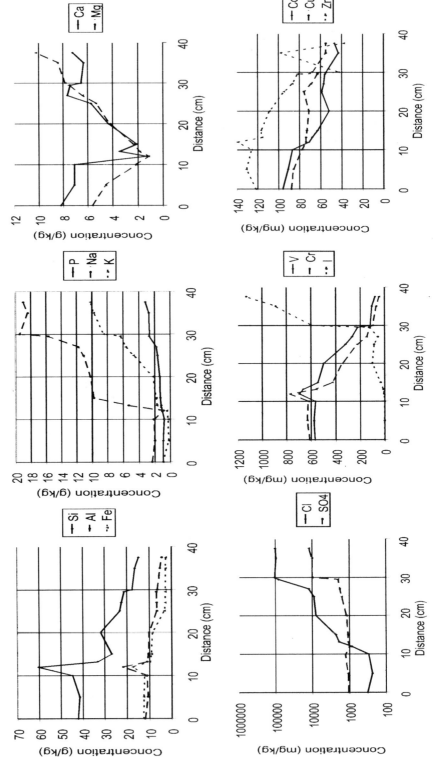

Fig. 5. Geochemical trends of major and minor elements through a salt crust layer: 0–12 cm: unaltered rock; 12–27 cm mixed layer; 27–38 cm outer salt crust.

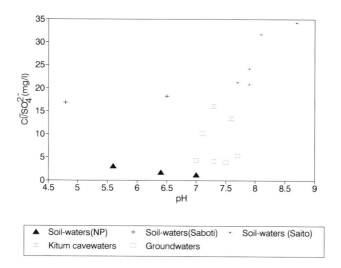

Fig. 6. Cl/SO$_4$ v. pH for all Mount Elgon groundwaters, cave waters and soil porewaters. NP, National Park.

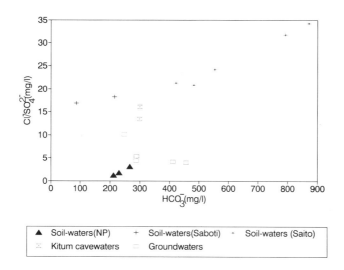

Fig. 7. Cl/SO$_4$ v. HCO$_3^-$ for all Mount Elgon groundwaters, cave waters and soil porewaters. NP, National Park.

at this depth in the aquifer. As would be anticipated, the cave waters have a much higher faecal coliform count than the groundwater (Table 4).

Discussion

The process of salt formation in the caves is likely to be complex and involve evaporation of the cave waters, re-solution of the evaporite salts and percolation into the wallrock, additional leaching of wallrock minerals, interaction with organic matter and further precipitation of evaporite salts. The formation of the salts is likely to occur through multiple cycles of dissolution–evaporation–reprecipitation and will be an important control on water chemistry. The multiple cycles and percolation of salt-rich cave waters into the rocks is reflected in the complex stratigraphy formed in the cave floor

Table 4. *Geochemistry of waters, Mount Elgon area*

	pH	TDS	HCO_3^-	F^-	Cl^-	I^-	SO_4^{2-}	NO_3^-	PO_4^{3-}	Na	Ca	K	Mg	SiO_2	Al	Fe	f.c.
Surface waters																	
Maji Chumbi	6.0	280	164	0.8	58	0.84	12.9	0.39	0.12	69	32	9.8	16.9	15	7.9	12.4	>250
Maji Rongai	7.3	494	421	1.3	34	0.45	8.6	0.27	0.09	110	56.5	7.8	12.8	26	1.8	10.8	22
Saito Dam	8.1	3690	5100	10.8	3650	19.9	171	8.9	0.82	3850	79.3	22.5	11.9	49.5	<0.1	<0.1	180
Soil porewaters – National Park, 150 m west of Kitum cave																	
0–20 cm	5.6	255	267	1.9	34	1.05	10.9	1.4	0.12	89	17.9	9.8	8.6	13.4	1.6	1.7	89
20–40 cm	6.4	179	230	2.2	22.9	0.81	12.8	0.56	0.05	106	26.8	5.9	11.3	15.4	0.9	1.2	63
40–60 cm	7.0	165	212	2.9	14.6	0.79	11.5	0.34	0.11	111	29.6	6.2	10.9	16.7	0.9	0.9	25
Soil porewaters – BH 67/3 Saboti village																	
0–20 cm	4.8	270	88	3.1	225	2.7	13.3	7.7	0.06	75	44	43	8.9	10	12.5	29.8	>250
80–100 cm	6.5	189	215	4.2	190	2.3	10.4	1.5	0.05	69	39	13	11.2	12	7.9	18.7	>250
Soil porewaters – salt-lick at Saito Dam																	
0–20 cm	8.7	850	872	2.59	349	16.9	10.2	3.98	0.29	445	85	38	75	20	5.6	3.3	112
20–40 cm	8.1	779	793	2.62	311	11.4	9.8	1.12	0.21	371	76.5	29.2	67.1	18.5	5.2	2.7	98
40–60 cm	7.9	529	554	2.58	288	8.9	11.9	1.59	0.18	272	66.3	24.3	58.4	32.7	4.3	3.4	54
60–80 cm	7.9	452	482	2.63	254	6.3	12.2	1.63	0.14	254	52.3	22.5	64.2	26.2	3.9	3.5	109
80–100 cm	7.7	387	422	2.8	230	4.5	10.8	1.22	0.16	326	69	21	63	27	5.7	4.1	78
Cave waters – Kitum cave																	
site K 2	7.3	298	300	2.5	138	0.97	15.6	3.65	0.37	265	59	11	36	21	3.7	3.1	165
site K 4	7.6	365	299	1.3	291	1.1	21.8	1.22	0.29	358	68.5	18.9	47.6	37.2	3.5	2.9	190
Groundwaters – BH 67/3 Saboti village																	
25 m depth	7.0	363	288	13.98	46	0.14	10.5	0.49	0.05	163	29.2	4.8	14.7	20.9	1.6	1.9	8
45 m depth	7.1	256	247	10.49	87	0.17	8.6	1.54	0.09	150	24.5	7.6	12.5	18	5	3.3	65
55 m depth	7.3	390	412	19.6	28	0.1	6.7	0.34	0.06	199	42.3	3.65	26.1	22	2.3	2.1	0
70 m depth	7.5	422	456	22.9	24.3	0.1	6.1	0.14	0.05	189	39.6	3.2	19.3	25.2	2.3	2.1	0
80 m depth	7.7	224	289	26.8	29	0.1	5.4	0.11	0.07	211	45	2.9	23	21	4.8	2.2	5

All concentrations in mg l^{-1} except faecal coliforms (f.c.) which are number in 100 ml water.

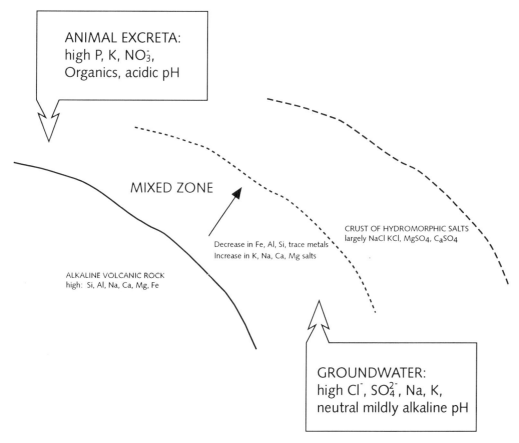

Fig. 8. Schematic diagram of salt crust stratigraphy.

salt crusts (Fig. 8). This stratigraphy of a mixed salt-wallrock minerals zone and an outer hydromorphic salt crust is reflected in the chemical changes observed through the crust with the more soluble salts precipitated in the outer crust (rich in Mg, Ca, K and Na).

Experimental studies of evaporite salt formation have shown that the salts will precipitate out in sequence from the least soluble, such as Ca-carbonates, to the most soluble such as Mg-sulphates, regardless of the total concentration of the brine (Harvie & Weare 1980). The salts developed in the Kitum cave follow this general sequence and show two similar assemblages although these have some major differences as well. The evaporite salts precipitated on rock debris on the floor of the Kitum cave are volumetrically greater and are more varied in mineral species present (21 as opposed to 10 on the cave walls) and chemistry of the salts (the floor salts contain nitrates, phosphates, NH_4^+, Fe and Al evaporite salts) than those on the cave walls. But apart from the presence of gypsum in the floor salts and the presence of anhydrite in the wall evaporite assemblage, most of the other major salts are present in both assemblages (calcite, epsomite, halite, hexahydrite, mirabilite, natron and polyhalite) suggesting a common mechanism of formation. However, the chemical differences between the two assemblages suggest that elements from additional sources are available to form evaporite salts on the rock debris on the floor of the cave. These sources have provided nitrate, phosphate, NH_4^+, Fe, Al and additional sulphate, Na, Mg and K. The additional sulphate, Na, Mg and K may be a product of a greater volume of cave water on the floor of the cave or, alternatively, cave waters may be additionally enriched by mineral–water interaction by percolating through unaltered rocks exposed to supergene leaching by elephant quarrying. Such a process could also enhance Fe and Al, although the concentrations of these elements in the cave waters are no different to borehole groundwater chemistry (Table 4).

The additional source of phosphate, nitrate and NH_4^+ is most likely from groundwater interaction with animal excreta particularly from bat guano, which is known to be rich in P and N (Ukumi 1990), and elephant excreta which is volumetrically more important in the Kitum cave. Despite the high nutrient content of bat guano it is only utilized as a nutrient source by a few species such as waterbuck (*Kobus defassa*, Ruppell) (Ukumi, 1990) because of the low pH (3.8–5) and high content of dissolved organic acids. The importance of animal excreta as an additional source of components for the evaporite salt crusts is reflected in observations by Sutcliffe (1973) who found a much lower volume of salts in caves not frequently visited by elephants in the Mount Elgon area. The common presence of zeolites in rocks which are frequently utilized as a nutrient source in the caves is unclear and possibly only a coincidence as the nutrients are more concentrated and soluble in the evaporites. However, the expandable nature of zeolites and their ability to sorb nutrients (Tsitsishvili *et al.* 1992; Bish & Guthrie 1993) could possibly contribute in the formation of the salt crusts or the use of their lithologies as a target of elephant quarrying. However, it must be remembered that those rock units which host zeolites have been hydrothermally altered and are mechanically softer than unaltered equivalent rock types.

The similarities in major nutrients between the salt-licks and cave salts (Table 3) suggest that no additional nutritional benefit is gained from utilization of the cave salts. Nevertheless from the hydrogeochemistry data it is clear that for some elements, such as P, a greater portion is in a labile form in the cave salts than in the soils at the salt-lick. The superiority of the cave salts may have more to do with location than nutritional benefit over surface salt-licks.

Conclusions

In the Kitum cave, Mount Elgon National Park, a suite of alkaline volcanic rocks is undergoing supergene leaching with the loss of mobile elements such as Na, K, Mg and Ca and consequent relative enrichment of more immobile elements such as Al, Fe and Si. The formation of hydromorphic salt crusts occurs through evaporation of groundwater enriched by mineral–water reactions with wallrock and leaching of organic matter in the caves. Multiple cycles of dissolution–evaporation–reprecipitation lead to a two-zone stratigraphy of salts on the wallrock.

From analysis of surface salt-licks and waters from throughout the national park it is apparent that nutrients are readily available. The surface salt-licks are very similar in composition to the cave salt crusts based on hydrogeochemistry and possibly only inferior in the availability of P. Consequently although the cave salts are a rich source of nutrients in a soluble form, utilization by elephants is more likely to be influenced by the physical environment of the caves, which provides a warm safe haven with available water where they can quarry the cave wallrock and salt crusts on the cave floor. The salts being more soluble than wallrock minerals would be a more efficient source of nutritional elements. However, any preference for the salt crusts may be related more to the ease of mining the soft salts over the harder wallrock salts.

This project was supported by The Natural History Museum and Wateraid. We acknowledge the contribution of Mike Ukumi, Jenny Saars, Hassim Mjengera, Paul Malisa and Ndensani Kimaro in the field. Earlier work by IR came under the Mount Elgon Elephant Caves Research Project, which is indebted to J. Blashford-Snell, M. Carter, C. Redmond, J. Simons and the Davey, Barnley, Hughes and Mills families for their help and support. M. Le Bas of Leicester University is thanked for advice and discussion. M. Gill, J. Maskall and M.Ramsay of Imperial College; J. Francis; A. Sutcliffe, P. Jenkins and M. Coe are thanked for discussion and review of the manuscript.

References

ANSAH, R. K. 1990. *A study of elephant behaviour in Mole National Park, Ghana*. Ghana Department of Game and Wildlife, Report, Accra.

BELL, R. H. V. 1982. The effect of soil nutrient availability on African ecosystems. *In:* HUNTLY, B. & WALKER, B. (eds) *Ecology of tropical savannahs*. Springer, Berlin, 193–216.

BEN-SHAHAR, R. & COE, M. 1992. The relationship between soil factors, grass nutrients and the foraging behaviour of wildebeest and zebra. *Oecologia*, **90**, 422–428.

BISH, D. L. & GUTHRIE, G. D. JR. 1993. Mineralogy of clay and zeolite dusts. *In:* GUTHRIE, G. D. & MOSSMAN, B. T. (eds) *Health Effects of Mineral Dusts*, Mineralogical Society of America, Reviews in Mineralogy, **28**, 139–184.

BOWELL, R. J. & ANSAH, R. 1993. Trace element budget in an African savannah ecosystem. *Biogeochemistry*, **20**, 103–126.

—— & —— 1994. Mineral status of soils and forage, Mole National Park, Ghana and implications for wildlife management. *Environmental Geochemistry and Health*, **16**, 41–58.

COWAN, L., MCTAGGART, M. & BRINK, V. C. 1949. Natural game licks in the Rocky Mountain National Park of Canada. *Journal of Mammals*, **30**, 379–387.

DALKE, P. D., BEEMAN, R. D., KINDEL, F. S., ROBEL, R. J. & WILLIAMS, T. R. 1965. Use of salt by elk in Idaho. *Journal of Wildlife Management*, **29**, 319–332.

DAVIES, K. A. 1952. *The building of Mount Elgon*, Geological Survey of Uganda, **7**.

DENTON, D. 1972. Instinct, Appetites and Medicine. *Australian and New Zealand Journal of Medicine*, **2**, 203–212.

DETHIER, V. G. 1977. The Taste of Salt. *American Scientist*, **65**, 744–751.

DOUGALL, H. W. & SHELDRICK, D. L. W. 1964. The chemical composition of a day's diet of an elephant. *East African Wildlife Journal*, **2**, 51–59.

EUGSTER, H. P. 1970. Chemistry and origin of Brines of Lake Magadi, Kenya. *Mineralogical Society of America Special Report*, **3**, 213–235.

FAO. 1961. *Calcium requirements: report on FAO/WHO meeting*. FAO, Nutrition Report, **30**.

FRENCH, M. H. 1945. Geophagia in animals. *East African Medical Journal*, **22**, 103–110; 152–161.

HARVIE, C. E. & WEARE, J. H. 1980. The prediction of mineral solubilities in natural waters: the Na–K–Mg–Ca–Cl–SO$_4$–H$_2$O system from zero to high concentrations at 25°C. *Geochimica et Cosmochimica Acta*, **44**, 981–997.

HENSHAW, J. & AYENI, J. 1971. Some aspects of big game utilization of mineral licks in Yankari game reserve, Nigeria. *East African Wildlife Journal*, **9**, 73–82.

JONES, B. F., EUGSTER, H. P. & RETTIG, S. L. 1977. Hydrochemistry of the Lake Magadi basin, Kenya. *Geochimica et Cosmochimica Acta*, **41**, 53–72.

LE BAS, M. J. 1977. *Carbonatite–Nephelinite Volcanism: An African Case History*. Wiley, London.

McCULLAGH, K. 1969a. The growth and nutrition of the african elephant. I: Seasonal variations in rate of growth and urinary excretion of hydroxyproline. *East African Journal of Widlife*, **7**, 85–90.

—— 1969b. The growth and nutrition of the african elephant. II: The chemical nature of the diet. *East African Journal of Widlife*, **7**, 91–97.

McNAUGHTON, S. J. 1990. Mineral nutrition and seasonal movements of African migratory ungulates. *Nature*, **345**, 613–615.

MASKALL, J. E. & THORNTON, I. 1989. The mineral status of Lake Nakuru National Park, Kenya: a reconnaissance survey. *African Journal of Ecology*, **27**, 191–200.

—— & —— 1991. Trace element geochemistry of soils and plants in Kenyan conservation areas and implications for wildlife nutrition. *Environmental Geochemistry and Health*, **13**, 93–107.

OLLIER, C. D. & HARROP, J. F. 1958. The Caves of Mount Elgon. *Uganda Journal*, **22**, 158–163.

PATTERSON, R. J., FRAPE, S. K., DYKES, L. S. & McLEOD, R. A. 1978. A coring and squeezing technique for the detailed study of subsurface water chemistry. *Canadian Journal of Earth Science*, **15**, 162–169.

REDMOND, I. 1982. Salt-mining elephants of Mount Elgon. *Wildlife*, **24**, 288–293.

—— 1991. Elephants Underground. *In:* ORENSTEIN, R. (ed.) *Elephants-Saving the Gentle Giants*. Bloomsbury, London, 128–130.

—— 1992 Erosion by elephants. *In:* SHOSHANI, J. (ed.) *Elephants, majestic creatures of the wild*. RD Press, Oxford, 128–130.

—— & SHOSHANI, J. 1987. Mount Elgon elephants. *Elephant Interest Group Newsletter*, **2**, 46–66.

SCHULKIN, J. 1991. *Sodium hunger: The search for a salty taste*. Cambridge University Press.

SEARLE, D. L. 1952. *Geology of the area northwest of Kitale township*. Geological Survey of Kenya, **64**.

SUTCLIFFE, A. J. 1973. Caves of the East African Rift Valley. *Transactions of the Cave Research Group of Great Britain*, **15**, 41–65.

TSITSISHVILI, G. V., ANDRONIKASHVILI, T. G., KIROV, G. N., FILIZOVA, L. D. 1992. *Natural Zeolites*. Ellis Horwood, Series in Inorganic Chemistry.

UKUMI, M. 1990. *A provisional report on bat ecology in Mount Elgon*. Kenya Wildlife Dept.

WEEKS, H. & KIRKPATRICK, C. M. 1976. Adaptations of White-tailed deer to naturally occurring sodium deficiencies. *Journal of Wildlife Management*, **40**, 610–625.

WEIR, J. S. 1969. Chemical properties on Kalahari sands of salt licks created by elephants. *Journal of Zoology*, **158**, 293–310.

—— 1972. Spatial distribution of elephants in African National Parks in relation to environmental sodium. *Oikos*, **23**, 1–13.

—— 1973. Exploitation of water soluble sodium by elephants. *East African Wildlife Journal*, **11**, 1–7.

—— & DAVISON, E. C. 1965. Daily occurrence of African game animals at waterholes during the dry season. *Zoology of Africa*, **1**, 353–368.

WYATT, J. R. & ELTRINGHAM, S. K. 1974. The daily activity of the elephant in the Rwenzori National Park, Uganda. *East African Wildlife Journal*, **12**, 273–289.

Biogeochemistry and metal biology

O. SELINUS,[1] A. FRANK[2] & V. GALGAN[3]

[1] *Geological Survey of Sweden, PO Box 670, S-75128 Uppsala, Sweden*
[2] *Centre for Metal Biology, University of Uppsala, ISV, PO Box 535, S-75121 Uppsala, Sweden*
[3] *Dept. of Chemistry, National Veterinary Institute, PO Box 7073, S-75007 Uppsala, Sweden*

Abstract: Environmental monitoring of metals on two trophic levels is presented: a biogeochemical technique developed at the Geological Survey of Sweden, using aquatic mosses and roots of aquatic plants, and a bioanalytical–chemical technique developed at the National Veterinary Institute based on organ tissues from the moose (*Alces alces* L.), a wild ruminant living in the Swedish forests.
The usefulness of the techniques is exemplified by monitoring of Cd in southern Sweden. The results of both monitoring systems are in close agreement. Together with analysis of crops and drinking water the results indicate a region with elevated Cd burden. Also the changing environment is monitored by the moose. Decreasing concentrations of essential and toxic metals (cations) and increasing molybdenum concentration (anion) were found in a strongly acidified region of Sweden by comparison with a reference material from 1982. pH increase of the environment of the moose is indicated, probably by liming. It resulted in severe copper and chromium deficiency of the moose and was suggested as the cause of a 'mysterious' disease in the moose.
Comparison of the techniques confirms the advantage of using metal-monitoring maps in interpreting biological data and also the predictive value of the monitoring maps in biological contexts.

The Centre for Metal Biology was established in 1993 on the initiative of Swedish politicians. The members of the Centre are Uppsala University, the Swedish University of Agricultural Sciences, the Geological Survey of Sweden, the National Veterinary Institute, the University Hospital at Uppsala and the municipality of Uppsala. The Centre was created as a scientific platform with the important tasks of performing research and informing various target groups such as politicians, physicians, veterinarians, teachers and students about the present state of knowledge in the field of heavy metals and diseases related to them in humans and animals.

In order to solve intricate biological/environmental problems in modern society, and to work towards explaining the interactions between metals, scientists working in different fields of biology, toxicology, nutrition and health, were brought together to facilitate interdisciplinary co-operation. The broad spectrum of participants, representing different scientific fields, enables the Centre to cover knowledge and research in many areas with impact on human and animal health. The Centre is expected to provide guidelines and to propose studies with environmental impact, including suggestions on solving, eliminating or counteracting problems identified. Such measures could be exemplified by detoxifying agents and agents for large-scale metal elimination from, for example, industrial waste water.

A model problem which was identified at an early stage was the study of the possible adverse effects of mercury from dental amalgams. Approaches to shed light upon this problem have been proposed and tried by physicians at the Amalgam Unit at the University Hospital. Knowledge emanating from the toxicological research within the centre proved to be of great value in this case and is an example of interdisciplinary co-operation.

Through the Centre for Metal Biology opportunities for co-operation between disciplines are thus greatly increased. The present paper is a result of such an interdisciplinary co-operation between the Geological Survey of Sweden and the National Veterinary Institute.

Environmental monitoring

Introduction

Contamination of the environment with hazardous compounds and elements of anthropogenic origin is of increasing concern because of the effect on the whole biosphere, i.e. the micro-flora and -fauna of soils, plants and higher life, including humans and animals. Use of different monitors at various trophic levels has been suggested to collect relevant information. Biological monitoring should be designed to obtain and make use of the optimum amount of available information by complementing existing environmental studies, or through the simultaneous collection of other environmental data (Wren 1983). Monitors have recently been defined as organisms in which changes in known characteristics can be measured to assess the extent of environmental contamination so that conclusions on the health implication for other species of the environment as a whole can be drawn (O'Brien et al. 1993). Monitors may provide information about the environmental concentration of essential as well as toxic metals of importance for life, displaying deficiency and toxicity, respectively.

Some elements may derive from environmental pollution while some enrichments may be natural. The metals may affect life on different trophic levels. Essential and toxic elements in bedrock or soils may become a direct risk for human and animal health and may be the underlying cause of both deficiency and toxicity (Crounce et al. 1983; Låg 1990, 1991). Aerial deposition of sulphuric and nitric oxides (acid rain) may influence weathering of bedrock and soils. The concentration of metals in upper soil layers may change as a result of mobilization and the metals become more available via plants to grazing animals. Elements that are easily mobilized are Ca, Mg, Mn, Al, Ni, Zn and Cd, and to a lesser extent Hg, Pb and Cu. When the buffering capacity of the soil is insufficient, acid rain may cause deficiencies in plants and via plants in herbivorous animals as a result of leaching and eluting of metals essential to plants and animals. Certain essential trace elements, such as Se and Mo, become less soluble in acidic environment and their availability to plants decreases. Changes in the uptake via plants may result in changed metal concentrations as well as changed relationships between metal concentrations in organ tissues, with severe consequences for grazing animals.

Cadmium

Cadmium found in soils, waters, plants and other environmental matrices not affected by pollution, can be regarded as natural. There is little difference among the igneous rocks in Cd content while among sedimentary rocks, the carbonaceous shales, formed under reducing conditions, contain the most Cd. The Zn/Cd ratio in terrestrial rocks is about 250 (Thornton 1986). In non-contaminated, non-cultivated soils, Cd concentration is largely governed by the amount of Cd in the parent material (Purves 1985; Adriano 1986). In an extensive survey of Swedish soils Andersson (1977) found an average of $0.22\,mg\,Cd\,kg^{-1}$ (0.03–2.3 mg/kg range) for both cultivated and non-cultivated Swedish soils. The soil geochemical mapping carried out by the Geological Survey of Sweden so far has analysed 10 000 samples from all parts of Sweden. The median value of these samples is $0.1\,mg\,Cd\,kg^{-1}$ in natural unweathered podzolic soils, while the 90th percentile is $c.\ 0.4\,mg\,kg^{-1}$, and the maximum value is $6.4\,mg\,kg^{-1}$ (M. Andersson, pers. comm.).

Cd is readily taken up by roots and is distributed throughout the plants. The uptake in plants can be both active and passive (Kabata-Pendias & Pendias 1992), however, these aspects will not be discussed in the present context. The amount of uptake is influenced by soil factors such as pH, cation exchange capacity, redox potential, phosphatic fertilization, organic matter, other metals and other factors. In general, there is a positive, almost linear correlation between the different Cd concentrations in the substrate and the resulting Cd concentration in the plant tissues (Adriano 1986).

In a *Filipendula ulmaria* meadow ecosystem Balsberg (1982) found that water solutions of different Cd concentrations added to the soil, < 10% of the total Cd in the ecosystem was retained in the plant biomass. Root concentrations exceeded those in the soil and were several times the concentration in above ground organs. The Cd concentration in various plant parts decreased in the order: new roots > old roots > rhizomes > stem leaf-stalks > stem leaf-blades > reproductive organs.

In the present paper two examples are given of monitoring metals, in this case especially cadmium, in the environment by use of (i) aquatic mosses and the roots of aquatic higher plants, a technique developed at the Geological Survey of Sweden; and (ii) organ tissues of the moose (*Alces alces* L.), a wild ruminant used for monitoring by the Chemistry Department of the National Veterinary Institute since 1980.

Aquatic roots and mosses as environmental monitors

In order to delineate the geochemical distribution of metals, the Geological Survey of Sweden started a nationwide mapping programme in 1980. The purpose of the programme is to compile a geochemical atlas of the entire country. The maps give a general outline of the distribution for heavy metals in the surficial environment. A new method is used, whereby metal concentrations are determined in organic material consisting of aquatic mosses and roots of aquatic higher plants. These are barrier-free with respect to trace metal uptake and reflect the metal concentrations in stream water (Brundin et al. 1987).

Aerial parts of many plant species do not generally respond to increasing metal concentrations in the growth medium because of physiological barriers between roots and above ground parts of plants. These barriers protect them from uptake of toxic levels of metals into the vital reproductive organs (Kabata-Pendias & Pendias 1992; Kovalevsky 1987). The roots and mosses respond closely to chemical variations in background levels related to different bedrock types in addition to effects of pollution (Brundin & Nairis 1972; Brundin et al. 1988; Selinus 1988, 1989; Nilsson & Selinus 1991). Variation of uptake with growing season and between plant species is of concern only for above earth parts of plants but not for the roots (Brundin et al. 1988).

Due to chemical weathering processes, the metal concentrations in the stream water reflect the chemical composition of the surrounding bedrock and soils. When the groundwater reaches surface waters some of the metals may precipitate. Enrichment takes place both in the roots and in iron- and manganese hydroxides. The exchange of metals between the water and the roots is a slow process whereby the influence from seasonal variations is of minor importance. One great advantage of using biogeochemical samples instead of water samples is that the biogeochemical samples provide integrated information of the metal contents in the water for a period of some years. Water samples suffer from seasonal and annual variations depending on, for example, precipitation. One great advantage of using biogeochemical samples instead of water samples is that the biogeochemical samples provide integrated information of the metal contents in the water for a period of some years. The biogeochemical samples also provide information on the time-related *bioavailable* metal contents in aquatic plants.

All sampling points are chosen in such a way that they each represent a drainage area (one sample every 6 km^2). All samples are analysed after ashing by X-ray fluorescence (XRF) for Al, As, Ba, Ca, Co, Cr, Cu, K, Mg, Mo, Nb, Na, Ni, P, Pb, Rb, S, Si, Sr, Ti, U, V, W, Y, Zn, Zr. Every fifth sample is also analysed by Atomic Absorption (AAS) for Hg, Se and Cd. After analysing the samples, all analytical results are normalized with respect to organic content and limonite content (Fe and Mn) using stepwise regression analysis (Selinus 1983).

The biogeochemical mapping programme now (1996) covers about 50% of the land area of Sweden (30 000 sample points), where about 75% of the population of Sweden lives. This means that the Geological Survey now has an extensive database of analyses for use in environmental and geomedical research. The samples are related to the periods in which they are sampled, which means that resampling and follow-up research will render invaluable monitoring information in the future. An existing sample bank with all biogeochemical samples could also be used for future environmental research. The geochemical mapping programme also includes geochemical soil surveys and bedrock surveys. The soil surveys provide information on the regional variation of major, minor and trace element distributions in the unaffected part of the till cover of the country. The bedrock surveys are used for separating the natural background of the metals from anthropogenic sources by means of multivariate statistical methods (Selinus & Esbensen 1995).

Cadmium levels in biogeochemical samples (roots and aquatic mosses) from southern Sweden are shown in Fig. 1 (Ressar & Ohlsson 1985). The contents are enhanced in the southernmost counties and along the west coast of Sweden. The latter distribution is derived mainly from transboundary atmospheric transport and deposition of anthropogenic origin. The contents of Cd in the southernmost part are, however, much higher, and the highest levels so far detected by the Geological Survey since 1982 are located in this region. This region is a densely populated farming region from which growing crops are distributed to the rest of Sweden. The important source of uptake of Cd for the non-smoking Swedish population is cereals. Therefore, a sampling programme of wheat in the affected areas was performed by the farmers' organizations after contact with the Geological Survey.

In 1989 samples of autumn wheat were taken in certain geographical regions, and in the Skåne region 54 samples were taken. The results showed that samples from Skåne had an average of 73 μg Cd kg^{-1} d.w., with several areas exceeding 100 μg Cd kg^{-1} d.w. In comparison, an area in central Sweden yielded on analysis only 29 μg Cd kg^{-1} d.w. on average in autumn wheat

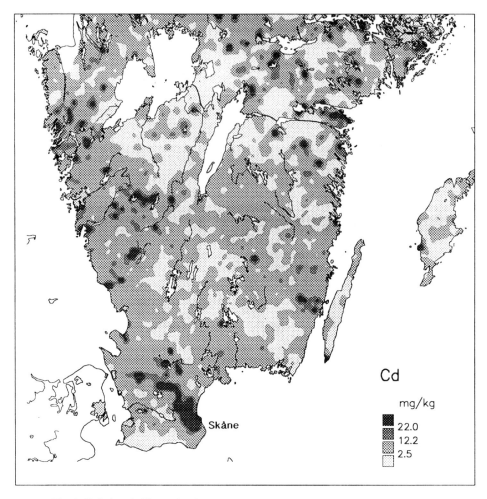

Fig. 1. Cadmium in biogeochemical samples (roots and mosses) from southern Sweden.

(A. Jonsson, pers. comm.). Japan is known to have the highest cadmium burden in the world, especially in home produced rice. It is noteworthy that the concentration of cadmium in polished rice was found to be 30–300 $\mu g\,kg^{-1}$, and an average value from 200 samples of polished rice was 66 $\mu g\,kg^{-1}$ w.w. (Friberg et al. 1974). The Cd contents in wheat from Skåne could therefore be a matter for concern. In this region drinking water is taken from many wells. In those that have been analysed for Cd, the results show an almost identical distribution of high Cd contents as depicted in the biogeochemical map. For drinking water, the WHO has set a limit of 5 $\mu g\,Cd\,l^{-1}$ (WHO 1984). In comparison, the wells in Skåne, within the region with high Cd burden, have average levels of about 400 $\mu g\,Cd\,l^{-1}$.

Several factors may interact with each other and contribute to the high Cd concentrations found in this region, for example, deposition of airborne Cd as well as acid rain from Eastern, Western and Central Europe. Cadmium may also originate from local sources, for example, from phosphate fertilizers used in agriculture or sewage sludge, or high contents of natural Cd derived from the sedimentary bedrock in Skåne. Research is now being carried out in order to investigate combinations of these factors.

The concurring results from biogeochemistry (Fig. 1) and analyses of crops and drinking water indicate, thus, a region of pronounced Cd contents. Similar results were found when monitoring was performed by analysing samples from a wild herbivorous ruminant, the moose.

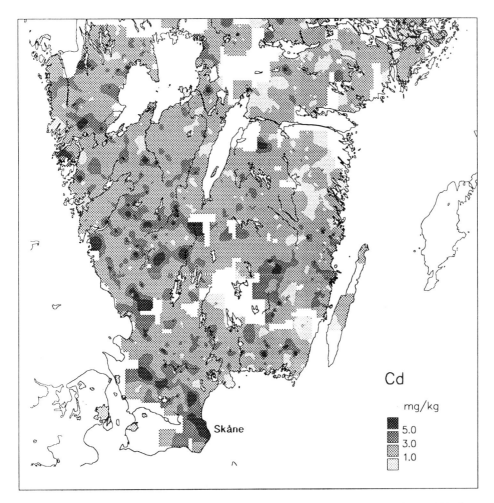

Fig. 2. Average yearly cadmium uptake in moose kidney in southern Sweden expressed in mg Cd kg^{-1} w.w.

The moose (*Alces alces* L.) as an environmental monitor

Cadmium in organ tissues of the moose

In the search for monitors among the wild Swedish fauna during 1973–1976 that could be used for cadmium, the moose (*Alces alces* L.) was found to be useful (Frank 1986). This large wild ruminant is found in most parts of Sweden. The moose is relatively sedentary, the range of migration within the territory seldom exceeds 50–80 km to some extent also depending on the density of the population. Tissues may be obtained through co-operation with the hunters' organization during regular hunting seasons in the autumn. Regional differences in cadmium burden were demonstrated by Frank et al. (1981). By far the highest burden was found in the region of Skåne (Mattsson et al. 1981; Frank 1982).

Prior to the hunting seasons in 1981 and 1982, standardized packages for sample collection were distributed to the hunters. All but one of the 24 counties of Sweden participated in sample collection. Tissues of liver and kidney, as well as the left mandible (for age determination by cutting the first molar teeth and microscopic examination) were collected from more than 4300 animals (Frank 1989). The samples were pretreated by automatic wet digestion (Frank 1976, 1988). Simultaneous analysis of 13 metals using a direct current plasma-atomic emission

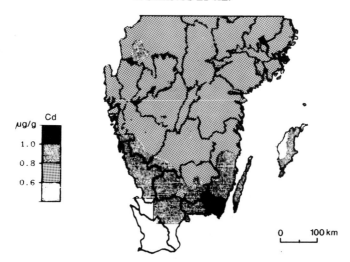

Fig. 3. Cadmium content (μg/g d.w.) in terrestrial moss samples collected in 1975. Adapted from Rühling & Skärby 1979. (The white area in southernmost Sweden (Skåne) is not sampled.)

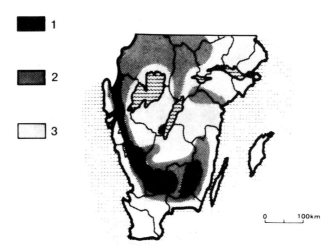

Fig. 4. pH values in the lakes of Sweden in the late 1970s. (1) pH < 5.0 all the year in at least 33% of all lakes; (2) pH < 5.5 some time during the year in at least 50% of all lakes; (3) less than 50% of all lakes acidified. (Adapted from Monitor 1981, Swedish Environmental Protection Agency.)

spectrometer (DCP-AES) was applied in determination of the tissue metal concentrations (Frank & Petersson 1983). The metals determined were Al, Ca, Cd, Co, Cr, Cu, Fe, Mg, Mn, Ni, Pb, V, and Zn.

At the time of sample collection, the main interest was focused on determination of the cadmium burden in the different geographical regions of Sweden. In addition, the concentrations of essential elements in liver and kidney gave increased knowledge of normal variations in the different regions, which is important information from a nutritional point of view. All these values represent *time-related reference values for moose in 1982*. Future investigations compared with reference values of the same geographical region would display changes in the environment.

The renal cadmium accumulation is age dependent. To some extent, the rate of accumulation depends on the total amount of cadmium present in the environment. However, influencing factors are the geochemical background, atmospheric transport and deposition of particulate matter, oxides of sulphur and nitrogen, acid rain, anthropogenic activities etc. – all seem to contribute to the flux of cadmium in the biosphere. Acidification mobilizes cadmium, and its availability for uptake via plants in herbivorous animals increases. Early investigations demonstrated very high Cd burdens in the south and southwest of the country. Once again, the highest values were found in the two counties making up the most southern province of Sweden (Frank & Petersson 1984; Frank et al. 1984).

A number of statistical models were tested to express the cadmium burden of a population of animals. Eventually a statistical model allowing comparison of the cadmium burden of different regions was developed and found most useful (Eriksson et al. 1990). It is presented in an abbreviated form as follows. A quotient is formed by dividing the cadmium concentration of the respective organs with the estimated age of the animal. From these quotients, median values are calculated for each age class and region, and the age classes are weighted for age distribution. The resulting constant is characteristic for the region in question and is the average yearly cadmium uptake in liver and kidney respectively at the time of sample collection. Using this model, the cadmium burdens for all the investigated counties were calculated.

A shaded map expressing the cadmium burden of southern Sweden is shown in Fig. 2. When this cadmium map was compared with the map of atmospheric deposition of cadmium, measured by moss analysis (Rühling & Skärby 1979) (Fig. 3), with the bedrock geological map and the acidification map of southern Sweden (Fig. 4), the best agreement was achieved by comparison with the acidification map. Comparison of Fig. 2 with the biogeochemical map based on analysis of roots of water plants (Fig. 1) shows close agreement in the region of Skåne.

Secondary copper deficiency, chromium deficiency and an earlier unknown disease of the moose

In connection with a new type of moose disease of unknown etiology in southwestern Sweden (Steen et al. 1989), reexamination of tissue metal concentrations was performed in 1988 and 1992. Decreased hepatic Cu concentrations in 1988 and 1992 (30% and 50% respectively) were observed in comparison with the corresponding value in 1982 (Frank et al. 1994). Decrease of other essential and toxic metal concentrations was noted in the liver such as Cr (80%), Fe, Zn, Pb and Cd. In the kidney, decreases of Ca, Mg, Mn indicate severe metabolic disturbances. The Cd concentration in the kidney decreased by about 30%.

Decreasing metal concentrations in organ tissues of the moose, after such a short period of 5–6 years, indicate changes in the concentration of metals via plants in the upper soil surface layer. Thus, the moose appears to be a sensitive monitor of environmental changes. In ruminants, the utilization and availability of Cu in feed is greatly influenced by interactions between Cu, Mo and S (Mills & Davis 1987). Increased Mo concentrations in respect to Cu in the feed causes secondary copper deficiency in ruminants with clinical signs in close agreement with the signs related to the unknown moose disease.

When liver from yearlings, collected in 1982 and 1992, was analyzed for Mo, increased hepatic Mo concentrations were found in northern and southern districts of Älvsborg County (24% and 21%, respectively) during the 10 years between sample collections (Frank et al. 1994).

When pH in soils decreases, Mo becomes less available to plants and its uptake diminishes. Nonetheless, the increased Mo uptake by plants in the strongly acidified region is difficult to explain. A small pH increase in the environment causes a marked increase of Mo uptake by plants (Mills & Davis 1987). Thus, the results evidently indicate elevated pH in the soils. The intensive liming of lakes, wetlands, fields and pastures in the western part of Sweden during the later 1980s, and to some extent the liming of forest areas during recent years, are suggested as possible explanations of increased hepatic Mo concentrations in this exceptionally acidified region. This side effect of liming and its harmful consequences for domestic and wild ruminants appear to have been overlooked.

A map showing the ratio of Cu/Mo was recently constructed on the basis of biogeochemical data provided by the Geological Survey of Sweden (to be published). This map demonstrates the risk areas of secondary copper deficiency as well as of copper toxicity in ruminants. The area of the former and the region for the highest prevalance of the unknown moose disease appear to coincide.

Conclusions

Two monitoring techniques were compared. Both techniques map the bioavailabilty of elements in the environment, however, on different trophic levels. The map based on roots of aquatic plants and aquatic mosses, and those of using organ tissues from the moose, a wild ruminant, appear to be remarkably similar, in the vicinity of Skåne, where a region of elevated Cd burden is identified.

The two examples given in the paper elucidate the usefulness of the mapping techniques in monitoring and detecting toxic metal burdens of regions (biospheres) as well as low prevalence of essential elements.

The maps, especially the detailed maps of biogeochemical data comprising about 30 elements, contribute to and facilitate interpretation of biological data and have predictive value in biological contexts. In addition, the moose seems to be a reliable monitor to detect rapid changes in the environment.

References

ADRIANO, D. C. 1986. *Trace elements in the terrestrial environment*. Springer-Verlag, 110–129.

ANDERSSON, A. 1977. Heavy metals in Swedish soils: on their retention, distribution and amounts. *Swedish Journal of Agricultural Research*, **7**, 7–20.

BALSBERG, A. M. 1982. Seasonal changes in concentration and distribution of supplied cadmium in a Filipendula ulmaria ecosystem. *Oikos*, **38**, 91–98.

BRUNDIN, N. H. & NAIRIS, B. 1972. Alternative sample types in regional geochemical prospecting. *Journal of Geochemical Exploration*, **1**, 7–46.

——, EK, J. I. & SELINUS, O. C. 1988. Biogeochemical studies of plants from stream banks in northern Sweden. *Journal of Geochemical Exploration*, **27**, 157–188.

CROUNCE, R. G., PORIES, W. J., BRAY, J. T. & MAUGER, R. L. 1983. Geochemistry and man: Health and disease. *In:* THORNTON, I. (ed.) *Applied environmental geochemistry*, Academic, London, 267–334.

ERIKSSON, O., FRANK, A., NORDKVIST, M. & PETERSSON, L. R. 1990. Heavy metals in reindeer and their forage plants. *Rangifer*, Special issue, **3**, 315–331.

FRANK, A. 1976. Automated wet ashing and multimetal determination in biological material by atomic absorption spectrometry. *Zeitschrift für Analytische Chemie*, **279**, 101–102.

—— 1982. Hotet från skyn. *Svensk Jakt*, **120**, 626–628 [in Swedish].

—— 1986. In search of biomonitors for cadmium. Cadmium content of wild Swedish fauna during 1973–1976. *Science of the Total Environment*, **57**, 57–65.

—— 1988. Semi-micro accessory to an automated wet digestion system for ashing small sample amounts. *In:* BRÄTTER, P. & SCHRAMEL, P. (eds) *Trace element analysis in medicine and biology*. Vol. 5. Walter de Gruyter, Berlin, 78–83.

—— 1989. Bly-och kadmiumhalt i lever och njure av älg samt koncentrationer av kobolt, koppar, mangan och zink i älglever. *Rapport till SNV* [in Swedish].

—— & PETERSSON, L. R. 1983. Selection of operating conditions and analytical procedure in multi-metal analysis of animal tissues by d c plasma-atomic emission spectroscopy. *Spectrochimica Acta*, **38 B**, 207–220.

—— & —— 1984. Assessment of the bioavailability of cadmium in the Swedish environment. *Zeitschrift für Analytische Chemie*, **317**, 652–653.

——, GALGAN, V. & PETERSSON, L. R. 1994. Secondary copper deficiency, chromium deficiency and trace element imbalance in the moose (*Alces alces* L.). (Effect of anthropogenic activity.) *Ambio*, **23**, 315–317.

——, PETERSSON, L. R. & MÖRNER, T. 1981. Bly-och kadmiumhalter i organ från alg, radjur och hare. (Lead and cadmium in tissues from elk (*Alces alces*), roe deer (*Capreolus capreolus*) and hares (*Lepus europaeus, Lepus timidus*).) *Svensk Veterinärtidn*, **33**, 151–156 [in Swedish with English summary].

——, —— & —— 1984. The moose (*Alces alces*) as an indicator of the bioavailability of cadmium in the Swedish environment. *In: Das freilebende Tier als Indicator für das Funktionszustand der Umwelt*. Proceedings of Symposium, Nov. 1984, Wien.

FRIBERG, L., PISCATOR, M., NORDBERG, G. F. & KJELLSTRÖM, T., 1974. *Cadmium in the Environment*, 2nd ed. CRC, Cleveland.

KABATA-PENDIAS, A. & PENDIAS, H. 1992. *Trace elements in soils and plants*. CRC, Cleveland, 67–72.

KOVALEVSKY, A. L. 1987. *Biogeochemical exploration for mineral deposits*. NU Science, Utrecht, 8–23.

LÅG, J. 1990. General survey of geomedicine. *In:* LÅG, J. (ed.) *Geomedicine*, CRC, Cleveland, 1–24.

—— 1991. Attention paid to selenium and cadmium since the 1950's. *In:* LÅG, J. (ed.) *Human and animal health in relation to circulation processes of selenium and cadmium*. Norwegian Academy of Science and Letters, 9–21.

MATTSSON, P., ALBANUS, L. & FRANK, A. 1981. Kadmium och vissa andra metaller i lever och njure från älg. (Cadmium and some other elements in liver and kidney from moose (*Alces alces*).) *Vår Föda*, **33**, 335–345 [in Swedish with English summary].

MILLS, C. F. & DAVIS, G. K. 1987. Molybdenum. *In:* MERTZ, W. (ed.) *Trace Elements in Human and Animal Nutrition Vol. 1*. Academic, New York, 429–457.

MONITOR. 1981. Swedish Environmental Protection Agency.

NILSSON, C. A. & SELINUS, O. 1991. Biogeochemical mapping of Sweden for Exploration and Environmental Research *In: 15th IGES Abstracts*, 41.

O'BRIEN, D. J., KANEENE, J. B. & POPPENGA, R. H. 1993. The use of mammals as sentinels for human exposure to toxic contaminants in the environment. *Environmental Health Perspectives*, **99**, 351–368.

PURVES, D. 1985. *Trace-element contamination of the environment*. Elsevier, Amsterdam, 22–28.

RESSAR, H. & OHLSSON, S. A. 1985. Geokemisk kartering. *SGU Rapporter och meddelanden*, **42**.

RÜHLING, A. & SKÄRBY, L. 1979. Landsomfattande kartering av regionala tungmetallhalter i mossa. *SNV PM 1191* [in Swedish with English summary].

SELINUS, O. 1983. Regression analysis applied to interpretation of geochemical data at the Geological Survey of Sweden. *In:* HOWARTH, R. J. (ed.) *Handbook of exploration geochemistry, part 2, Statistics and data analysis in Geochemical prospecting*, Elsevier, Amsterdam, 293–302.

—— 1988. Biogeochemical mapping of Sweden for geomedical and environmental research. *In:* THORNTON, I. (ed.) *Proceedings of the 2nd symposium on geochemistry and health*. Science Reviews Ltd, 13–20.

—— 1989. Heavy metals and health – results of the biogeochemical mapping programme of Sweden. *EUG V*, Strasbourg.

—— & ESBENSEN, K. 1995 Separating anthropogenic from natural anomalies in environmental geochemistry. *Journal of Geochemical Exploration*, **55**, 55–66.

STÉEN, M., FRANK, A., BERGSTEN, M. & REHBINDER, C. 1989. En ny sjukdomsbild hos älg. *Svensk Veterinärtidn*, **41**, 73–77 [in Swedish].

THORNTON, I. 1986. Geochemistry of cadmium. *In:* MISLIN, H. & RAVERA, O. (eds) *Cadmium in the environment*. Birkhäuser, Basels, 8.

WHO. 1984. *Guidelines for drinking water quality, vol 1, recommendations 1984*. WHO, Geneva.

WREN, C. D. 1983. *Review of the occurrence and toxicity of metals in wild animals*. Ref no KN107-2-4609, Canadian Wildlife Service, 100 Gamelin Bvd., Hull, Quebec, 158.

Groundwater geochemistry and health: an overview

W. M. EDMUNDS & P. L. SMEDLEY

British Geological Survey, Wallingford, OX10 8BB, UK

Abstract: The natural geological and geochemical environment, in addition to providing beneficial mineral content and bioessential elements to groundwaters, may also give rise to undesirable or toxic properties through a deficiency or an excess of various elements. In this paper the controls on the release of toxic elements are considered together with the geochemical conditions that give rise to excess and deficiency. Many studies have pointed towards an inverse correlation between water hardness and cardiovascular disease and the associated mobility of metals in soft waters may also be a contributory factor. Under acidic conditions Al and Be as well as other metals may be released, whilst changes in redox conditions as well as pH will affect the mobility of Fe, Mn and As in particular. Some potentially toxic elements such as Ba and F are usually held at acceptable concentrations in groundwaters by the respective solubility controls of barite and fluorite; quality problems arise from these elements when groundwaters contain respectively low sulphate or low calcium concentrations. Deficiencies of Se and I (as well as F) in groundwater are related to the low geochemical abundance of these elements in certain environments. In the context of groundwater pollution it is important first to define the natural baseline concentrations of key elements of health importance.

The provision of safe drinking water as well as water of acceptable quality remain prime targets for both advanced and developing countries. By far the greatest water quality problem in developing countries is the prevalence of water-borne diseases, especially gastro-enteritis which is related to faecal pollution and inadequate hygiene (Tebbutt 1983). Such problems are usually related to poor well siting and construction as well as to insufficient water and water distribution. Pollution from agricultural chemicals, for example nitrate and pesticides, once considered a problem restricted to rich nations, is now also a rapidly growing problem in developing countries.

In addition to the anthropogenic sources, the natural baseline geochemistry of groundwaters and surface waters resulting from interaction with rocks, also creates widespread health and acceptability problems in many parts of the world, mainly on a regional scale. The concentrations of naturally occurring mobile elements often exceed those recommended as the maximum for potable waters, and/or their concentrations may exceed limits of general acceptability for domestic use.

These natural problems of groundwater have been exacerbated during the past two decades by the widespread installation of rural groundwater supplies. Well or borehole water, replacing otherwise unhygienic surface supplies, has all too often been developed in places where specific geochemical conditions may have led to excessive concentrations of toxic or undesirable elements (As, Cr, Fe, Mn, Sb, Al or F).

In the present paper, those elements considered by international organizations to be toxic in natural waters are reviewed with regard to their mobility under various hydrogeochemical conditions. Elements which may give rise to health problems due to their deficiency in water supplies are also considered, as are those elements such as iron which present problems of acceptability in water supplies. It is not possible to review the vast numbers of studies worldwide comprehensively, but rather a selective overview of the current problems is presented using examples mainly from developing countries as a key to future needs, especially work of a remedial or preventive nature.

Essential and non-essential elements

Nearly all natural waters contain traces of most of the chemical elements but often at extremely low or unquantifiable concentrations. The typical abundance of the elements in natural waters at pH 7 is summarized in Fig. 1. Nine major species (HCO_3, Na, Ca, SO_4, Cl, NO_3, Mg, K and Si) invariably make up over 99% of the solute content of natural waters. The abundance of minor and trace elements, under 1% of the total, can change significantly from the concentrations shown depending on geochemical conditions. In particular a pH decrease of one unit may lead to an increase of more than one order of magnitude in the concentration of certain metals. A change from oxidizing to reducing conditions may have a similar effect on elements such as iron.

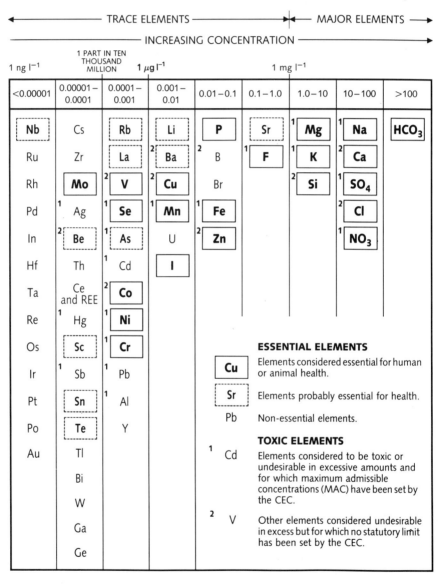

Fig. 1. Major and trace elements in groundwater and their significance in terms of health. Concentrations shown are those typical of dilute oxygenated groundwater at pH 7. Elements outlined are those considered to be essential (or probably essential) for health. Those elements which have guideline or statutory limits set by the CEC or WHO (see Tables 1, 2) are indicated.

Table 1. Chemicals of health significance in drinking water

	WHO (1993) guideline maxima (mg l^{-1})	CEC (1980) guideline values (mg l^{-1})	CEC (1980) max. admissible concentrations (mg l^{-1})
Antimony (Sb)	0.005 (P)		0.010
Arsenic (As)	0.01 (P)		0.050
Barium (Ba)	0.7	0.1	
Beryllium (Be)	NAD		
Boron (B)	0.3	1.0	
Cadmium (Cd)	0.003		0.005
Chromium (Cr)	0.05 (P)		0.050
Copper (Cu)	2 (P)	0.1	
Fluoride (F)	1.5		1.5*
Lead (Pb)	0.01		0.05
Manganese (Mn)	0.5 (P)	0.02	0.05
Mercury (Hg) total	0.001		0.001
Molybdenum (Mo)	0.07		
Nickel (Ni)	0.02		0.05
Nitrate (NO$_3$)	50	25	50
Selenium (Se)	0.01		0.01
Uranium (U)	NAD		

Data from WHO (1993) with values from CEC (1980) where appropriate. P, provisional value; NAD, no adequate data to permit recommendation of a health-based guideline value; * climatic conditions, volume of water consumed and intake from other sources should be considered when setting national standards.

Table 2. Substances in drinking water that may give rise to complaints from consumers

	WHO (1993) guideline maxima (mg l^{-1})	CEC (1980) guideline values (mg l^{-1})	CEC (1980) max. admissible concentrations (mg l^{-1})
Aluminium (Al)	0.2	0.05	0.2
Ammonium (NH$_4$)	1.5	0.05	0.5
Calcium (Ca)		100	
Chloride (Cl)	250	25	
Sulphide (H$_2$S)	0.05		
Iron (Fe)	0.3	0.05	0.2
Magnesium (Mg)		30	50
Potassium (K)		10	12
Sodium (Na)	200	20	150*
Sulphate (SO$_4$)	250	25	250
Zinc (Zn)	3	0.1	5

Data from WHO (1993) compared where appropriate with CEC limits. *More than 80% compliance over a period of 3 years.

Those elements currently considered to be essential for human health and metabolism (Moynahan 1979) are also indicated in Fig. 1, together with elements which are considered to have harmful effects if present in water supplies above certain limits (CEC 1980; World Health Organisation 1993). The currently agreed limits for inorganic constituents of significance in relation to health are summarized in Table 1. It should be noted that for some elements (e.g. Be, U) no agreed limit has been reached despite their known health effects. Some bioessential elements (e.g. F and Se) may also have a harmful effect if present above certain limits. Those elements that may give rise to acceptability problems are summarized in Table 2; some of the limits quoted in both tables have qualifications placed on them for which reference to the original text is recommended.

Geochemical baseline conditions

The composition of surface waters and shallow groundwaters will closely reflect the local geology. Reactions between rainwater and bedrock over a timescale of days or months during percolation, followed by emergence as springs or as inputs to the water table, give the groundwater its essential mineral character. During this process the physical properties of the bedrock and hence mechanisms of flow (i.e. by intergranular or via fractures) will be of importance. The extent of reaction with the host rock will be controlled by the residence times of the water and the primary mineralogy of the aquifer. In this context the initial concentrations of CO_2 in the soil will determine the amount of reaction of carbonate or silicate minerals that takes place in the aquifer. The geological map may be used to highlight the distribution of carbonate (infinite buffering), from non-carbonate (base-poor) terrains which offer poor buffering capacity and can give rise to acid waters and mobilization of metals harmful to health (Lucas & Cowell 1984; Edmunds & Kinniburgh 1986).

In contrast to shallow environments where the baseline chemistry closely mirrors the surface geology, deeper groundwaters can undergo significant changes in the baseline conditions along flow lines with increasing residence times (Edmunds et al. 1987). It is therefore necessary to consider the possible changes taking place not only between areas with different geology but also the sequential changes taking place within an aquifer. Classic examples of health effects relating to differences in water chemistry along flow gradients in aquifers may be found. In the confined Lincolnshire Limestone aquifer in the UK (Lamont 1959), for example, a zonation in dental health could be identified in relation to groundwater use. Once recognized as a fluoride-related problem, blending of high and low F-waters was adopted as a means of eradicating fluorosis and of providing an optimum F concentration in water supplies.

Health and acceptability problems are really only important for a small number of elements and it is convenient to consider these under categories relating to geochemical controls.

Natural water quality problems

Water hardness

It has long been suspected that a causal link exists between water hardness (i.e. principally dissolved Ca and Mg) and cardiovascular disease (Gardner 1976; NAS 1977; Masironi 1979) although water is only one of many factors that might be involved (Shaper et al. 1981). In Britain, detailed studies were carried out during the 1960s and 1970s, especially in towns which had experienced changes in the hardness of their water supplies (Lacey & Shaper 1984). These results supported the hypothesis of a weak causal relationship between drinking water hardness and mortality, especially in men, from cardiovascular disease. More recently the British Committee on Medical Aspects of Food Policy (COMA 1994) as part of a wider review of the nutritional aspects of cardiovascular disease, also found a weak inverse relationship between water hardness and cardiovascular disease mortality, but noted that the size of the effect was small and most clearly seen at water hardness levels below 170 $mg\,l^{-1}$ (as $CaCO_3$). It was also pointed out (by COMA) that further studies since the 1970s had not altered the balance of evidence for the association and that the explanation for the association remained unknown. Other studies reviewed by Foster (1992) have shown that the relationship between water hardness and heart disease is not straightforward and that the other factors involved (e.g. diet, exercise, smoking) are likely to be more important. On balance, it is probable that hardness may only be a general pointer towards other agents connected with the disease. Several hypotheses for the link with water have been proposed including: (1) potential for Ca and/or Mg to protect against some forms of cardiovascular disease; (2) that some trace elements, more prevalent in hard water, may be beneficial; (3) many metals are more soluble in soft water and hence may promote cardiovascular disease.

From the geochemical viewpoint it is considered that more studies are needed that investigate the ratio of Mg to Ca in natural waters in relation to cardiovascular disease, rather than total hardness alone as a factor. Elevated sodium in drinking water is also implicated in contributing to high blood pressure (Calabrese & Tuthill 1981) and it seems important that further work in this area should consider not only sodium (and total mineralization) but also the ratio of Na to other competing ions, especially potassium.

Fewer studies have been carried out in developing countries but Dissanayake et al. (1982) for example found a similar negative correlation between water hardness and various forms of cardiovascular disease and leukaemia in Sri Lanka. Links between water hardness and endemic goitre have also been suggested (Day & Powell-Jackson 1972).

Acid water and mobilization of toxic metals

Acid groundwater may result either from natural processes such as flow through non-carbonate rocks (e.g. granite), from pyrite oxidation or from pollution (acid rain). Acid water is in itself not thought to be a health risk but since many minerals are more soluble in acid water, toxic trace metals (e.g. Al, Be, Cd, Pb) may be present in higher concentrations. Most acid waters are base-poor, soft waters and their Ca and Mg deficiencies may also be implicated in various forms of heart disease referred to above.

Acid groundwaters are common in many parts of Africa, Asia and South America where granitic basement rocks with little acid-buffering capacity compose large areas of the land mass. High trace-metal contents of such waters are likely as a result. Acidic water is, for example, documented in south-western Ghana where Langanegger (1991) found that 45% of groundwaters had pH below 6.5. This is supported by recent work (Smedley et al. 1996). Two elements in particular, Al and Be, are worth highlighting as potentially problematic in acid waters.

Aluminium (Al). The WHO maximum recommended concentration for Al in drinking water is 0.2 mg l^{-1} (WHO 1993). Al is a major element in alumino-silicate minerals and is therefore a common constituent of most rocks. The solubility of Al is strongly pH-dependent and significant environmental concentrations are only found below pH 5.5 where the increasing concentrations are related to the solubility of microcrystalline gibbsite (Bache 1986). The presence of inorganic ligands, notably F and SO$_4$ also increase the solubility of aluminium (May et al. 1979). At pH greater than 5, it is unlikely that labile, monomeric forms of aluminium will be present in natural waters although colloidal aluminium and other aluminosilicate colloids and particles may contribute to total aluminium in waters. It is customary to measure total aluminium in water on samples filtered through 0.45 μm filters. This will give an early warning of high aluminium problems, although analysis of the monomeric forms (Driscoll 1984) is also needed for health and environmental studies where thermodynamic treatment of aluminium solubility is desirable.

The occurrence of high Al in drinking water has been linked to the development of Alzheimer's disease (e.g. Martyn et al. 1989). Recent studies related to Al toxicity have been carried out in relation to acidification by acid rain (Vogt 1986), although concern has also been expressed in relation to the quality of water released from water treatment plants where aluminium salts are used to clarify water by coagulation of organics, particulate matter and bacteria. Aluminium in drinking water forms only a small part of the total daily intake. It is more likely to be toxic if present in the labile, monomeric form in which it may make a disproportionate contribution to the amount absorbed from the gastro-intestinal tract. The greater bioavailability of Al in drinking water may therefore render it a relatively more harmful source than food (Martyn et al. 1989). The role of aluminium in Alzheimer's disease (let alone the role of Al in water) remains a controversial subject. The WHO guideline of 0.2 mg l^{-1} is based on a compromise between reducing aesthetic problems and retaining the efficacy of water treatment; they conclude that it is not presently possible to derive a health-based guideline.

It is likely that high aluminium occurs in groundwater in a number of developing countries where acid waters are generated. However, any health effects are unlikely to have been reported due to the absence of long-term records and the overriding influence of other diseases.

Beryllium (Be). Beryllium is known to be toxic at industrial exposure levels (Griffits et al. 1977), yet to date only limited data have been reported on its occurrence and toxicity in natural waters and elsewhere in the environment. Of particular concern is the possibility that beryllium might be mobilized along with aluminium under conditions of increasing acidity (Vesely et al. 1989). Beryllium is substituted as a trace component in the silicate lattice of some rock-forming minerals. It is also present as the minerals beryl and bertrandite (Be$_4$Si$_2$O$_7$(OH)$_2$) and is concentrated in residual deposits of silicic volcanic rocks (Brookins 1988). It is especially concentrated in acidic waters, being present as dissolved Be^{2+} at pH < 5.5 (e.g. Edmunds & Trafford 1993) but may also be soluble as Be(OH)$_x$ complexes at higher pH. Beryllium solubility may also be enhanced by the formation of F-complexes. No WHO guide level has been set for Be in drinking water because of insufficient toxicological data and on the assumption that its concentrations in drinking waters must be very low (WHO 1993).

Redox-related controls

The complete reduction of oxygen in an aquifer is accompanied by a sharp decrease in the redox potential. The mobility of some species (NO$_3$, Se, U for example) is favoured under oxidizing conditions whilst others (especially Fe) have increased mobility under reducing conditions. A redox boundary will be found at shallow depths especially in soils or aquifers high in organic matter or sulphide minerals which will act as the main substrate (electron donor) for the reduction of oxygen but in organic-deficient sediments, oxidizing conditions may persist for thousands of years.

Nitrate. Nitrate is often a major concern in developed countries as a result of well documented concerns about its potential health problems, including links with methaemoglobinaemia and stomach cancers. It is not discussed in detail here because excessive nitrate concentrations are related mainly to pollution and have been comprehensively reviewed elsewhere (Foster *et al.* 1982; Chilton *et al.* 1994). Two points are emphasized, however, in the context of this review on geochemical controls. In many arid regions the natural baseline concentrations may well exceed the recommended limits due to natural rather than anthropogenic factors, for example fixation by leguminous plants (Edmunds 1994) or microorganisms (Barnes *et al.* 1992). Secondly, concentrations of NO_3 will persist under oxidizing conditions but *in-situ* denitrification will rapidly take place once oxygen has reacted completely if electron donors are available in the system.

Iron and manganese. Under reducing conditions concentrations of dissolved Fe and Mn can reach several $mg\,l^{-1}$, although much may be in colloidal rather than truly dissolved form. The solubility of Fe and Mn is also greater at low pH. Manganese is an essential element and is readily absorbed. There is limited evidence that it may be toxic at high concentrations (Loranger *et al.* 1994) and a provisional limit of $0.5\,mg\,l^{-1}$ has been set (WHO 1993). Water with high Mn and/or Fe concentrations is usually unpalatable in terms of taste, odour, staining of laundry and discoloration of food (Gale & Smedley 1989). These aesthetic problems may be exacerbated by the presence of Fe bacteria (e.g. *Thiobacillus ferro-oxidans, Gallionella*) which obtain energy from the oxidation of Fe(II) and are responsible for biofouling of aquifers in some Fe-rich areas.

Most Fe problems relate to dissolution of Fe-bearing minerals but problems may also arise from corrosion of ferrous casing, pumps and pipework in supply boreholes. Iron especially is a very common problem in groundwater globally. The problem is well documented in developing countries where communities may be poorly equipped to treat affected supplies. Iron problems are reported for example in parts of India, Ghana (Pelig-Ba *et al.* 1991), Thailand (Ramnarong 1991) and Sri Lanka, Malaysia, Vietnam, Indonesia (Lawrence & Foster 1991), South Africa (Chibi 1991) and the former Soviet Union (Kraynov & Solomin 1982).

Arsenic. Arsenic is toxic and carcinogenic. Hyperpigmentation, depigmentation, keratosis and peripheral vascular disorders are the most commonly reported symptoms of chronic arsenic exposure (Matisoff *et al.* 1982; Chen *et al.* 1994; Morton & Dunette 1994), but skin cancer and a number of internal cancers can also result. Toxicity depends on the form of As ingested, notably the oxidation state and whether in organic or inorganic form. Reduced forms of As are apparently more toxic than oxidized forms, with the order of toxicity from greatest to least being arsine, organo-arsine compounds, arsenite and oxides, arsenate, arsonium, native arsenic (e.g. Welch *et al.* 1988) although there is some evidence for *in-vivo* reduction of arsenate species (Vahter & Envall 1983). Arsenic intake by humans is probably greater from food (e.g. seafood) than from drinking water; however, that present in fish is overwhelmingly present as organic forms of low toxicity. Drinking water therefore represents by far the greatest hazard since the species present in groundwater are predominantly the more toxic inorganic forms (Ferguson & Gavis, 1972; Smedley *et al.* 1996). WHO has recently reduced its recommended limit for As in drinking water from $50\,\mu g\,l^{-1}$ to $10\,\mu g\,l^{-1}$ in response to evidence from toxicological studies. Regulatory bodies such as the EU and US-EPA are currently considering similar revision of the maximum value.

The crustal average As concentration is $2\,mg\,kg^{-1}$ (Tebbutt 1983). Arsenic occurs as a trace element in many rocks and minerals but is especially concentrated in sulphide minerals such as orpiment (As_2S_3), arsenopyrite (FeAsS), realgar (AsS) and enargite (Cu_3AsS_4). Coal may contain about $2000\,mg\,kg^{-1}$ As (Onishi 1969) and phosphorite may also be enriched.

Arsenic species in aqueous systems consist principally of arsenite ($H_nAsO_3^{3-n}$) and arsenate ($H_nAsO_4^{3-n}$) oxyanions. These species are highly soluble over a wide range of Eh and pH conditions. However, under reducing conditions in the presence of sulphide, As mobility is reduced due to precipitation as orpiment, realgar or arsenopyrite (although at low pH, the aqueous species $HAsS_2$ may be present). Biomethylation of As may also take place resulting in the production of monomethylarsonic acid (MMAA) and dimethylarsinic acid (DMAA). In natural waters these are usually rare compared to the organic forms but may be present in relatively high concentrations in organic-rich waters.

Arsenic is strongly sorbed onto, or coprecipitated with, ferric hydroxide (Fe(OH)$_3$), the arsenate forms (5+ oxidation state) being more strongly sorbed than the arsenite forms (3+ oxidation state). This results in potentially much greater concentrations of dissolved As under reducing conditions, not only because of the lower sorption affinity but also because Fe(OH)$_3$ is more soluble at low Eh. Many studies of groundwater and sediments have detected correlations between As concentration and both Fe and Eh (e.g. Matisoff et al. 1982; Belzile 1988; Varsányi et al. 1991). However, As may also be present in oxidizing waters, particularly in groundwater environments where oxidation of sulphide minerals is occurring (Smedley et al. 1995, 1996). Arsenic is also readily sorbed onto aluminium hydroxide (Al(OH)$_3$), except at low pH where Al is itself stable in dissolved ionic form.

Occurrences of high As in drinking water are relatively rare. Most recorded cases are associated with sources of natural sulphide minerals, most notably pyrite and arsenopyrite, and these are often exacerbated by sulphide-mining activity. In the latter case the high As may be generated by oxidation *in situ* or by the processing and disposal of mine wastes. Incidences of high As have been noticed particularly in Taiwan (Tseng et al. 1968), S. America (Zaldivar 1974; Henriquez & Gischler 1980), Mexico (Cebrian et al. 1994), Ghana (Smedley et al. 1996), India (Chatterjee et al. 1995; Das et al. 1995) and Thailand (Ramnarong 1991).

Element deficiencies related to geology

Some water-related health problems are created by element deficiencies rather than excesses. Such diseases are most apparent in rural communities where water and food are locally derived and little exotic produce is consumed. In 'developed' societies, such deficiencies have normally diminished due to broader diet, wider provenance of foodstuffs and dietary supplements. It is possible to delineate large areas containing element deficiencies which are closely related to the local geology and/or geographical location. Three elements (Se, F, I) have well documented deficiency-related health problems, although Se and F also give rise to disease if present above threshold values defined earlier.

Selenium. Trace concentrations of Se are essential in the diet of humans and animals (e.g. Oldfield 1972) and Se deficiency may promote a health problem. Symptoms include muscular degeneration, impeded growth, fertility disorders, anaemia and liver disease (Låg 1984; Peereboom 1985). However, at high ingested concentrations of 10 mg day^{-1} or greater, other problems such as gastro-intestinal ailments, skin discoloration and tooth decay may occur (Tebbutt 1983). Selenium toxicity in American Indians has been reported by Beath (1962). The WHO recommended limit for Se in drinking water is 10 μg l^{-1}, but concentrations in natural water rarely exceed 1 μg l^{-1}.

The geochemistry of Se is similar to that of sulphur. It occurs naturally in four oxidation states: 2−, 0, 4+ and 6+. In its 2− state, Se occurs as H$_2$Se, a highly toxic and reactive gas which readily oxidizes in the presence of oxygen. In elemental form (Se^0), Se is insoluble and therefore non-toxic. The element occurs in the 4+ oxidation state as inorganic selenite (SeO$_3^{2-}$) which is highly toxic. However, under reducing and acidic conditions selenite is readily reduced to elemental Se (NRC 1976; Howard 1977). Oxidizing and alkaline conditions favour the stability of the 6+ form, selenate (SeO$_4^{2-}$), which is highly soluble. Selenium mobility should therefore be greater in oxidizing aquifers, although its dissolved concentration may be limited by the fact that it readily sorbs onto ferric hydroxide which precipitates under such conditions (Howard 1977).

Selenium has a strong affinity for organic matter and is readily incorporated into sulphide minerals. It is therefore often associated with sulphide-bearing hydrothermal veins and is present in relatively high concentrations in U deposits (Naftz & Rice 1989). It may also form the mineral ferroselite (FeSe$_2$) if present in sufficiently high concentrations.

Few studies of Se in drinking water in developing countries have been carried out, but Iyengar & Gopal-Ayengar (1988), for example, cited the incidence of endemic Keshan disease, a chronic cardiomyopathy thought to be related to Se deficiency, in many parts of China. The disease afflicted several thousand people, principally in hilly and mountainous districts (altitude > 1600 m). There is also some evidence that remoteness from the sea may lead to Se deficiency (Låg 1984).

Iodine. The association of I deficiency in the human diet with endemic goitre has long been recognized. Goitre results from enlargement of the thyroid in order to compensate for I deficiency in hormone production. The condition has been recognized in many areas all over the world (e.g. Kelly & Sneddon 1960).

Iodine is not a major element in minerals and does not enter readily into the crystal lattice. It is, however, chalcophile and may be found in

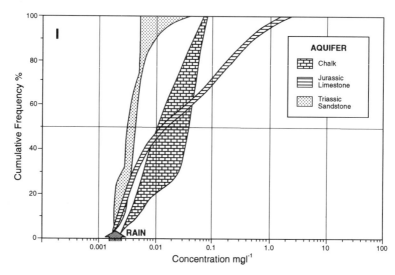

Fig. 2. Iodine concentrations in groundwater from UK aquifers (Chalk, Jurassic limestones, Triassic sandstones). Note that only carbonate rocks contribute I from water–rock interaction, over and above rainfall inputs.

higher concentrations in association with organic carbon (e.g. Fuge & Johnson 1986). Iodine is also readily adsorbed onto Fe and Al oxides (Whitehead 1984).

The principal natural source of I is sea water (mean value 58 μg l^{-1}; Fuge & Johnson 1986), but additional sources are formation waters, fluid inclusions and volcanic emanations. Iodine concentrations in the environment are increased by man's activities. I is used in herbicides, fungicides, sterilants, detergents, pharmaceuticals and the food industry. Iodine is also released into the environment from fossil fuel combustion, car exhausts and from sewage. The geochemical cycle of I involves volatilization to atmospheric I (as iodine gas, I_2 or as methyl iodide, CH_3I), atmospheric transport and subsequent loss to the biosphere and lithosphere as wet and dry deposition (e.g. Whitehead 1984; Fuge & Johnson 1986; Fuge 1989). Iodine in rainfall over coastal areas is therefore generally higher (1.5–2.5 μg l^{-1}; Whitehead 1984) than over inland areas (1 μg l^{-1} or less; Fuge 1989).

Soils generally have higher concentrations of I than their parent rocks, especially shallow soils, presumably owing to the addition of I from the atmosphere (Fuge & Johnson 1986). Whitehead (1979) found a range of 0.5–98.2 mg kg^{-1} (dry weight) in surface soils from the UK, the highest being in fen peat and the lowest in podzolic sands poor in organic carbon. Soil I concentration has in some places been enhanced by addition of seaweed as a fertilizer (Whitehead 1984). High natural baseline I concentrations may be found in carbonate aquifers probably derived from the oxidation of organic matter (Edmunds et al. 1989). This is well illustrated for the Chalk aquifer of the London basin (Fig. 2) where the median concentration is 32 μg l^{-1} and the I/Cl ratio is 5.84 × 10^{-4} which is about four times higher than for non-carbonate aquifers in the UK.

About 20% of the daily I requirement of humans is likely to come from drinking water, the remaining 80% being derived from food (Fuge 1987). Dairy products, meat and fish are especially enriched in I, as is iodized salt where available. Since drinking water is a relatively minor I source, links between concentrations in water and occurrence of endemic goitre must be relatively tenuous. Nonetheless, they can serve as an indicator of I levels in the local environment (e.g. soils, local vegetation) and will therefore be useful for the determination of local health risk.

Total I concentrations in drinking waters range between 0.01 and c. 70 μg l^{-1}, depending on location, topography and rainfall pattern (mean river water content 5 μg l^{-1}; Fuge & Johnson 1986; Fuge 1989). Concentrations much below this mean value are frequently associated with the occurrence of goitres. Day & Powell-Jackson (1972) for example reported concentrations of < 1 μg l^{-1} in goitrous areas

Table 3. *Impact of fluoride in drinking water on health (from Dissanayake 1991)*

Concentration of fluoride (mg l^{-1})	Impact on health
Nil	Limited growth and fertility
0–0.5	Dental caries
0.5–1.5	Promotes dental health resulting in healthy teeth, prevents tooth decay
1.5–4.0	Dental fluorosis (mottling of teeth)
4.0–10.0	Dental fluorosis, skeletal fluorosis (pain in back and neck bones)
> 10.0	Crippling fluorosis

of Nepal. Kelly and Sneddon (1960) produced maps of the distribution of endemic goitre and found that almost all countries regardless of climate, race or wealth had some recorded evidence of the problem although for reasons stated above the problem today tends to be restricted to rural areas of developing countries (Wilson 1953; Coble et al. 1968; Mahadeva et al. 1968; Kambal et al. 1969; Fuge & Johnson 1986; Rosenthal & Mates 1986). It has frequently been associated with mountainous areas, especially the Alps, Himalayas and Andes and regions distant from the coast (Fuge 1987).

There is therefore a very clear association between the geochemical occurrence of iodine, especially in natural waters, and the incidence of goitre. Remediation through dietary supplement is generally effective such that goitre now remains a political rather than a geochemical problem.

Mineral-saturation control

The upper limits of solubility for some elements are naturally maintained due to saturation with certain minerals. Health-related problems usually emerge when abnormally low concentrations of associated ions allow the concentrations of the harmful element to increase. These increases can be predicted from and described by the relevant solubility product (K_s). Mineral saturation exerts a strong control on F and Ba concentrations.

Fluorine. An extensive literature exists on the occurrence of F in natural waters, both in industrialized and developing countries. This is because it is a fairly common trace element and its health effects have been recognized in many parts of the world. At low concentrations of less than about 0.5 mg l^{-1} total F, dental caries may result, whilst at higher concentrations chronic exposure can result in dental fluorosis (mottled enamel) or skeletal fluorosis (e.g. Rajagopal & Tobin 1991). Concentrations above which these become problematic are around 2 mg l^{-1} and 4 mg l^{-1} respectively (Table 3), although poor nutrition is also recognized as an important contributory factor. The effects are permanent and incurable. High F concentrations in drinking water have also been linked with cancer (Marshall 1990). The WHO recommended limit for F in drinking water is 1.5 mg l^{-1}.

During the 1950s in the USA and Europe, it was found that introduction of F in toothpaste and fluoridation of public water supplies to a concentration of about 1 mg l^{-1} reduced the incidence of dental caries by more than 50% (Diesendorf 1986). The benefits of water fluoridation in recent years have been much less pronounced, probably as a result of long-term use of topical F, increased dietary F, improved dental health education and reduced sugar intake.

The average crustal abundance of F is 300 mg kg^{-1} (Tebbutt 1983). Fluorite (CaF_2) is the most common F-bearing mineral but it is also present in apatite ($Ca_5(Cl,F,OH)(PO_4)_3$) and in trace quantities in amphibole, mica, sphene and pyroxene. Fluoride occurrence is

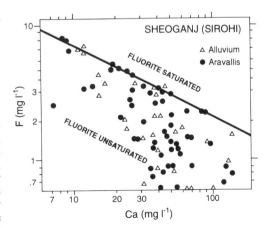

Fig. 3. Fluoride concentration in groundwaters from Rajasthan, India showing solubility control by fluorite (from Handa 1975).

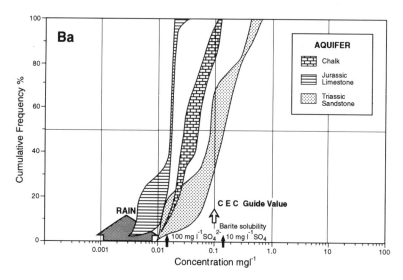

Fig. 4. Barium concentrations in groundwater from UK aquifers (Chalk, Jurassic limestone, Triassic sandstone). The Ba concentrations in rain and in groundwaters at saturation with 10 and 100 mg l^{-1} SO$_4$ are also shown. Note that highest concentrations of Ba occur in sandstone aquifers.

commonly associated with volcanic activity (being especially high in volcanic glasses), geothermal fluids and granitic rocks. Thermal, high pH waters have especially high concentrations.

The principal form of F in water is as free dissolved F$^-$ but at low pH, the species HF^0 may be stabilized (at pH 3.5 this may be the dominant species; Hem 1985). F readily forms complexes with Al, Be, Fe^{3+}, B and Si. Concentrations of F in water are limited by fluorite solubility, such that in the presence of 10^{-3} M Ca, F should be limited to 3.1 mg l^{-1} (Hem 1985). It is therefore the absence of Ca in solution which allows higher concentrations of F to be stable (Fig. 3). High F concentrations may therefore be expected in groundwater in Ca-poor aquifers and in areas where F minerals (or F-substituted minerals e.g. biotite) are common. Fluoride concentration will also increase in groundwaters where cation exchange of Ca for Na takes place.

In many developing countries high F concentrations have been reported in association with rift zones, volcanic rocks and granitic (Ca-poor) basement rocks (Bugaisa 1971; Kilham & Hecky 1973; Hadwen 1975; Nanyaro et al. 1984). High F concentrations are noted in the Kenyan part of the African Rift Valley (Ockerse 1953; Gaciri & Davies 1993) and in Uganda (Møller et al. 1970) where incidences of dental fluorosis have been linked with concentrations of F up to 3 mg l^{-1}.

Fluoride problems have also received much attention in Asia. Teotia et al. (1981) note that endemic fluorosis affects nearly one million people in India, the high concentrations of dissolved fluoride in drinking water resulting from dissolution of fluorite, apatite, francolite and topaz in the bedrocks. Handa (1975) noted the general negative correlation between F and Ca concentration in Indian groundwater (Fig. 3). Dissanayake (1991) found concentrations of F in the Dry Zone of Sri Lanka up to 10 mg l^{-1} associated with dental fluorosis and possibly skeletal fluorosis. In the Wet Zone, intensive rainfall and the long-term leaching of F from rocks is probably responsible for low groundwater F concentrations. Here, the incidence of dental caries is reported to be high. Excessive F concentrations in water have also been found in Algeria and Kenya (Tjook 1983), Turkey (Pekdeger et al. 1992), South America (Lloyd & Helmer 1991), Ghana (UN 1988; Amoah 1990; Smedley et al. 1995) and Ivory Coast (especially in granitic areas; Akiti et al. 1990), Thailand (Ramnarong 1991; Table 4) and China (Zhaoli et al. 1989).

Barium. Barium occurs as a minor element in many rock types but is most abundant in acid igneous rocks. It is readily released during water–rock interaction but its solubility is controlled by the solubility of barite. Thus concentrations of barium in natural waters should be inversely proportional to the sulphate concentrations. High barium concentrations are to be anticipated mainly where sulphate reduction has occurred and where SO_4 concentrations are less than $10\,mg\,l^{-1}$. Barium has a possible association with cardiovascular disease (Brenniman et al. 1981; WHO 1993) and a guideline maximum value for drinking waters of $0.7\,mg\,l^{-1}$ is given by WHO.

The concentration of Ba rarely exceeds $1\,mg\,l^{-1}$ in natural waters (Edmunds et al. 1989). Median concentrations in waters unsaturated with respect to barite tend to be highest in groundwaters from non-carbonate aquifers (Fig. 4) and this reflects the lower geochemical abundance in carbonate rocks and the higher Ba in silicate minerals such as K-feldspar.

Other elements

Lead. Lead is present as a major element in galena (PbS) and is a common constituent in hydrothermal mineral veins. Its average crustal abundance is $16\,mg\,kg^{-1}$. Pb is also produced from smelting, motor-vehicle exhaust fumes and from corrosion of lead pipework. Lead solubility is controlled principally by $PbCO_3$ and low-alkalinity, low-pH waters can have higher Pb concentrations (Hem 1985). The WHO maximum recommended concentration for Pb has recently been reduced from $50\,\mu g\,l^{-1}$ to $10\,\mu g\,l^{-1}$ because of concerns about chronic toxicity, although this concentration is rarely exceeded in natural waters. There is an extensive literature on lead in the environment derived from leaded petrol and from lead pipes in soft water areas.

Lead is a cumulative poison, initiating tiredness, irritability, anaemia, behavioural changes and impairment of intellectual functions in affected patients (Tebbutt 1983). Ramnarong (1991) cites a case of lead poisoning in Thailand, where five out of ten patients died in 1979. Water from the local well was found to contain $53.5\,mg\,l^{-1}$ Pb and soils contained $0.13-4.92$ $mg\,kg^{-1}$ as a result of pollution from leachate derived from a local refuse dump. Pelig-Ba et al. (1991) also reported relatively high natural Pb concentrations (around $0.15\,mg\,l^{-1}$) in acid water from granitic terrains in Ghana.

Cadmium. Cadmium occurrence in the environment is from both natural and human sources. It is usually associated with zinc ores and may be present in volcanic emissions and released from vegetation (Robards & Worsford 1991). Environmental levels are greatly enhanced by industrial operations as Cd is commonly used as a pigment, in paint, plastics, ceramics and glass manufacture, in metal fabrication and finishing. It is also released from smelting of copper ores and from sewage sludge (Nicholson et al. 1983). Cadmium is an acute toxin, producing symptoms such as giddiness, vomiting, respiratory difficulties, cramps and loss of consciousness at high doses. Chronic exposure to the metal can lead to anaemia, anosmia (loss of sense of smell), cardiovascular diseases, renal problems and hypertension (Mielke et al. 1991, Robards & Worsford 1991). There is also evidence that increased Cd ingestion can promote Cu and Zn deficiency in humans, both necessary elements in metabolic processes (Petering et al. 1971). Cd may also be a carcinogen (Tebbutt 1983).

Exposure of humans to Cd is likely to be greatest from food intake and inhalation. Drinking water should have lower Cd concentrations unless water sources are affected by volcanic exhalations, landfill leachate, or mine waters. Evidence of endemic Cd poisoning was, for example, found in Toyonia Prefecture, Japan. This resulted from the consumption of rice grown in irrigation water contaminated by local mining effluent (Robards & Worsfold 1991).

The WHO limit for Cd in drinking water is $5\,\mu g\,l^{-1}$ but Nicholson et al. (1983) detected renal damage in seabirds as a result of exposure to both Cd and Hg at concentrations below this limit. Cadmium solubility is limited by $CdCO_3$ (Hem 1985) and is therefore found in higher concentrations at low pH. Cd may also be sorbed onto organic substances such as humic and fulvic acids and hence organic-rich waters may have higher Cd concentrations given a local Cd source. Few data are available on Cd in drinking water in developing countries. Concentrations are generally expected to be low, but acid waters and especially those close to mines and sewage effluents may have higher concentrations.

Conclusions

Not all elements relevant to human health have been considered in detail in this paper. Other potentially harmful dissolved constituents in some groundwaters include U, Sb, Th, CN, Hg, Ni and Cr. It is clear that health relates not only to excesses of trace elements in drinking water supplies, but may also relate to deficiencies (e.g. Se). With some elements health depends upon a delicate balance between the two (e.g. I, F, Se). The relationships between trace elements in water and health are very complex. Water is not their only dietary source and often relationships may be masked by the effects of other elements. Competition between different elements in the body for example, can either exacerbate health problems or effect some form of protection (e.g. the protective effects of Fe, Cu and Zn on Cd toxicity; Underwood 1979). Water may, however, be a useful indicator of the local environmental levels of trace elements (e.g. in food, soils, rocks, atmosphere). The links between excesses or deficiencies of particular trace elements and health are likely to be more noticeable in developing countries especially in rural areas because of a much greater dependence on water and food of local provenance.

Trace element mobility is dependent upon physico-chemical conditions and interaction of other chemical constituents but varies with each individual element. Al, Be, Pb, Cd, Fe and Mn for example are preferentially mobilized under acidic conditions. Arsenic (along with Fe and Mn) may be more soluble under reducing conditions, whilst Se and U are more mobile in oxidizing environments. Fluoride is most mobile in alkaline conditions, given low dissolved Ca concentrations and I may be a largely conservative element, depending mainly on I concentrations in local input sources. Deficiency in both I and Se have been observed in regions remote from the sea, particularly at high altitudes. Microbiological processes also exert an important influence on trace element speciation. Disregarding the additional effects of industry and agriculture on trace element content of groundwater, acid-mobile elements should be concentrated in mining effluent, especially those associated with pyrite oxidation, and in hydrothermal and geothermal areas. Likewise, As, Se and U should have elevated concentrations in groundwater of hydrothermal and geothermal areas. High F concentrations have been reported in zones of extensional tectonism and volcanism (e.g. the East African Rift) and in association with F-bearing hydrothermal mineral veins.

Further geochemical work on natural baseline variations in the chemistry of surface and groundwaters is still required for those elements for which few data currently exist. In addition there is a need to convey more information on natural baseline conditions to those working in water agencies who may wrongly assign anomalous inorganic concentrations to pollution origins. Geochemical studies have sometimes been 'ahead' of related epidemiological and especially clinical studies and it is recognized that it is much easier to define a particular geochemical province for a given element than to establish the relationships with health effects. Water of high quality is nevertheless of such a high priority worldwide that hydrogeochemical databases are an essential component in epidemiological studies.

From the practical point of view, some conclusions may be drawn in relation to studies connected with rural water schemes. The problem of arsenic which is mainly related to sulphide oxidation can be avoided or minimized by preventing oxidation through good management practice (e.g. avoiding excessive drawdown). The other main area of concern, fluoride, is difficult to manage, although groundwater with shorter residence times in general should be more likely to contain lower levels of fluoride. Those responsible for development schemes must be aware of these geochemical provinces likely to give rise to problems. Furthermore, there is a tendency in many aid schemes to specify a certain depth and design and then to complete many boreholes to this specification. If concentrations of iron and manganese are too high, villagers will abandon the wells. Therefore, it is most important that a pilot investigation of the water quality is carried out before large wellfield projects are undertaken.

This paper has benefited from reviews by R. Fuge and one anonymous reviewer; it is published with the approval of the Director, British Geological Survey (Natural Environment Research Council).

References

AKITI, T. T., ASSOUMA, D., BOUKARI, M. & KABORE, F. 1990. *L'hydrogéologie de l'Afrique de l'Ouest*. 2nd ed. Ministère de la Cooperation et du Développement.

AMOAH, C. 1990. Problems of maintenance of water quality in arid and semi-arid regions of West Africa. *In:* STOUT, G. E. & DEMISSIE, M. (eds) *The State-of-the-Art of Hydrology and Hydrogeology in the Arid and Semi-arid Areas of Africa*. Proceedings of the Sahel Forum, Burkina Faso, UNESCO, 723–735.

BACHE, B. W. 1986. Aluminium mobilisation in soils and waters. *Journal of the Geological Society, London*, **143**, 699–706.

BARNES, C. J., JACOBSON, G. & SMITH, G. D. 1992. The origin of high-nitrate groundwaters in the Australian arid zone. *Journal of Hydrology*, **137**, 181–197.

BEATH, A. 1962. Selenium poisons Indians. *Science News Letter*, **81**, 254.

BELZILE, N. 1988. The fate of arsenic in sediments of the Laurentian Trough. *Geochimica et Cosmochimica Acta*, **52**, 2293–2302.

BRENNIMAN, G. R., KOJOLA, W. H., LEVEY, P. S., CARNOW, B. W. & NAMETAKA, T. 1981. High barium levels in public drinking water and its association with elevated blood pressure. *Archives of Environmental Health*, **36**, 28–32.

BROOKINS, D. G. 1988. *Eh-pH Diagrams for Geochem-*

istry. Springer, New York.
BUGAISA, S. L. 1971. Significance of fluorine in Tanzania drinking water. *In: Proceedings of the Conference on Rural Water Supply in East Africa, Dar-es-Salaam, 1971*, 107–113.
CALABRESE, E. J. & TUTHILL, R. W. 1981. The influence of elevated levels of sodium in drinking-water on elementary and high-school students in Massachusetts. *Science of the Total Environment*, **18**, 117–133.
CEBRIÁN, M. E., ALBORES, A., GARCÍA-VARGAS, G., DEL RAZO, L. M. & OSTROSKY-WEGMAN, P. 1994. Chronic arsenic poisoning in humans: the case of Mexico. *In:* NRIAGU, J. O. (ed.) *Arsenic in the environment, part II: Human Health and Ecosystem Effects*. Wiley, New York.
CEC (Commission of the European Community). 1980. *Directive relating to the Quality of Water intended for human consumption (80/778/EEC)*. CEC, Brussels.
CHATTERJEE, A., DAS, D., MANDAL, B. K., CHOWDHURY, T. R., SAMANTA, G. & CHAKRABORTI, D. 1995. Arsenic in groundwater in six districts of West Bengal: the biggest arsenic calamity in the world. Part 1. Arsenic species in drinking water and urine of affected people. *Analyst*, **120**, 643–650.
CHEN, S.-L., DZENG, S. R., YANG, M-H., CHIU, K-H., SHIEH, G-M. & WAI, C. M. 1994. Arsenic species in groundwaters of the Blackfoot disease area, Taiwan. *Environmental Science Technology*, **28**, 877–881.
CHIBI, C. 1991. Design and performance of a community-level iron removal plant. *Waterlines*, **10**, 9–10.
CHILTON, P. J., LAWRENCE, A. R. & STUART, M. E. 1994. The impact of tropical agriculture on groundwater quality. *In:* NASH, H. & MCCALL, G. J. H. (eds) *Groundwater Quality*. Chapman & Hall, London, 113–122.
COBLE, Y., DAVIS, J., SCHULBER, T. A., HETA, F. & AWAD, A. Y. 1968. Goiter and iodine deficiency in Egyptian oases. *American Journal of Clinical Nutrition*, **21**, 277–283.
COMA. 1994. *Nutritional Aspects of Cardiovascular Disease*. No. 46, Committee on Medical Aspects of Food Policy, HMSO, London.
DAS, D., CHATTERJEE, A., MANDAL, B. K., SAMANTA, G. & CHAKRABORTI, D. 1995. Arsenic in groundwater in six districts of West Bengal, India: the biggest arsenic calamity in the world. Part 2. Arsenic concentration in drinking water, hair, nails, urine, skin-scale and liver tissue (biopsy) of the affected people. *Analyst*, **120**, 917–924.
DAY, T. K. & POWELL-JACKSON, P. R. 1972. Fluoride, water hardness and endemic goitre. *The Lancet*, **1**, 1135–1138.
DIESENDORF, M. 1986. The mystery of declining tooth decay. *Nature*, **322**, 125–129.
DISSANAYAKE, C. B. 1991. The fluoride problem in the groundwater of Sri Lanka – environmental management and health. *International Journal of Environmental Studies*, **38**, 137–156.
——, SENARATNE, A. & WEERASOORIYA, V. R. 1982. Geochemistry of well water and cardiovascular diseases in Sri Lanka. *International Journal of Environmental Studies*, **19**, 195–203.
DRISCOLL, C. T. 1984. A procedure for the fractionation of aqueous aluminium in dilute acidic waters. *International Journal of Environmental Analytical Chemistry*, **16**, 267–283.
EDMUNDS, W. M. 1994. Characterisation of groundwaters in semi-arid and arid zones using minor elements. *In:* NASH, H. G. & MCCALL, G. J. H. (eds) *Groundwater Quality*, Chapman & Hall, London, 19–30.
—— & KINNIBURGH, D. G. 1986. The susceptibility of U.K. groundwaters to acidic deposition. *Journal of the Geological Society, London*, **143**, 707–720.
—— & TRAFFORD, J. M. 1993. Beryllium in river baseflow, shallow groundwaters and major aquifers in the U.K. *Applied Geochemistry*, Suppl. Issue 2, 223–233.
——, COOK, J. M., KINNIBURGH, D. G, MILES, D. L. & TRAFFORD, J. M. 1989. Trace-element occurrence in British groundwaters. BGS, Research Report, SD/89/3.
——, ——, DARLING, W. G. KINNIBURGH, D. G., MILES, D. L., BATH, A. H., MORGAN-JONES, M. & ANDREWS, J. N. 1987. Baseline geochemical conditions in the Chalk aquifer, Berkshire, U.K: a basis for groundwater quality management. *Applied Geochemistry*, **2**, 251–274.
FERGUSON, J. F. & GAVIS, J. 1972. A review of the arsenic cycle in natural waters. *Water Research*, **6**, 1259–1274.
FOSTER, H. D. 1992. *Health, disease and the environment*. Belhaven, London.
FOSTER, S. S. D., CRIPPS, A. C. & SMITH-CARINGTON, A. 1982. Nitrate leaching to groundwater. *Philosophical Transactions of the Royal Society of London*, **296**, 477–489.
FUGE, R. 1987. Iodine in the environment: its distribution and relationship to human health. *In:* HEMPHILL, D. D. (ed.) *Trace Substances in the Environment*, University of Missouri, 74–87.
—— 1989. Iodine in waters: possible links with endemic goitre. *Applied Geochemistry*, **4**, 203–208.
—— & JOHNSON, C. C. 1986. The geochemistry of iodine: a review. *Environmental Geochemistry and Health*, **8**, 31–54.
GACIRI, S. J. & DAVIES, T. C. 1993. The occurrence and geochemistry of fluoride in some natural waters of Kenya. *Journal of Hydrology*, **143**, 395–412.
GARDNER, M. J. 1976. Soft water and heart disease. *In:* LENIHAN, J. & FLETCHER, W. W. (eds) *Environment and Man*, Blackie, Glasgow, 116–135.
GALE, I. N. & SMEDLEY, P. L. 1989. Iron in groundwater – a survey of the extent and nature of the problems and methods of removal. BGS, Technical Report, WD/89/29.
GRIFFITS, W. R., ALLAWAY, W. H. & GROTH, D. H. 1977. *Beryllium. Geochemistry and the Environment*, Vol. II. National Academy of Science, Washington DC, 7–10.
HADWEN, P. 1975. *Fluoride in groundwater in Ethiopia*. Geological Survey of Ethiopia. unpublished re-

port.

HANDA, B. K. 1975. Geochemistry and genesis of fluoride-containing ground waters in India. *Groundwater*, **13**, 275–281.

HEM, J. D. 1985. *Study and interpretation of the chemical characteristics of natural water.* US Geological Survey Water-Supply Paper, **2254**.

HENRIQUEZ, H. & GISCHLER, C. 1980. *Introduction of assessment and control of natural arsenic and associated contaminants in the south Andean countries.* Montevideo, ROSTLAC.

HOWARD, J. H. 1977. Geochemistry of selenium: formation of ferroselite and selenium behaviour in the vicinity of oxidising sulphide and uranium deposits. *Geochimica et Cosmochimica Acta*, **41**, 1665–1678.

IYENGAR, G. V. & GOPAL-AYENGAR, A. R. 1988. Human health and trace elements including effects on high-altitude populations. *Ambio*, **17**, 31–35.

KAMBAL, A., RAHMAN, I. A., GREIG, W. R., GRAY, H. W. & MCGILL, E. M. 1969. Endemic goitre in Sudan. *The Lancet*, **1**, 233–235.

KELLY, F. C. & SNEDDON, W. W. 1960. *Endemic Goitre.* WHO Monograph Series **44**.

KILHAM, P. & HECKY, R. E. 1973. Fluoride: geochemical and ecological significance in East African waters and sediments. *Limnology Oceanography*, **18**, 932–945.

KRAYNOV, S. R. & SOLOMIN, G. A. 1982. Iron-bearing subsurface waters and the problem of water supply for household use and drinking. *International Geology Review*, **25**, 601–607.

LACEY, R. F. & SHAPER, A. G. 1984. Changes in water hardness and cardiovascular death rates. *International Journal of Epidemiology*, **13**, 18–24.

LÅG, J. 1984. A comparison of selenium deficiency in Scandinavia and China. *Ambio*, **13**, 286–287.

LAMONT, P. 1959. A soft water zone in the Lincolnshire Limestone. *Journal of British Water Association*, **41**, 48–71.

LANGANEGGER, O. 1991. Groundwater quality in rural areas of Western Africa. In: *The State-of-the-Art of Hydrology and Hydrogeology in the Arid and Semi-Arid Areas of Africa.* Proceedings of the Sahel Forum, Burkina Faso. UNESCO, 575–584.

LAWRENCE, A. R. & FOSTER, S. S. D. 1991. *Urban groundwater exploitation and aquifer pollution vulnerability in southern and southeastern Asia.* British Geological Survey, Technical Report, WD/91/74F.

LLOYD, B. & HELMER, R. 1991. *Surveillance of Drinking Water Quality in Rural Areas.* Longman, New York.

LORANGER, S., BIBEAU, B. C. & ZAYED, J. 1994. Le manganèse dans l'eau potable et sa contribution à l'exposition humaine. *Revue de l' Epidémiologie et Santé Publique*, **42**, 315–321.

LUCAS, A. E. & COWELL, D. W. 1984. Regional assessment of sensitivity to acidic deposition for eastern Canada. In: BRICKER, O. P. (ed.) *Geological aspects of acidic deposition.* Acid Precipitation Series, **7**, Ann Arbor Science (Butterworths), 113–119.

MAHADEVA, K., SENEVIRATNE, D. A., JAYATILLEKE, D.
B., SENTHE SHANMUGANATHAN, S., PREMACHANDRA, P. & NAGARAJAH, M. 1968. Further studies on the problem of endemic goitre in Ceylon. *British Journal of Nutrition*, **22**, 527–534.

MARSHALL, E. 1990. The fluoride debate: one more time. *Science*, **247**, 276–277.

MARTYN, C. N., BARKER, D. J. P., OSMOND, C., HARRIS, E. C., EDWARDSON, J. A. & LACEY, R. F. 1989. Geophysical relation between Alzheimers disease and aluminium in drinking water. *The Lancet*, **1**, 59–62.

MASIRONI, R. 1979. Geochemistry and cardiovascular diseases. *Philosophical Transactions of the Royal Society of London*, **B288**, 193–203.

MATISOFF, G., KHOUREY, C. J., HALL, J. F., VARNES, A. W. & STRAIN, W. H. 1982. The nature and source of arsenic in northeastern Ohio groundwater. *Groundwater*, **20**, 446–456.

MAY, H. M., HELMKE, P. A. & JACKSON, M. L. 1979. Gibbsite solubility and thermodynamic properties of hydroxy-aluminium ions in aqueous solution at 25°C. *Geochimica et Cosmochimica Acta*, **43**, 861–868.

MIELKE, H. W., ADAMS, J. L., CHANEY, R. L., MIELKE, P. W. & RAVIKUMAR, V. C. 1991. The pattern of cadmium in the environment of five Minnesota cities. *Environmental Geochemistry and Health*, **13**, 29–34.

MØLLER, I. J., PINDBORG, J. J., GEDALIA, I. & ROED-PETERSEN, B. 1970. The prevalence of dental fluorosis in the people of Uganda. *Archives of Oral Biology*, **15**, 213–225.

MORTON, W. E & DUNNETTE, D. A. 1994. Health effects of environmental arsenic. In: NRIAGU, J. O. (ed.) *Arsenic in the Environment, Part II: Human Health and Ecosystem Effects*, Wiley, New York, 159–172.

MOYNAHAN, E. J. 1979. Trace elements in man. *Philosophical Transactions of the Royal Society of London*, **B288**, 65–79.

NAFTZ, D. L. & RICE, J. A. 1989. Geochemical processes controlling selenium in groundwater after mining, Powder River Basin, Wyoming, USA. *Applied Geochemistry*, **4**, 565–575.

NANYARO, J. T., ASWATHANARAYANA, U., MUNGURE, J. S. & LAHERMO, P. W. 1984. A geochemical model for the abnormal fluoride concentrations in waters in parts of northern Tanzania. *Journal of African Earth Sciences*, **2**, 129–140.

NAS (National Academy of Sciences). 1977. *Drinking water and health.* Safe Drinking Water Committee, US National Research Council. Washington DC.

NICHOLSON, J. K., KENDALL, M. D. & OSBORN, D. 1983. Cadmium and mercury nephrotoxicity. *Nature*, **304**, 633–635.

NRC (National Research Council). 1976. *Selenium.* National Academy of Sciences. Washington.

OCKERSE, T. 1953. Chronic endemic dental fluorosis in Kenya, East Africa. *British Dental Journal*, **95**, 57–60.

OLDFIELD, J. E. 1972. Selenium deficiency in soils and its effect on animal health. *Geological Society of America, Special Paper*, **140**, 57–63.

ONISHI, H. 1969. Arsenic. In: CORRENS, C. W., SHAW, D. M., TUREKIAN, K. K. & ZEMANN, J. (eds) Handbook of Geochemistry, II-3. Springer, Berlin. 33B1–33M1.

PEEREBOOM, J. W. C. 1985. General aspects of trace elements and health. Science of the Total Environment, 42, 1–27.

PEKDEGER, A., ÖZGÜR, N. & SCHNEIDER, H-J. 1992. Hydrogeochemistry of fluorine in shallow aqueous systems of the Gölcük area, SW Turkey. In: KHARAKA, Y. K. & MAEST, A. S. (eds) Proc. 7th Intl. Conference on Water–Rock Interaction, Utah, 1992. 821–824.

PELIG-BA, K. B., BINEY, C. A. & ANTWI, L. A. 1991. Trace metal concentrations in borehole waters from the Upper Regions and the Accra Plains of Ghana. Water, Air and Soil Pollution, 59, 333–345.

PETERING, H. G., JOHNSON, M. A. & STEMMER, K. L. 1971. Studies of zinc metabolism in the rat. Archives of Environmental Health, 23, 93–101.

RAJAGOPAL, R. & TOBIN, G. 1991. Fluoride in drinking water: a survey of expert opinions. Environmental Geochemistry and Health, 13, 3–13.

RAMNARONG, V. 1991. Groundwater quality monitoring and management in Thailand. Proceedings of the Expert Group Meeting on Groundwater Monitoring in Asia and the Pacific, Bangkok.

ROBARDS, K. & WORSFOLD, P. 1991. Cadmium: toxicology and analysis: a review. Analyst, 116, 549–568.

ROSENTHAL, E. & MATES, A. 1986. Iodine concentrations in groundwater of northern Israel and their relations to the occurrence of goiter. Applied Geochemistry, 1, 591–600.

SHAPER, A. J., POCOCK, S. J., WALKER, M., COHEN, N. M., WALE, C. J. & THOMSOM, A. G. 1981. British Regional Heart Study: cardiovascular risk factors in middle-aged men in 24 towns. British Medicine, 283, 179–186.

SMEDLEY, P. L., EDMUNDS, W. M. & PELIG-BA, K. 1996. Mobility of arsenic in groundwater in the Obuasi gold-mining area of Ghana: some implications for human health. This volume.

——, ——, WEST, J. M., GARDNER, S. J. & PELIG-BA, K. 1995. Vulnerability of shallow groundwater quality due to natural geochemical environment: 2: Health problems related to groundwater in the Obuasi and Bolgatanga areas, Ghana. British Geological Survey, Technical Report, WC/95/43.

TEBBUTT, T. H. Y. 1983. Relationship Between Natural Water Quality and Health. UNESCO, Paris.

TEOTIA, S. P. S., TEOTIA, M. & SINGH, R. K. 1981. Hydro-geochemical aspects of endemic skeletal fluorosis in India – an epidemiological study. Fluoride, 14, 69–74.

TJOOK, T. K. 1983. Defluoridation of water supplies. Waterlines, 2, 26–27.

TSENG, W. P., CHU, H. M., HOW, S. W., FONG, J. M., LIN, C. S. & YEN, S. 1968. Prevalence of skin cancer in an endemic area of chronic arsenism in Taiwan. Journal of the National Cancer Institute, 40, 453–463.

UN. 1988. Groundwater in north and west Africa. Natural Resources/Water Series, 18, United Nations, New York.

UNDERWOOD, E. J. 1979. Trace elements and health: an overview. Philosophical Transactions of the Royal Society of London, B288, 5–14.

VAHTER, M. & ENVALL, J. 1983. In vivo reduction of arsenic in mice and rabbits. Environmental Research, 32, 14–24.

VARSANYI, I., FODRE, Z. & BARTHA, A. 1991. Arsenic in drinking water and mortality in the Southern Great Plain, Hungary. Environmental Geochemistry and Health, 13, 14–22.

VESELY, J., BENES, P. & SEVAK, K. 1989. Occurrence and speciation of beryllium in acidified fresh waters. Water Research, 23, 711–717.

VOGT, T. 1986. Water Quality and Health. Study of a Possible Relation Between Aluminium in Drinking Water and Dementia. Sosiale Og Okonomiske Studier 61, Staistisk Sentralbyra, Oslo-Kongsvinger, 60–63.

WELCH, A. H., LICO, M. S. & HUGHES, J. L. 1988. Arsenic in groundwater of the Western United States. Groundwater, 26, 333–347.

WHITEHEAD, D. C. 1979. Iodine in the UK environment with particular reference to agriculture. Journal of Applied Ecology, 16, 269–279.

—— 1984. The distribution and transformations of iodine in the environment. Environment International, 10, 321–339.

WHO (World Health Organisation). 1993. Guidelines for drinking water quality. Geneva.

WILSON, D. C. 1953. Goitre in Ceylon and Nigeria. British Journal of Nutrition, 8, 90–99.

ZALDIVAR, R. 1974. Arsenic contamination of drinking water and foodstuffs causing endemic and chronic poisoning. Beitrage Zur Pathologie, 151, 384–400.

ZHAOLI, S., MI, Z. & MINGGAO, T. 1989. The characteristics of fluorine in groundwater of North China and the significance of fluorite–water interaction to fluorine transportation. In: Proceedings of the 6th International Symposium on Water–Rock Interaction, Malvern, UK. Balkema, Rotterdam, 801–804.

Biogeochemical factors affecting groundwater quality in central Tanzania

R. J. BOWELL,[1] S. McELDOWNEY,[2] A. WARREN,[3] B. MATHEW[4] & M. BWANKUZO[5]

[1] *Steffen, Robertson and Kirsten (UK) Ltd, Cardiff CF1 3BX, UK*
[2] *School of Biological and Health Sciences, University of Westminster, 115 New Cavendish St., London W1M 8JS, UK*
[3] *Department of Zoology, The Natural History Museum, Cromwell Road, London SW7 5BD, UK*
[4] *FACO, c/o BHC Harare, Zimbabwe*
[5] *Maji, Dodoma PO Box 90, Tanzania*

Abstract: Analysis of groundwaters from the Makutuapora aquifer in the Dodoma region of central Tanzania has revealed a relationship between mineral–water interactions, water chemistry, bedrock geology, and microbiology. Groundwaters were slightly alkaline (pH 6–7.8) and essentially Na-Ca-HCO_3-Cl, with minor K, Mg, F, and SO_4^{2-}. Variations in water chemistry, particularly Ca/Na and Mg/Ca, ratios are related to the progressive alteration of feldspars and ferromagnesium minerals. The constant Na/Ca and Mg/Ca ratios noticed over mature aquifers and wells indicates that a steady-state is attained between aluminosilicates and groundwater. While erratic Fe/Mg and Na/K ratios denote a more open system or rather a greater diversity in minerals hosting these elements participating in mineral–water reactions. In places total concentrations of Fe, Mn, and Al can each exceed 1 mg l^{-1} with most of the metal held in particulate form (> 0.45 µm). The increase in metals suggests an imbalance in the steady-state reactions between magmatic minerals and leachate, possibly related to microbial activity. Fifty percent of the groundwaters were contaminated by significant numbers of thermotolerant coliforms indicating considerable risk of contamination by faecal pathogens. Numbers of faecal coliforms were positively correlated with K, Na, NO_3^-, PO_4^{3-} and BOD. Groundwater chemistry also affected the activity of the indigenous microbial community. Microbial biomass appeared to be unaffected by differences in groundwater chemistry. The numbers of selected physiological bacterial types (e.g. organisms contributing to the nitrogen and sulphur cycles) and the range of protist morphotypes, isolated from the tropical groundwater systems, were broadly similar to those found in temperate groundwater. Total concentration of metals such as Al, Fe, Co and Mn certainly exceed levels at which these metals could be considered toxic although if these metals are present in non-labile forms (as suggested by other studies) then the potential toxicity would be negligible. At present the major concerns for health are high seasonal salinities in the groundwaters and high faecal contamination.

Groundwater makes up over 95% of the world's available freshwater resources and is the main source of drinking water for a large percentage of the world's population. Historically groundwater has been considered a safe source of water protected by the soil layer which removes pollutants as the water percolates downwards. In many developing countries groundwater is not treated or even monitored prior to consumption. If these groundwaters contain high concentrations of potentially toxic elements (PTEs) or faecal contaminants there may be serious health implications.

Current understanding of the biogeochemical processes occurring in subsurface environments is based largely on studies carried out in temperate regions. Little is known about these processes in tropical regions where environmental conditions are thought to play an important role in determining the quality of groundwater; rocks weather more quickly and leaching is more intense than in temperate zones (Trescases 1992). Recent research has highlighted the high concentrations of potentially toxic elements in African groundwaters (Ogbukagu 1984; Lahermo et al. 1991; McFarlane & Bowden 1992; Bowell et al. 1994).

A wide range of microbial types contributes to

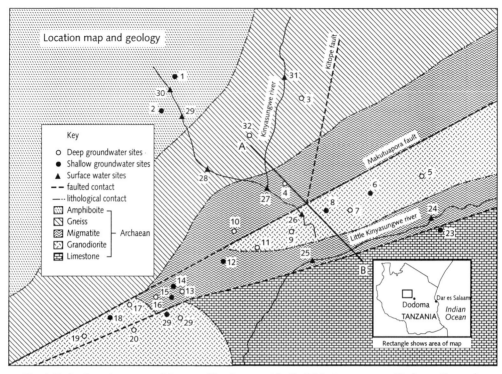

Fig. 1. Bedrock geology map of the crystalline basement in the Makutuapora area, Dodoma, Tanzania.

the biogeochemical cycles through oxidation/reduction reactions, alkylation and dealkylation processes, fossilization and solubilization, or mineralization reactions. Some microorganisms contribute to all the elemental cycles, e.g. heterotrophic bacteria are involved in the degradation and mineralization of organic matter, while others contribute significantly to one cycle, e.g. ammonia oxidizing bacteria to the nitrogen cycle. The presence of diverse and complex microbial communities in soils (Smith 1982), and shallow and deep groundwater systems is well established (Sinclair & Ghiorse 1989; Johnson & Wood 1992). It is likely that many different microbial populations within these communities contribute to element transformations and cycling in groundwater systems. This may have implications for the fate of PTEs in groundwater. Contamination of groundwater by faecal pathogens has considerable implications for human populations using the groundwater resource. The longevity of enteric pathogens in groundwater systems is determined by environmental factors including geochemistry and the indigenous microbial community.

Recent studies have shown that eukaryotic microorganisms (protists) are also widely distributed in subsurface sediments (Sinclair & Ghiorse 1989; Novarino et al. 1994). Protists are important bacterial predators in aquatic and terrestrial environments and may play a significant role in the population dynamics of groundwater bacteria. They have also been used in bioassays for a range of organic and inorganic elements and compounds. Consequently it may be possible to use protists as indicators of bacterial activity and to monitor any variations in groundwater chemistry which will influence biogeochemical processes. The aim of this research was to study the relationships between water chemistry and microbiological activity in deep (collected below 30 m, hosted by unweathered bedrock) and shallow (above 30 m, hosted in the weathered rocks and soils) groundwaters of the main aquifer in the Dodoma region of Tanzania at Makutuapora (Fig. 1). An assessment is made of possible advantages in using groundwater for drinking water rather than the traditional supplies of potable water available at the surface.

Methodology

Groundwater, sediment, soil and rock samples were collected in two field seasons in 1992:

April–May (wet season) and October–November (dry season). Thirty-two sites were sampled comprising 8 surface water sites, 11 shallow groundwater sites (34 samples from various depths) and 13 deep groundwater sites (22 samples from various depths). Sites designated as deep groundwater holes are those which draw water from below 30 m. Above this the boreholes were generally cased with clay or PVC lining. Exceptions to this were three holes which had yet to be cased and from which samples could be taken from the water-table to a depth of between 75 and 100 m. The shallow groundwater sites were designated as those which drained the weathering profile and were generally drilled by a WATERAID Landrover-mounted soil auger. None of these holes was lined.

Groundwater samples were collected by one of three methods: (1) From pre-drilled concrete-lined boreholes using hand-, wind or diescl-operated pumps; (2) from traditional hand-dug wells; (3) from freshly-drilled aquifer material collected either by hand augering or percussion drilling.

In some cases a stainless steel sampler was employed for collecting water at known depths down the water column. In the monitoring study of two newly drilled boreholes (sites 2 and 17) sampling was conducted at the surface during the first five days of pumping. For microbiological sampling all equipment was sterilized beforehand whenever possible.

For each sample of water three sub-samples were taken; unfiltered and unmodified (natural water); unfiltered and acidified with 10% (v/v) HNO_3; filtered through a 0.45 μm Durapore membrane filter and acidified with 10% (v/v) HNO_3. Field measurements of Eh, pH, electrical conductivity (EC), temperature and biological oxygen demand (BOD) were made at each site using OAKTON field instruments.

Mineralogy and lithogeochemistry

Mineral identification was carried out by optical and scanning electron microscopy and confirmed by X-ray diffraction (XRD) and Fourier transform infra-red spectroscopy. Chemical analysis of the rocks and soils was carried out by inductively coupled plasma atomic emission spectrometry (ICPAES, Fisons ARL3410 Mini-torch). Elements were extracted from 5 g of sediment by digestion with 15 ml of HNO_3 (70% v/v) 15 ml of HF (40% v/v) and 15 ml of perchloric acid (70% v/v). Precision of the technique was checked against known standards (Williams et al. 1993).

Hydrogeochemistry

All waters were analysed by ion-chromatography (Dionex-300) using an AS4A-AMMS column with Na_2CO_3 (1.8 mM) eluent at a flow rate of 2.5 ml min^{-1} for anion analysis, a CS 12 column with methane sulphonic acid eluent (20 mM) at a flow rate of 2 ml min^{-1} for group I/II cation analyses. A pulsed electrochemical detector in conductivity mode was used for detection. Transition metal analysis was accomplished with a Dionex CS 5 column with pyridine-di-carboxylic acid eluent and 4(2-pyridylazo) resorcinol post-column derivitization and measurement by a variable wavelength detector in the range 520–530 nm.

Microbiology

Bacterial enumeration was performed on representative samples (4) of selected soils and borehole sediments (deep 30–100 m; shallow 0–30 m). One gram (wet weight) of each sample was suspended in phosphate buffered saline (pH 7.3) and prepared as a dilution series (to 10^{-6}). Bacterial numbers were then determined as follows.

(i) Total viable count (TVC) for aerobic heterotrophs. One ml volumes of sample dilutions were grown on peptone yeast extract agar at 26°C for 7 days before counting. The results were expressed as colony forming units (CFU) g^{-1} (wet weight) soil.

(ii) Denitrifying bacteria. Replicate (5) MPN (Most Probable Number) tubes containing nutrient broth supplemented with 10 mM KNO_3 were inoculated with 1 ml of sample dilutions and incubated at 20°C to 22°C for 7 days. Gas production indicated the presence of denitrifying bacteria.

(iii) Ammonia-oxidizing bacteria. One ml volumes of diluted sediment/soil samples were inoculated into Alexander and Clarke medium containing 0.05% (w/v) $(NH_4)_2SO_4$. Five 200 μl volumes of each sample were transferred to MPN microtitre plates which were incubated in a humid environment at 20–22°C for 14 days. Griess-Ilosvay reagent was added to determine the presence of nitrite.

(iv) Nitrite-oxidizing bacteria. Microtitre plates were prepared and incubated as described above using Alexander and Clarke medium containing 0.006 g KNO_2 per 1 l medium. The presence of nitrite was examined (above).

(v) Sulphate reducing bacteria (SRB). Replicate (5) sediment/soil sample dilutions were prepared in 5 ml volumes of anoxic Baars medium supplemented with yeast extract and reducing agents. The tubes were incubated anaerobically for 14 days at 20°C to 22°C before determining the presence of SRBs by the formation of black coloration.

The numbers of bacteria, determined by MPN methods, were expressed as numbers per gram wet weight of sediment/soil.

Faecal coliforms were enumerated in the field using a DelAgua water testing kit (DelAgua Ltd, University of Surrey, Guildford, UK). Immediately upon collection, water samples (100 ml, 50 ml or 10 ml) were passed through a 0.45 μm sterile filter membrane. Each membrane was placed onto a pad containing Membrane Lauryl Sulphate Broth, and incubated for 14 to 18 hours at 44°C. Colony forming units of thermotolerant faecal coliform bacteria were then counted and expressed as number per ml of sample.

Bacterial biomass and activity were determined by the analysis of biochemical markers. Biomarkers used for biomass determination were DNA and lipid phosphate (a cell membrane component). Adenylates (ATP, ADP and AMP) and RNA were used as indicators of activity. Samples were freeze dried and the biomarkers were extracted by sonication in a suspension of chloroform, methanol and 10 mM phosphate buffer (pH 7.4) in a ratio of 1 : 2 : 0.8 for 2 hours at room temperature. The phases were then separated and reduced to dryness by rotary evaporation. The upper aqueous phase (containing DNA, RNA and adenylates) was resuspended in 18 MΩ water and the chloroform phase (containing lipid phosphate) was filtered and resuspended in chloroform.

The aqueous phases were mixed with 10 mg ml^{-1} polyvinyl polypyrollidone (PVPP) and centrifuged to remove humic contaminants from the extracts. Nucleic acids were extracted from the supernatant by adding buffered chloroform (pH 8). The phases were separated and the nucleic acid precipitated from the aqueous phase, and resuspended in 10 mM phosphate buffer (pH 7.4). The DNA content of each sample was quantified after binding with Hoechst 33258 by fluorescence detection at 350 nm excitation and 450 nm emission. Total nucleic acids were measured by absorbance at 260 nm and RNA concentration was then derived by difference. DNA and RNA content in the samples were expressed as ng g^{-1} dry weight sediment/soil.

AMP, ADP and ATP extracted from the aqueous phase were derivatized using chloracetaldehyde to produce the fluorescent N6 etheno derivatives. The derivatized samples were separated by ion pairing on HPLC and quantified by fluorescence at an excitation of 280 nm and an emission of 400 nm. After peak area integration, AMP, ADP and ATP sample content was expressed as p mole g^{-1} dry weight sediment/soil. The energy charge was also calculated as (ATP + 0.5 ADP)/(ATP + ADP + AMP).

Lipid phosphate present in the chloroform fraction after soil extraction was quantified using persulphate digestion and the phosphomolybdate reaction followed by reaction with malachite green. Absorbance was measured at 615 nm. Lipid phosphate soil content was expressed as n moles g^{-1} dry weight soil/sediment.

Samples were screened for the presence of protists. One gram of sediment or 1 ml of groundwater was transferred into Erdschreiber's Soil Extract medium in a sterile Petri dish and incubated at 20°C for 4–7 days. Dishes were examined daily by light microscopy for the presence of protists. Uniprotistan cultures were obtained either by micromanipulation of individual cells with a micropipette or by the 'dilution to extinction' method.

Geology of the Dodoma region

The lithologies of the Dodoma area are part of the Archaean–Lower Proterozoic Tanzanian shield (Cahen & Snelling 1984) which extends north beneath Lake Victoria and into southern Uganda and Kenya. In the Dodoma area the lithologies consist of migmatitic gneisses, amphibolites, feldspathic quartzites, siliceous marbles, garnet–kyanite schists, and quartz–feldspathic gneisses, metamorphosed to amphibolite and granulite facies. They represent an ancient granitoid–supracrustal terrane dated at 2850 and 2600 Ma (MADINI 1992).

Due to the extreme weathering which has occurred in central East Africa, a thick weathering mantle has been developed above the bedrock. This varies in thickness from usually 30 m but up to 40 m on the plains to less than 1 m on steep slopes. The bulk change to bedrock occurs as the coherent, dense, rock composed largely of anhydrous or weakly hydrated minerals are altered to friable units with a lower density consisting mainly of microcrystalline strongly hydrated phyllosilicates, oxides and hydroxides. The major mineralogical variations in the weathering profile are shown in Fig. 2. The main geochemical changes in the conversion of bedrock to saprolite and soils is the loss of elements such as Ca, Mg, Rb and retention of more immobile elements such as Ti, Fe and Al (Table 1). On plateaus a laterite ferricrete is present and consists largely of Fe and Al oxides

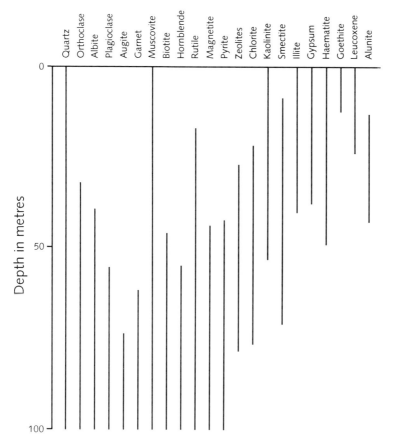

Fig. 2. Schematic representation of mineralogical variations with depth of the Makutuapora core samples.

and hydroxides. The soils in the area are largely ferrasols and vertisols but andosols and nitosols are present on peneplains (Bowell et al. 1995).

From mass balance reactions (Brimhall & Dietrich 1987; Brimhall et al. 1988) of the Dodoma weathering profiles, element behaviour can be quantified (Table 2). Elements such as Ca, Na, and K show a strong depletion as intensity of weathering increases. This is a reflection of the instability of primary silicate minerals (Velbel 1989, 1992). Aluminium is enriched in the clay-rich saprolite and surface soils. Similarly Si is retained in the saprolite and clay-rich soils but is lost from the surface soils. Iron is enriched during weathering with a larger enrichment in the ferricrete horizon, or cuirasse (over 600% v. average hypogene concentration) at the top of the mottled clay zone, possibly as a result of iron hydrolysis at the water table, or 'ferrolysis' (Mann 1983). However, the mass of added iron required to account for the enrichment in the overlying soils is much too great to be an *in situ* product. Given the sluggish mobility of iron in the oxide zone (Mann, 1983), some of the added iron must be due to mechanical concentration, possibly related to profile denudation. Trace elements such as Mn are largely depleted in the profile, although a Mn enrichment is observed in the zone of ferrolysis (ferricrete, Table 2) and is probably related to the same mechanism as Fe enrichment (Mann 1983). Like Zr, Nb is largely immobile but the strain indicators based on Nb v. Zr reveal a collapse zone in the soils.

The Dodoma region is similar to other areas of crystalline basement rocks in Africa, with groundwater occurring in two forms (Farquharson & Bullock 1992; Wright 1992). Superficial deposits above the granitic basement are composed of thin sand and gravel which retain infiltration water in its downward movement. These formations are shallow and are similar to palaeodambos identified elsewhere (Wright 1992). The recharge of these aquifers is depen-

Table 1. Geochemistry of protolith, saprolite and soils from the Makutuapora area, Dodoma, Tanzania

Soil	SiO$_2$	Al$_2$O$_3$	Fe$_2$O$_3$	TiO$_2$	MnO	Na$_2$O	K$_2$O	CaO	MgO	P$_2$O$_5$	Zr	Nb	V	Cr	Zn	Cu	Rb
vertisol surface	44.90	28.80	14.50	0.98	0.08	1.25	1.56	0.08	0.05	0.29	349	115	41.3	39	12	10	3.2
vertisol 0.5 m	46.45	28.30	13.90	0.81	0.09	1.32	1.65	0.19	0.09	0.21	345	114	38.7	36	12	10	3.2
vertisol 1.0 m	47.21	27.60	14.00	0.79	0.08	1.31	1.67	0.23	0.09	0.20	346	104	39.5	36	14	12	4.6
ferralsol surface	26.06	27.55	36.90	1.24	0.05	0.08	0.09	0.04	0.03	0.16	511	45	115	123	8	7	1.2
ferralsol 0.5 m	24.45	28.65	38.65	1.29	0.05	0.11	0.09	0.04	0.04	0.13	510	37	109	86	14	9	1.8
ferralsol 1.0 m	23.64	29.55	38.95	1.31	0.08	0.12	0.14	0.04	0.04	0.13	510	28	107	100	23	10	1.9
ferricrete	16.89	30.87	39.20	1.76	0.08	0.14	0.23	0.06	0.07	0.20	508	30	121	139	27	13	3.8
clay zone 10 m	62.03	18.23	12.95	0.68	0.16	1.48	1.51	0.23	0.08	0.15	345	98	30.8	40	16	12	11
clay zone 15 m	63.59	17.96	11.23	0.65	0.18	1.91	1.56	0.26	0.09	0.19	340	87	35.9	34	18	15	12
saprolite 10 m	65.61	15.91	10.98	0.51	0.21	2.11	2.11	0.27	0.10	0.16	340	96	34.2	33	20	21	15
saprolite 20 m	65.11	15.65	9.85	0.48	0.38	3.58	2.69	0.28	0.11	0.18	343	99	32.3	29	21	22	16
saprolite 25 m	66.04	15.22	8.55	0.36	0.42	4.23	2.96	0.31	0.12	0.20	340	105	29.3	27	21	24	19
saprolite 30 m	67.57	14.95	5.32	0.37	0.46	5.46	3.71	0.38	0.14	0.21	342	110	30.6	33	25	26	25
granodiorite	66.01	12.55	2.61	0.37	0.45	6.10	3.95	0.49	0.16	0.23	340	112	28.9	28	28	29	65
gneiss	59.55	25.93	2.71	0.38	0.50	5.23	2.55	0.33	0.16	0.21	506	108	112	39	27	14	71
pyx-gneiss	50.96	20.22	10.70	0.93	0.69	1.29	2.11	5.12	5.39	0.19	80	10.9	244	191	63	335	36
quartzite	89.00	2.69	2.18	0.18	0.11	0.48	1.26	1.30	1.21	0.39	255	24	65	62	10	12	16
marble	16.23	2.31	2.29	0.05	0.12	0.58	0.19	24.22	20.83	0.18	150	18	19	12	59	34	19
schist	56.23	24.23	4.21	0.25	0.63	3.88	4.68	1.96	1.81	0.23	187	39	36	78	98	89	46

Oxide concentration in wt% and element concentration in mg/kg. All analyses by ICPAES.

Table 2. *Element mass balance* in Dodoma weathering profiles (all rock types)*

Element	Saprolite	Clay zone	Ferricrete	Soil
SiO_2	98	112	65	82
Fe_2O_3	139	180	438	374
Al_2O_3	109	148	108	180
TiO_2	103	105	114	151
MnO	85	70	108	43
CaO	88	33	6	1
MgO	65	19	2	1
Na_2O	62	10	3	4
K_2O	81	52	15	3
P_2O_5	77	39	28	22
Zr	100	100	100	100
Nb	105	100	97	180
Cr	109	167	355	305
V	110	119	276	198
Cu	93	86	17	7
Rb	65	12	2	1
Sr	61	15	3	1
Ba	78	54	22	11

*Mass balances calculated using equations of Brimhall & Dietrich (1987) and Brimhall *et al.* (1988). T% of protolith concentration, assuming no variation in Zr concentration.

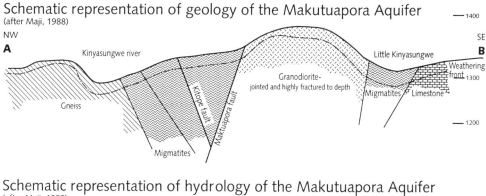

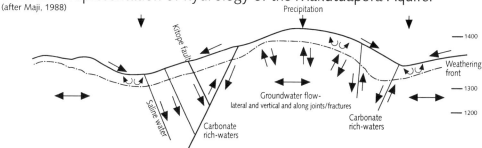

Fig. 3. Schematic cross-section and hydrological section of the Makutuapora aquifer (after MAJI 1988). (**a**) Geology: A–B represents position of cross-section as shown in Fig. 1; (**b**) hydrology: arrows show direction of flow.

Table 3. *Geochemistry of deep groundwaters (analysed by ICPAES and ion chromatography, n = 22)*

	Wet season		Dry season	
	Min.	Max.	Min.	Max.
pH	5.5	8.2	5.0	8.5
EC	200	2750	262	43200
Eh	198	622	211	708
Temp. °C	19	42	20	42
TDS	50	1450	43	1670
DOC	< 0.1	0.2	< 0.1	0.15
BOD	0	2.2	0	1.2
FC	0	88	0	50
HCO_3^-	23	320	36	350
SO_4^{2-}	1.9	22	5.9	39
NO_3^-	< 0.01	0.18	0.09	0.31
PO_4^{3-}	0.1	0.42	0.25	0.53
Li	< 0.01	1.6	0.23	2.21
Na	93	231	148	265
K	2	16	12	24
Mg	16	40	22	44
Ca	23	91	43	116
Sr	0.27	2.18	0.18	2.97
Ba	0.09	0.42	0.05	0.23
B	0.88	1.13	0.57	0.98
Al	0.12	33.95	1.51	21.40
Si	16.95	42.85	12.34	19.58
Mn	0.013	0.221	0.012	0.162
Fe	0.11	1.21	0.19	0.65
Cu	< 0.01	0.19	< 0.01	0.06
Zn	< 0.01	0.39	< 0.01	0.30

All anion and element concentrations in mg l^{-1}.
EC, electrical conductivity in μS cm^{-1}; Eh, redox potential in mV; TDS, total dissolved solids in mg l^{-1}; DOC, dissolved organic content in mg l^{-1} (technique of Ertel *et al.* 1986); BOD, biological oxygen demand, a measure of biological activity, in mg l^{-1}; FC, faecal coliform counts expressed as number of coliforms in 100 ml of water.
Elements analyzed below detection: Br, I (< 0.1 mg l^{-1}), Be, Sc, Zr, V, Cr, Mo, Co, As (< 0.01 mg l^{-1}).

dent on infiltration rates and is subject to large seasonal fluctuations. Due to the variable morphology of the region these aquifers have a wide distribution. They give a low yield (4.5 l min^{-1} per well on average) and are exploited by means of manual hand pumps. The major aquifers in the region occur within fractured crystalline rocks, granites and their metamorphosed equivalents, such as at Makutuapora. From a study of drill core from Makutuapora, a schematic cross-section was constructed (Fig. 3a). Most drill sites which hit water were closely associated with the major Kitoe fault system or within subsidiary fractures (Fig. 1).

Hydrogeochemistry

The physicochemical characteristics of groundwater and some surface waters from the Makutuapora aquifer are shown in Tables 3–5 for wet and dry season sampling. In general surface waters were weakly alkaline, at ambient temperature and had a high total dissolved solid concentration. The particulate solid load was slightly greater in the wet season than in the dry season, but the dissolved load was higher in the dry season. Most of the deep groundwaters were above pH 7 and were essentially Na–Ca–HCO_3^-–Cl, with minor K, Mg, F and SO_4^{2-}. In the dry season the higher salinity reduces water quality and would be a major factor in determining the use of water resources from individual wells as well as the whole aquifer. Most groundwaters had a lower range of total dissolved solid (TDS) content and metal content than surface water and shallow groundwaters.

Seasonal variations existed in the wells for trace element concentration depending upon whether or not the element was associated

Table 4. *Geochemistry of shallow groundwaters (analysed by ICPAES and ion chromatography, n = 34)*

	Wet season		Dry season	
	Min.	Max.	Min.	Max.
pH	4.6	6.9	4.7	6.5
EC	157	680	187	822
Eh	221	565	228	608
Temp. °C	15	29	19	33
TDS	60	720	73	895
DOC	0.15	0.39	0.23	0.45
BOD	0.4	6.9	0	6.2
FC	0	128	0	50
HCO_3^-	45	346	50	225
F^-	< 0.1	1.2	< 0.1	2.3
Cl^-	10.6	178	12.9	212
SO_4^{2-}	0.9	16.2	0.8	15.8
NO_3^-	< 0.01	0.39	0.05	0.49
PO_4^{2-}	0.1	0.56	0.36	0.95
Li	< 0.01	0.98	0.15	0.89
Na	36	295	42	312
K	10.5	21	13	26.50
Mg	12	72	16	81
Ca	20	63	23	68
Sr	0.15	0.96	0.15	1.01
Ba	0.05	0.31	0.06	0.23
B	0.65	0.89	0.58	0.72
Al	3.97	34.56	3.21	25.97
Si	9.5	28.7	6.58	30.10
Mn	0.016	0.079	0.011	0.060
Fe	0.39	1.98	0.16	0.79
Cu	< 0.01	0.25	< 0.01	0.16
Zn	< 0.01	0.46	< 0.01	0.39

All anion and element concentrations in mg l^{-1}.
EC, electrical conductivity in $\mu S\ cm^{-1}$; Eh, redox potential in mV; TDS, total dissolved solids in mg l^{-1}; DOC, dissolved organic content in mg l^{-1} (technique of Ertel *et al*. 1986); BOD, biological oxygen demand, a measure of biological activity, in mg l^{-1}; FC, faecal coliform counts expressed as number of coliforms in 100 ml of water.
Elements analyzed but below detection: Br, I (< 0.1 mg l^{-1}), Be, Sc, Zr, V, Cr, Mo, Co, As (< 0.01 mg l^{-1}).

largely with the dissolved or particulate fraction. For Fe, Al, Mn and SiO_2, the dominant fraction was in the particulate load and concentrations were higher in the wet season than in the dry season (Table 4). Other trace metals, such as V, Cr, Co, Pb, and Mo were all below the analytical detection limit of 0.01 mg l^{-1} in both bulk sample and filtrate. The subsurface or shallow groundwaters were more acidic with pH as low as 4.6, probably due to high levels of dissolved organic acids. An attempt was made to analyse dissolved organic acids although the rapid degradation of the acids and their labile nature made the analyses unreliable. Carboxylic and phenolic acids were identified and if these were similar to the organic acids in the original waters then they could represent a potential mechanism for Fe, Mn, Zn and Al mobility (Shotyk 1984).

From correlation coefficients a number of relationships can be observed in the Makutuapora groundwaters (Table 6). Bicarbonate content is positively correlated to pH, Na, Ca and Mg and inversely correlated to Al, Fe and sulphate. A positive correlation occurs between the anions nitrate, phosphate, chloride and sulphate. This correlation is probably biased by the shallow groundwaters which will reflect rainfall recharge, the main source of chloride and sulphate, to a greater extent than deeper groundwaters. Shallow groundwaters will also be more contaminated by soil leachates, reflected in the nitrate and phosphate content probably arising from decaying organic matter and leaching of fertilizers. The possible contamination of groundwater by faecal matter and biological waste is reflected by the positive correlation between biological oxygen demand (BOD), thermotolerant faecal coliforms (FC) with

Table 5. *Geochemistry of surface waters (analysed by ICPAES and ion chromatography, n = 8)*

	Wet season		Dry season	
	Min.	Max.	Min.	Max.
pH	7.5	8.0	7.0	8.0
EC	705	1900	700	2000
Eh	125	225	195	286
Temp. °C	11	22	18	38
TDS	270	960	295	1080
DOC	0.53	0.75	0.51	0.66
FC	112	> 250	100	> 250
HCO_3^-	250	652	165	655
F^-	< 0.1	0.2	< 0.1	0.3
Cl^-	11.9	16.5	12.6	59.5
SO_4^{2-}	1.8	30	2.2	33.5
NO_3^-	0.09	0.69	0.18	0.86
PO_4^{3-}	0.81	1.29	0.98	1.98
Li	< 0.01	0.51	< 0.01	0.63
Na	115	178	169	226
K	18.90	23.20	19.50	24.50
Mg	10	68.50	28	72
Ca	26	119	30	136
Sr	0.78	0.83	0.77	0.82
Ba	0.12	0.22	0.15	0.24
B	0.61	0.64	0.52	0.55
Al	< 0.01	1.30	< 0.01	1.0
Si	8.70	11.20	5.80	7.90
Mn	0.017	0.31	0.029	0.196
Fe	0.11	0.56	< 0.01	0.39
Cu	< 0.01		< 0.01	
Zn	< 0.01		< 0.01	

All anion and element concentrations in $mg\,l^{-1}$.
EC, electrical conductivity in $\mu S\,cm^{-1}$; Eh, redox potential in mV; TDS, total dissolved solids in $mg\,l^{-1}$; DOC, dissolved organic content in $mg\,l^{-1}$ (technique of Ertel *et al.* 1986); FC, faecal coliform counts expressed as number of coliforms in 100 ml of water.
Elements analyzed but below detection: Br, I ($< 0.1\,mg\,l^{-1}$), Be, Sc, Zr, V, Cr, Mo, Co, As ($< 0.01\,mg\,l^{-1}$).

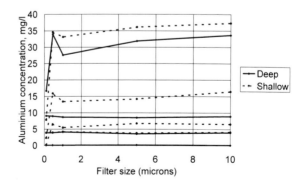

Fig. 4. Distribution of Al concentration with filter pore size after filtration of Makutuapora groundwaters.

Table 6. *Correlation matrices for Makutuapora groundwaters (all groundwaters)*

	pH	alk	NO_3^-	PO_4^{3-}	SO_4^{2-}	Cl^-	BOD	FC	Na	K	Ca	Mg	Al	Mn	Fe
pH	1.000														
alk	0.765	1.000													
NO_3^-	0.034	0.180	1.000												
PO_4^{3-}	0.129	0.186	0.321	1.000											
SO_4^{2-}	−0.411	−0.379	0.251	0.236	1.000										
Cl^-	−0.143	0.192	0.311	0.192	0.419	1.000									
BOD	−0.066	0.018	0.562	0.505	−0.119	−0.078	1.000								
FC	−0.032	0.012	0.505	0.479	−0.205	−0.082	0.765	1.000							
Na	0.019	0.615	0.211	0.211	0.046	0.298	0.231	0.297	1.000						
K	0.024	0.:24	0.369	0.176	0.036	0.196	0.363	0.289	0.311	1.000					
Ca	−0.238	0.795	0.211	−0.231	0.110	0.158	0.150	0.212	0.389	0.175	1.000				
Mg	−0.241	0.699	0.059	−0.142	0.150	0.127	0.137	0.188	0.212	0.163	0.721	1.000			
Al	−0.569	−0.279	0.021	−0.561	0.213	0.159	0.119	0.154	−0.129	0.113	0.119	0.121	1.000		
Mn	−0.512	−0.196	−0.119	−0.126	0.105	0.131	0.158	0.139	−0.071	0.069	0.182	0.198	0.318	1.000	
Fe	−0.611	−0.312	−0.126	−0.685	0.200	0.172	0.167	0.127	−0.095	0.093	0.169	0.198	0.495	0.621	1.000

Pearson correlation coefficients calculated using TECHBASE software package. Number of samples = 56; $r95\%$ = 0.238.
BOD, biological oxygen demand; FC, faecal coliform.

Table 7. *Proportion of total metal concentration in < 0.45 μm filtrate*

Sample	Depth (m)	pH	Fe (%)	Al (%)	Mn (%)
4 deep groundwater	100	7.8	1.3	0.5	2.0
3 deep groundwater	93	7.5	1.9	1.3	13.6
4 deep groundwater	89	8.0	nd	nd	0.1
3 deep groundwater	75	7.6	1.6	1.2	6.5
7 deep groundwater	70	7.9	nd	nd	0.4
19 deep groundwater	65	7.0	6.7	4.9	30.0
9 deep groundwater	62	7.7	0.9	1.8	6.4
11 deep groundwater	50	7.5	1.5	1.4	16.5
17 deep groundwater	45	7.6	1.6	1.2	9.6
5 deep groundwater	40	7.3	2.3	1.9	20.7
5 deep groundwater	36	7.4	2.2	2.2	22.8
21 deep groundwater	30	6.8	11.9	6.2	33.8
2 shallow groundwater	25	5.8	19.2	8.9	46.8
6 shallow groundwater	20	4.9	37.8	12.7	57.0
8 shallow groundwater	15	6.0	14.6	7.4	45.4
14 shallow groundwater	8	5.7	22.4	9.3	47.7
8 shallow groundwater	5	5.6	19.5	10.8	48.7
24 surface water	0	7.0	5.6	3.2	26.6
30 surface water	0	7.7	0.8	1.1	4.3
28 surface water	0	8.0	nd	0.1	0.1

nd, not detected. All values expressed as a % of total metal concentration in groundwater.

nitrate and phosphate. Faecal coliform numbers are also positively correlated to Na and K, and BOD is positively correlated to K (Table 6) (see below).

Aluminium, Fe and Mn are all positively correlated to each other suggesting they are influenced by similar mechanisms in the groundwaters. Additionally all metals are inversely correlated to pH as would be anticipated since mobility is enhanced at low pH (Jones et al. 1974; Jeffries & Hendershot 1981; Lindsay & Walthall 1981). However, none of these elements is significantly positively correlated with any anion (Table 6) so mobilization may be as hydroxide, organically-bound species (Shotyk 1984; Smith et al. 1996) or by microbial activity (see below). Both Al and Fe are strongly positively correlated with phosphate, as high dissolved Al and Fe in groundwaters would lead to Fe/Al-oxide precipitation and phosphate retention through surface adsorption processes (Chesworth et al. 1989). In acidic shallow groundwaters total concentrations of Al, Fe and Mn can exceed $1\,mg\,l^{-1}$, the majority of which is retained by the 0.45 μm filter and is considered to be colloidal (Table 7). From the filtration of Al into several fractions it is clear that much of the groundwater Al is held in particulate forms (Fig. 4). The transport and mobilization of Al, Fe and Mn can be inferred from the chemistry and mineralogy of sediments collected from Dodoma water pipes and Makutuapora water storage tanks (Table 8). Sediments are composed of evaporite salts, amorphous material, clay minerals (from the wall of the pipes) and Al–Fe hydroxides and sulphates. Additionally, other trace metals are also mobilized based on the chemistry of the sediments (Table 8). It is not possible to infer what proportion of these elements has been leached away from the clay pipes and what has migrated in the waters from the aquifer.

From three boreholes (9, 11 and 22, all in granodiorite lithology) sampling was possible from the water-table to 98 m as the holes were unlined (waters represent both deep, > 30 m, and shallow, < 30 m, groundwaters). The hydrochemistry of the three holes is shown in Fig. 5. Levels of Al, Fe, and Mn decreased with depth in the aquifer, with the highest concentration at the weathering front, in the ferricrete at c. 20 m depth (Figs 5e, 5h, 5f respectively). At this point pH was lowest (Fig. 5a) promoting dissolution and mobilization of the metals. Silica concentration was relatively static throughout the water column with a maximum at 85 m depth (possibly a density effect within the brine). Like Na, Ca and sulphate alkalinity, Mg and B also showed no systematic variation with depth. As would be expected biological oxygen demand and dissolved oxygen decreased with depth, although the apparently high values below 75 m depth agreed with the surprising observation of active protists deep in the aquifer.

Table 8. *Mineralogy and geochemistry of pipe precipitates from Makutuapora aquifer, Dodoma, Tanzania*

Site	Mineralogy	SiO$_2$	Fe$_2$O$_3$	Al$_2$O$_3$	CaO	MgO	Na$_2$O	K$_2$O	SO$_4$	Mn	Co	Cu	Zn
storage tank	goethite, quartz, kaolinite, jarosite-alunite	36.9	21.5	20.5	0.9	0.49	1.58	2.32	3.4	23450	79	112	76
clay pipe 1.5 km	goethite, copiapite, aluminite, manganite, halite, gypsum, quartz, sylvite, natron	21.2	23.9	23.9	1.7	0.58	1.96	3.20	8.8	32250	96	36	97
clay pipe 5 km	goethite, halite, gypsum, sylvite, illite, natron-trona, gibbsite	8.9	12.9	18.6	16.44	2.92	8.22	2.12	12.8	120	19	15	25

Mineralogy by XRD and geochemistry by ICPAES (oxides shown in wt% and elements in mg/kg).

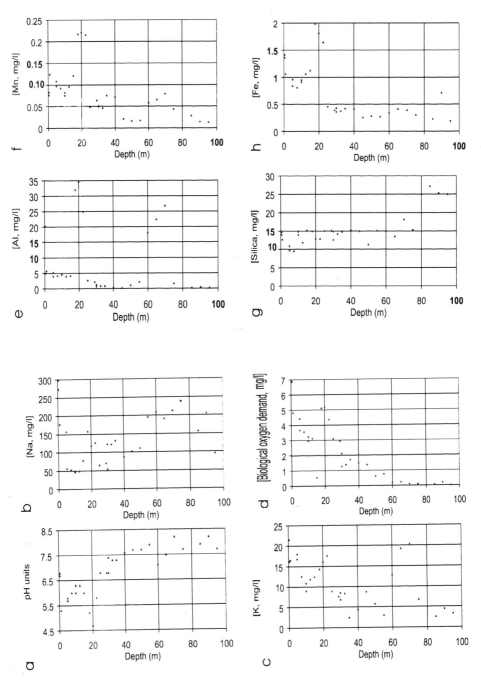

Fig. 5. Hydrochemical and biological oxygen demand variations with depth from sites 9, 11 and 22.

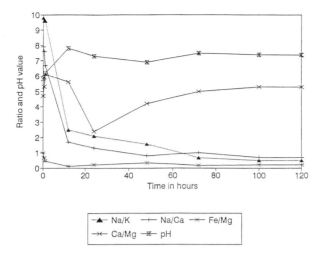

Fig. 6. Change in element ratios and pH with time from site 17 hosted by granodiorite.

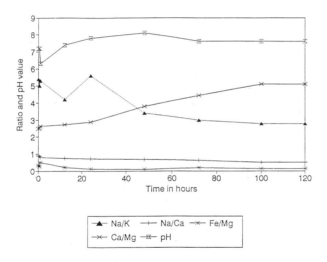

Fig. 7. Change in element ratios and pH with time from site 2 hosted by amphibolite.

Intriguingly, this apparent increase correlates to a relative increase in sulphate, chloride and K (Fig. 5) at a depth of 70–80 m in the groundwaters.

During the course of fieldwork two new boreholes were sampled immediately following initial contact with the aquifer over a 5 day period. The first hole was in the granodiorite (site 17) and the second in the amphibolite (site 2). At the initial stage of water release at site 17, fluid chemistry was dominated by Na and bicarbonate (not shown), but after 12 hours K and Ca increased in concentration and this can be observed in the variation in Na/K and Na/Ca ratios (Fig. 6). The initial Na/K ratio was 9.8 while the initial Na/Ca ratio was greater than 5 but this decreased after 12 hours to 2.5 for Na/K and less than 2 for Na/Ca. After 100 hours of continuous pumping the ratios stabilized at Na/K = 0.5 and Na/Ca = 1 (Fig. 6). No stability was observed in the Ca/Mg or Fe/Mg ratio. Over the same period the level of total Fe dropped from $18.2 \, \text{mg} \, l^{-1}$ to less than $0.2 \, \text{mg} \, l^{-1}$ and pH increased from 5.8 to 7.4 (Fig. 6). These ratios were also recorded when the well was resampled five months later. This high initial Fe level may explain the poor taste often experienced with newly drilled wells. Another newly drilled well was monitored from an initial point (0.5 hours after water was hit) but with amphibolite as the host rock. Here the fluid chemistry was dominated by Ca and bicarbonate over the whole 100 hour period of continuous pumping and also in samples collected 5 months later. The Na/Ca ratio over a 100 hour period decreased from 0.9 to 0.5 while the Ca/Mg ratio increased from 2.5 to 3.8 in the first 48 hours (Fig. 7). The Fe/Mg ratio was even more erratic than in the granodiorite well and pH was largely constant varying from an initial 7.2 to 7.6 60 hours later (Fig. 7). The stabilization of the ratios occurred after about 100 hours in these wells.

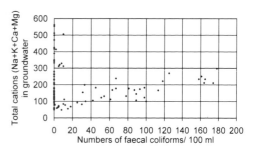

Fig. 8. Total alkalis v. faecal coliform counts for Makutuapora groundwaters.

Microbiology of Dodoma aquifers

Significant numbers of thermotolerant faecal coliforms (> 10 per 100 ml) were present at 50% of the sites studied. This indicates excessive faecal contamination of groundwater sites, particularly in shallow aquifers. The numbers of faecal coliforms showed positive correlations with NO_3^- (r = 0.505), PO_4^{3-} (r = 0.479), Na (r = 0.297), K (r = 0.289) and BOD (r = 0.765) in the Makutuapora groundwaters (Table 6). There was a relatively complex relationship between the total content of the cations Na, Ca, K and Mg and numbers of faecal coliforms in the Makutuapora groundwaters. Faecal coliform numbers of between 1 and 80 per 100 ml groundwater were present over the whole range of cation concentrations found. The highest numbers of faecal coliforms (100–180 per 100 ml groundwater), however, were only found to be present in samples with the higher cation content (<300 mg/l, Fig. 8). The numbers of bacteria capable of aerobic heterotrophic growth (i.e. utilizing organic carbon as a carbon and energy source) in the surface and subsurface borehole soils were in the range 9.0×10^5 to 2.3×10^7 CFU g^{-1} wet weight soil and was not clearly related to any geochemical characteristics of the groundwater systems.

Bacteria involved in the cycling of nitrogen were isolated from all the sampling sites. Dentrifying bacteria were most abundant with 2.0×10^2 to $1.8 \times 10^5 g^{-1}$ wet weight groundwater material. Nitrifying bacteria were also present at all sample sites, including both ammonia oxidizing bacteria (in general $< 2.0 \times 10^0 g^{-1}$ wet weight soil), and nitrite oxidizing bacteria (in general $<2.0 \times 10^0 g^{-1}$ wet weight soil).

Sulphate reducing bacteria, which contribute to the cycling of sulphur, were present in low numbers between 2.0×10^0 to $1.4 \times 10^3 g^{-1}$ wet weight soils at all sample sites studied.

The biomass measured by lipid phosphate determination was similar for all the sample sites, 171 and 172 nMole g^{-1} dry weight (Table 9). This suggests a far more constant microbial community size than demonstrated through culture techniques, which varied by several orders of magnitude between sites (above). The amount of DNA, another measure of biomass, also varied with sample. There was a range of DNA content from 158–418 ng g^{-1} dry weight sample (Table 9). The amount of biomass measured by lipid phosphate or DNA did not appear to be related to the geochemical characteristics of the groundwater system and was independent of depth.

RNA, an indicator of microbial activity, showed an exceptionally wide variation between sites (< 1–2827 ng g^{-1} dry weight sample). The amounts of individual adenylates (ATP, ADP, AMP) fluctuated with sample as did the total adenylates (Table 9). In a number of samples ATP was below the levels of detection. The calculated EC ratios for the Tanzanian aquifer (only calculated for those sites with measurable ATP) were all below 0.5. Again there was no apparent relationship between the depth the sample was taken from and activity. Some relationships did exist between geochemical characteristics in the groundwater and microbial activity by RNA and adenylate measurement. Microbial activity measured by total adenylates (Fig. 9a) and RNA (Fig. 9b) tended to be highest at the lowest groundwater concentration of the halides F^- and Cl^-. There was a positive relationship between metals, particularly Al, and total adenylates reaching a plateau at adenylate concentrations above 10 p mole per gram dry weight sample (Fig. 10a). A positive relationship between RNA and all three metals

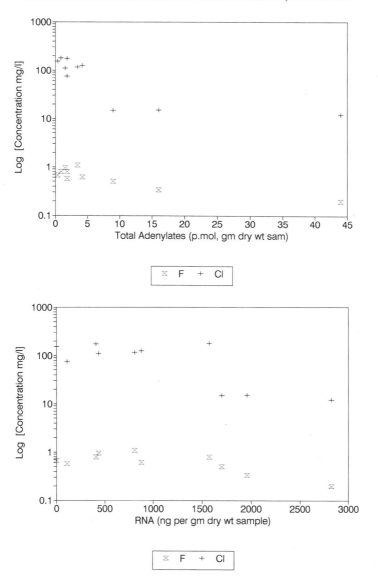

Fig. 9. Halides F⁻ and Cl⁻ v. microbial activity measured by biomarkers for Makutuapora groundwaters; (**a**). total adenylates v. F and Cl; (**b**) RNA v. F and Cl.

was also found (Fig. 10b). It seems clear that there was a relationship between microbial activity in groundwater and the concentration of Al, Mn and Fe.

Protists were isolated in enrichment cultures from 22 out of 32 sites suggesting that they are widespread in the subsurface throughout the region. Representatives of three main morphological groups were present: flagellates, ciliates and amoebae. Of these, flagellates were by far the most abundant. All protists which could be recognized are known to be bacterivorous.

Discussion

Microbiological health of Dodoma groundwaters

The microbiological health of groundwater is controlled by a variety of biological and abiotic parameters. These factors include the frequency

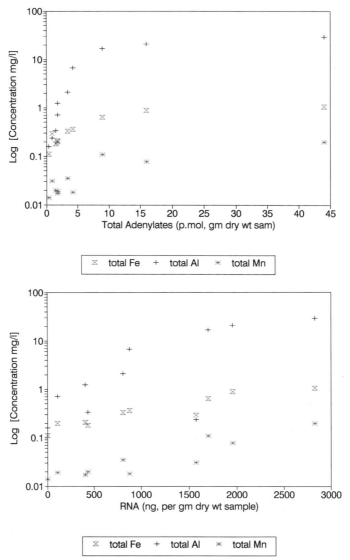

Fig. 10. Total metal concentration v. microbial activity measured by biomarkers for Makutuapora groundwaters; (a) total adenylates v. Al, Fe and Mn; (b) RNA versus Al, Fe and Mn.

of faecal contamination of groundwater from human or animal sources and the longevity of enteric pathogens in groundwater. The survival of microbial pathogens is influenced by the abiotic conditions in the groundwater and the biomass and activity of the indigenous microbial community.

The faecal pollution indicated by the presence of thermotolerant coliforms in the Tanzanian aquifers is likely to have originated from both animal and human sources. Animal sources may be particularly important as the land in the region is heavily grazed. The presence of these indicators of faecal pollution suggests a considerable risk of contamination of these aquifers by enteric pathogens.

Among the abiotic factors that may influence the survival of thermotolerant coliforms and faecal pathogens, are the geochemical characteristics of the aquifer. There was a positive correlation between the numbers of thermotolerant faecal coliforms and the concentrations of NO_3^- and PO_4^{3-}. It is possible that higher soil concentrations of these anions have a protective role aiding survival of thermotolerant coliforms in soils. The longevity of enteric pathogens

Table 9. *Microbial biomass and activity measured by biomarkers at selected sites in the Makutuapora area*

Sample	Lipid phosphate (in moles)*	DNA (ng)*	RNA (ng)*	AMP (p mole)*	ADP (p mole)*	ATP (p mole)*	Total adenylates (p mole)*	Energy charge†
2 shallow	172	220	2827	30.4	8.5	5.1	44.0	0.21
7 deep	172	235	1701	4.3	4.0	0.7	9.0	0.30
8 shallow	171	276	806	2.7	0.7	nd‡	3.4	nc§
10 deep	172	165	407	1.4	0.4	nd	1.8	nc
10 deep	171	418	878	3.4	0.8	nd	4.2	nc
14 shallow	172	320	1959	5.6	6.2	4.2	16.0	0.46
14 shallow	172	193	436	0.6	0.4	0.5	1.5	0.47
15 shallow	171	305	110	0.9	0.9	nd	1.8	nc
15 shallow	172	158	< 1	0.4	nd	nd	0.4	nc
23 shallow	172	181	1573	0.5	0.4	nd	0.9	nc

*Measurements per g/dry weight sediment; †energy charge calculated as (ATP+0.5 ADP)/(ATP+AMP=ADP); ‡nd, not detected; §nc, not calculated.
Samples taken at the same site, e.g. 10, 14 and 15, were from different depths in the profile.

may, also, be increased in this way. A positive correlation was also observed between numbers of thermotolerant faecal coliforms and BOD (Table 6). The higher concentrations of BOD are likely to be due to elevated concentrations of organic matter, arising from faecal contamination, acting as a source of nutrients for the growth of indigenous microorganisms (Harvey et al. 1984; Madsen et al. 1991).

The biomass and activity of the soil microbial community are also likely to affect the longevity of thermotolerant coliforms and enteric pathogens in the groundwater systems, since the indigenous community tends to eliminate foreign microorganisms. The microbial biomass measured by lipid phosphate was remarkably constant between samples and with depth (Table 9). Lipid phosphate is considered a relatively reliable measure of biomass due to its rapid turnover and loss on cell death (White et al. 1979). Measures of DNA can also be taken to provide an estimate of biomass (Paul & Meyers 1982) and have been correlated with direct bacterial counts (McCoy & Olson 1985). The use of DNA as a measure of biomass is slightly complicated, however, by the increase in DNA in a cell just prior to division during cell growth (Neidhardt et al. 1990). This may result in a rise in DNA measured but no actual increase in cell numbers, although cell mass will increase, until after cell division has occurred. In addition, DNA may be released from microorganisms and other soil or surface organisms on their death. This DNA, which is not cell associated, may persist for a period in the environment. Any free DNA in the aquifer would influence estimates of biomass by DNA measures. Given that the lipid phosphate remains relatively constant between aquifer samples in the Tanzanian study, the higher values of DNA in some samples i.e. $> 300 \, ng \, g^{-1}$ dry weight sediment (Table 9) compared to others, may indicate that the bacteria at these sites were actively dividing or that there was variation in the amount of free DNA with sample.

The amount of RNA (Karl et al. 1981) and adenylates (Karl & Holm-Hansen 1978) present in samples gives information on the physiological status of the bacterial community. High RNA levels are found in active cells and drop rapidly at the onset of starvation or other stress (Harder et al. 1984). There is a clear divergence in RNA content with sample indicating considerable fluctuations in microbial activity (Table 9). ATP is a precursor to cell nucleic acids (DNA and RNA), as well as providing energy for general cell biosynthesis (Harder et al. 1984). Levels of adenylates fluctuate considerably with nutritional and physiological conditions and provide information on the physiological status and activity of the microbial community. There were variations in the levels of individual and total adenylates with sample (Table 9). The highest values of ATP and total adenylates were in general found in those samples with the largest RNA content, confirming these samples as potentially the most microbially active. EC (see methods) is a measure of the physiological state of the organisms within a sample and is independent of population size (Karl & Holm-Hansen 1978). EC could only be calculated for four of the Tanzanian samples studied since ATP in all the remaining samples was below levels of detection. The calculated EC values were < 0.5 suggesting relatively low activity in these samples. There is some discrepancy between this and the RNA results. It is possible that EC determinations are in fact influenced by the role of ATP as a precursor for the nucleic acids. In natural environments ATP and other adenylates may have a rapid turnover rate in cells resulting in the detection of only low concentrations.

The environmental characteristics of the groundwater system are likely to be an important determinant of the activity of the microbial community. It is often difficult to find correlations between geochemical factors and microbial growth and activity since many different factors interact to determine levels of bacterial activity. Even so, a slight inverse relationship was observed between microbial activity (biomarker data) and concentrations of the halides F^- and Cl^- in the groundwater (Fig. 9). Data on interactions between halides and microorganisms are relatively limited. A number of reactions, however, have been described. These include microbial halide binding (Bors et al. 1984), oxidation/reduction reactions (Tsungai & Sase 1969; Gozlan & Margalith 1973), and alkylation reactions (Harper & Hamilton 1988; Manley & Dastoor 1988). In addition, the halides are highly efficient disinfectants which may partly explain the inverse relationship between activity and F^- and Cl^-. Positive relationships between microbial activity, measured by either RNA or adenylate biomarkers, and the concentration of Al, Mn and Fe were found (Fig. 10a, b). Manganese and Fe can be oxidized by certain heterotrophic and autotrophic bacteria as part of their normal physiological processes, although the latter organisms are unlikely to be present in these tropical groundwaters. In addition, microorganisms can reduce oxidized forms of these metals (Beveridge & Doyle 1989; McEldowney et al. 1993). Iron and Mn are also required by microorganisms as trace elements i.e. micro-nutrients, and therefore their availablity may have an impact on microbial activity. Aluminium has no known microbiological function. Microbial activity may, however, affect the fate of Al in tropical groundwater (see below).

Eukaryotic microbes (protists) were found to be widespread throughout the Makutuapora aquifer. Until relatively recently protists were thought to be absent in subsurface environments. However, evidence is now accumulating that protists comprise a substantial part of many subsurface microbial populations (Sinclair & Ghiorse 1987, 1989; Kinner et al. 1992; Novarino et al. 1994). Previous studies of subsurface protists have been confined almost exclusively to temperate regions. This is thought to be the first record of protists in tropical groundwaters.

Recent research indicates that the vast majority of groundwater protists have a strong predilection for surfaces and are usually found attached to, or closely associated with, the aquifer sediment (Harvey et al. 1992; Novarino et al. 1994). Few protists are found free-swimming in groundwaters. Collection of subsurface protists therefore requires that cores of aquifer material are taken. Access to freshly-collected aquifer material was available at only a few sites in the Makutuapora aquifer. Sampling procedures mostly consisted of pumping groundwater to the surface. Therefore, it is assumed that the protists isolated during the present study represented only a small proportion of those present in the aquifer.

The vast majority of the protists isolated from the Makutuapora aquifer were heterotrophic. This is consistent with previous studies carried out in temperate regions (Sinclair & Ghiorse 1987; Novarino et al. 1994). Furthermore, all the protists identified during the present study were bacterivorous. Consquently, these may directly influence the levels of bacterial activity in the subsurface environment thereby indirectly affecting biogeochemical processes.

Mineral–water interactions

The major changes in the mineral assemblages are the loss of feldspar, pyroxene, biotite, and hornblende with the formation of zeolites (natrolite), hydrated phyllosilicates (kaolinite, sericite, vermiculite, montmorillonite and chlorite), oxides and hydroxides (goethite, gibbsite, hematite, and leucoxene). These changes are a function of mineral–water interactions in the bedrock aquifer and tropical weathering in the upper layers of the weathering profiles. Silicate minerals such as feldspar are being altered to hydrated forms, such as albite or microcline to kaolinite. This can also be inferred from the systematic changes in Na/K and Na/Ca ratios for newly drilled wells. Initially there is leaching of alkalis and alteration of oligoclase to albite and paragonite with time. In the amphibolites a similar variation in pumped water chemistry from a freshly drilled aquifer was noticed by variation in Na/Ca ratio, related to the alteration of feldspars and in the Fe/Mg and Ca/Mg ratio related to the alteration of ferromagnesium silicates augite and bronzite. The lack of constancy in the Fe/Mg ratio could be due to dissolution of ilmenite–magnetite as well as pyroxenes. The constant Na/Ca and Ca/Mg ratios noticed in established wells (those drilled prior to sampling seasons) indicate that a steady-state is attained between aluminosilicates in the wallrock and leachate chemistry. These steady states are attained instantly on geological time scales, within the space of a few hundred hours.

Biogeochemistry of Makutuapora waters

The biomass and activity of the microbial community has an impact, not only on the survival of enteric pathogens, but also on mobility, form and fate of organic and inorganic compounds in groundwater. This has health implications when water is extracted for drinking. There is a paucity of information available on the biomass and activity of microbial communities in tropical groundwater systems (see above), and the microbial community structure of soils and sediments associated with tropical aquifers.

This study demonstrated that the heterotrophic bacterial population present in the Tanzanian surface and subsurface soils associated with boreholes is towards the upper end of the range reported for aerobic heterotrophs isolated from shallow (Balkwill & Ghiorse 1985) and deep (Albrechtsen & Winding 1992; Fredrickson et al. 1989) groundwater in nontropical regions. Heterotrophic bacteria play an important role in the degradation of organic compounds.

Bacteria involved in the nitrogen biogeochemical cycle were present at all sites examined. Denitrifying bacteria represented a significant fraction of the heterotrophic bacterial population. Denitrifying bacteria are commonly found to be present and active in groundwater systems in temperate environments (Francis et al. 1989; Johnson & Wood 1992). Nitrifying bacteria, both ammonia oxidizing bacteria and nitrite oxidizing bacteria, were present at the Tanzanian sites. It has previously been found that the population density of nitrifying bacteria in temperate groundwater sediments is low, e.g. $10\,g^{-1}$, even when ammonia is not limiting. This has been attributed to competition for oxygen with aerobic heterotrophs (Fredrickson et al. 1989). Ammonia oxidizing bacteria are unable to dominate over heterotrophs particularly at available organic carbon levels below $1–2\,mg\,l^{-1}$ (McCarthy et al. 1981). The heterotrophic numbers isolated from the Tanzanian aquifer were high and the explanation for the low numbers of nitrifying bacteria may be the result of adverse competition for oxygen.

Sulphate reducing bacteria (SRB) have been isolated from shallow and deep aquifers, their numbers often increasing with depth (Jones et al. 1989; Johnson & Wood 1992). Their distribution and numbers have previously been related to the clay content of sediment, and have been difficult to correlate with groundwater sulphate concentration (Jones et al. 1989). Similarly, there was no significant correlation between groundwater sulphate concentration and the numbers of sulphate reducing bacteria at Makutuapora. In fact, the number of SRBs was low at most sites.

There is an apparent contradiction in the variation in isolated microbial numbers with sample and the relatively constant estimate of biomass shown in the lipid phosphate determinations. This discrepancy is probably simply related to the inevitably selective nature of culture media and chosen growth conditions failing to permit growth of many types of bacteria.

The presence of Fe and Al minerals in mineral precipitates in the feeder pipes between Dodoma and Makutuapora suggest that some mechanism exists for mobility of both metals. Similarly high levels of Mn and other metals occur in the pipe precipitates as well. A similar phenomenon observed in Wales has also been reported by Fuge et al. (1992) and likewise used to support the mobilization of metals in groundwater. Previous studies in temperate regions have shown that in carbon-limited habitats such as the subsurface, organic contamination of faecal origin often results in a significant increase in microbial activity (Harvey et al. 1984; Madsen et al. 1991). Increased microbial activity may in turn influence the mobility of certain elements in the aquifer through a variety of mechanisms discussed previously. Elevated levels of dissolved organic matter released by microbial activity in the groundwaters may also explain the observed relationships between elevated metal concentration and high BOD in some groundwaters.

Most of the aluminium was associated with particles in the size range $0.1–10.0\,\mu m$, which is colloidal either mineralogical or bacterial. The source of the aluminium may have been from the congruent dissolution of kaolinite in the upper soils. The dissolution of aluminium may also have been assisted by indigenous microorganisms, a phenomenon previously reported by McFarlane & Heydeman (1984) and by McFarlane & Bowden (1992). There is, however, little information on interactions between bacteria and aluminium in the environment. The importance of microorganisms in the mobilization of aluminium in deeply weathered profiles of the African surface has been highlighted by McFarlane (1987) and McFarlane & Bowden (1992). In these studies it was found that aluminium was leached from kaolinite by the action of indigenous populations of microorganisms, and furthermore that the aluminium remained mobile within the groundwaters, probably as a result of microbially-mediated organic complexing. Bacteria can accumulate metals at their surface and internally (Beveridge & Doyle 1989). Internal metal accumulation by bacteria is normally an energy dependent process and increases with time. Even potentially toxic metals with no physiological function can be taken up internally. Cell surface sorption of metals is a physicochemical process and tends to be relatively rapid, sometimes reaching equilibrium within minutes of metal exposure (McEldowney et al. 1993). Aluminium at the levels present in Makutuapora groundwater could be considered 'toxic' (WHO 1971; Connery 1990). Microorganisms within these groundwaters may assist in attaining these conditions and further, may be involved in the transportation and deposition of the metals. However, more recent work suggested that in tropical waters much of the Al is present in a non-labile form of relatively low potential toxicity (Smith et al. 1996).

Conclusions

In this study the geochemistry and microbiology of groundwaters and surface waters of the Makutuapora aquifer have been studied. Groundwater is essentially Na-Ca-Cl-HCO_3^- with minor K, Mg, F and SO_4^{2-}. Water chemistry is largely influenced by mineral–water interactions and less so by microbial activity, although the concentrations of Al, Mn and Fe in groundwater may be related in part to microbial activity.

The biochemical marker analysis suggests that the bacterial community is of comparable size between the different study sites, but that the physiological status of the community varies. This is undoubtedly related to variations in the physicochemical and nutrient conditions encountered by the bacteria at the sites. The numbers of nitrifying and denitrifying bacteria and SRBs isolated from the tropical groundwater systems were broadly comparable with numbers found to be present in temperate systems. By contrast, numbers of heterotrophs appeared relatively high in these tropical systems. In common with temperate groundwaters, there were no apparent correlations between the geochemical characteristics of the tropical groundwater and the numbers of different physiological types. There was considerable contamination of the groundwaters by thermotolerant coliforms suggesting a significant health risk from enteric pathogens in these groundwaters. The survival of the thermotolerant coliforms appeared to be linked with some of the geochemical characteristics of the groundwaters. Heterotrophic protists were widespread throughout the aquifer and are likely to influence bacterial activity by their predatory behaviour.

Metal content of the groundwaters can exceed $1\,mg\,l^{-1}$ of Al, Mn and Fe. These metals are present largely in a particulate, possibly colloidal form. The increase in metals suggests an imbalance in the steady-state reactions between groundwaters and magmatic minerals. This imbalance could in part be in response to microbial activity.

At present only weak correlations can be drawn from the field data which could be coincidental rather than casual and experimental work will be required to prove these possible relationships and to aid modelling of the biogeochemical cycles in the Makutuapora groundwaters.

This project was supported through a grant from The Natural History Museum, London and Wateraid, and supported in the field by the Tanzanian Ministry of Water, Energy, and Minerals and Overseas Development Administration. M. Yunusu, L. H. Rugeiyamu, I. Westbury, B. Kindoro, E. Wright, S. Kalli, R. Mtonga, A. Yates and E. H. Mwangimba are thanked for assistance and discussion in Tanzania. Laboratory assistance at The Natural History Museum by V. K. Din, T. Greenwood and J. Francis, at Southampton by N. H. Morley and at Imperial College by M. Gill is acknowledged. At the University of Westminster laboratory assistance was provided by D. Hard, P. Jackson, D. McNulty and S. Pickett.

References

ALBRECHSTEN, H-J. & WINDING, A. 1992. Microbial biomass and activity in subsurface sediments from Vezen, Denmark. *Microbial Ecology*, **23**, 303–317.

BALKWILL, D. L. & GHIORSE, W. C. 1985. Characterization of subsurface bacteria associated with two shallow aquifers in Oklahoma. *Applied and Environmental Microbiology*, **50**, 580–588.

BEVERIDGE, T. J. & DOYLE, R. J. 1989. *Metal Ions and Bacteria*. Wiley, New York.

BORS, J., MASTENS, R. & KUHN, W. 1984. Investigations on the influence of micro-organisms on the translocation of radio-iodide in soil. *In:* BONNYS-VAN GELDER, E. & KIRCHMAN, R. (eds) *Role of Microorganisms on the Behaviour of Radionuclides in Aquatic and Terrestrial Systems and their Transfer to Man*, Proc. Workshop Int. Union of Radioecologists, Brussels, 25–27 April 1984, 219–227.

BOWELL, R. J., MCELDOWNEY, S. & WARREN, A. 1995. *Biogeochemical Factors Affecting Water Quality in Tanzanian Waters*. Final report of NHM-Water-

Aid-University of Westminster project, 1992/93 (unpublished).
——, MORLEY, N. H. & DIN, V. K. 1994. Arsenic speciation in porewaters, Ashanti, Ghana. *Applied Geochemistry*, 9, 15–23.
BRIMHALL, G. H. & DIETRICH, W. E. 1987. Constitutive mass balance relations between chemical composition, volume, density, porosity, and strain in metasomatic hydrochemical systems: results of weathering and pedogenesis. *Geochimica et Cosmochimica Acta*, 51, 567–587.
——, LEWIS, C. J., AGUE, J. J., DIETRICH, W. E., HAMPEL, J., TEAGUE, T. & RIX, P. 1988. Metal enrichment in bauxites by deposition of chemically mature aeolian dust. *Nature*, 333, 819–824.
CAHEN, L. & SNELLING, N. J. 1984. *The Geochronology and Evolution of Africa*. Clarendon, Oxford.
CHESWORTH, W., VAN STRAATEN, P. & SEMOKA, J. M. R. 1989. Agrogeology in East Africa. *African Journal of Earth Sciences*, 9, 352–362.
CONNERY, J. 1990. Summary report of workshop on aluminium and health, Oslo, May 2–5 1988. *Environmental Geochemistry and Health*, 12, 179–196.
ERTEL, J. R., HEDGES, J. I., DEVOL, A. H., RICHEY, J. E. & RIBEIRO, M. de N. G., 1986. Dissolved humic substances of the Amazon river system. *Limnology and Oceanography*, 31, 739–754.
FARQUHARSON, F. A. & BULLOCK, A. 1992. The hydrology of basement complex regions of Africa with particular reference to southern Africa. *In:* WRIGHT, E. P. & BURGESS, W. P. (eds) *Hydrogeology of Crystalline Basement Aquifers in Africa*, Geological Society, London, Special Publication, 66, 59–76.
FRANCIS, A. J., SLATER, J. M. & DODGE, C. J. 1989. Denitrification in deep subsurface sediments. *Geomicrobiology Journal*, 7, 103–116.
FREDRICKSON, J. K., GARLAND, T. R., HICKS, R. J., THOMAS, J. M., LI, S. W. & MCFADDEN, K. M. 1989. Lithotrophic and heterotrophic bacteria in deep subsurface sediments and their relation to sediment properties. *Geomicrobiology Journal*, 7, 53–66.
FUGE, R. F., PEARCE, N. J. G. & PERKINS, W. T. 1992. Unusual sources of aluminium and heavy metals in potable water. *Environmental Geochemistry and Health*, 14, 15–18.
GOZLAN, R. S. & MARGALITH, P. 1973. Iodide oxidation by a marine bacterium. *Journal of Applied Bacteriology*, 36, 407–417.
HARDER, W., DIJKHUIZEN, L. & VELDKAMP, H. 1984. Environmental regulation of microbial metabolism. *In:* KELLY, D. P. & CARR, N. G. (eds) *The Microbe 1984. II Prokaryotes and Eukaryotes*. Cambridge University Press, Cambridge.
HARPER, D. B. & HAMILTON, J. T. G. 1988. Biogenesis of halomethanes by fungi. *In:* CRAIG, P. J. & GLOCKING, F. (eds) *The Biological Alkylation of Heavy Elements*. Royal Society of Chemistry, Special Publication, 66, 197–200.
HARVEY, R. W., SMITH, R. L. & GEORGE, L. 1984. Effects of organic contamination upon microbial distributions and heterotrophic uptake in a Cape Cod, Mass., aquifer. *Applied and Environmental Microbiology*, 48, 1197–1202.
——, KINNER, N. E., BUNN, A. L. & MACDONALD, D. 1992. Transport of protozoa through an organically contaminated sandy aquifer. *In: First International Conference on Groundwater Ecology, US EPA*. American Research Association, 111–118.
JEFFRIES, D. S. & HENDERSHOT, W. H. 1981. Aluminium geochemistry at the catchment scale in watersheds influenced by acidic precipitation. *In:* SPOISTO, G. S. (ed.) *The Environmental Chemistry of Aluminium*, CRC, Boca Raton, 279–302.
JOHNSON, A. C. & WOOD, M. 1992. Microbial potential of sandy aquifer material in the London Basin. *Geomicrobiology Journal*, 10, 1–13.
JONES, B. F., KENNEDY, V. C. & ZELLUREGE, G. W. 1974. Comparison of observed and calculated concentrations of dissolved Al and Fe in stream water. *Water Resources Research*, 10, 791–793.
JONES, R. E., BEEMAN, R. E. & SUFLITA, J. M. 1989. Anaerobic metabolic processes in deep terrestrial subsurface. *Geomicrobiology Journal*, 7, 117–130.
KARL, D. M. & HOLM-HANSEN, O. 1978. Methodology and measurement of adenylate energy charge ratios in environmental samples. *Marine Biology*, 48, 185–197.
——, WINN, C. D. & WONG, D. C. L. 1981. RNA synthesis as a measure of microbial growth in aquatic environments: evaluation, verification, and optimization of methods. *Marine Biology*, 64, 13–21.
KINNER, N. E., BUNN, A. L., HARVEY, R. W., WARREN, A. & MEEKER, L. D. 1992. Preliminary evaluation of the relationships among protozoa, bacteria and chemical parameters in sewage contaminated groundwater at Otis Air Base, Massachusetts. *In:* MULLARD, G. E. & ARONSON, D. A. (eds) *USGS Toxic Substances Hydrology Program Proc. Tech. Mtg.* WRI Report 91-4034, 148–151.
LAHERMO, P., SANDSTROM, H. & MALISA, M. 1991. The occurrence and geochemistry of fluoride in natural waters in Finland and East Africa with reference to geomedial implications. *Journal of Geochemical Exploration*, 4, 65–79.
LINDSAY, W. L. & WALTHALL, P. M. 1981. The solubility of aluminium in soils. *In:* SPOISTO, G. S. (ed.), *The Environmental Chemistry of Aluminium*, CRC, Boca Raton, 221–240.
MCCARTHY, P. L., REINHARD, M. & RITTMANN, B. E. 1981. Trace organics in groundwater. *Environmental Science and Technology*, 15, 40–51.
MCCOY, W. F. & OLSON, B. 1985. Fluorometric determination of the DNA concentration in municipal drinking water. *Applied Environmental Microbiology*, 49, 811–817.
MCELDOWNEY, S., HARDMAN, D. J. & WAITE, S. 1993. *Pollution: Ecology and Biotreatment*. Longman Scientific & Technical, Harlow, 261–289.
MCFARLANE, M. J. 1987. The key role of microorganisms in the process of bauxitisation. *Modern Geology*, 11, 325–344.
—— & BOWDEN, D. J. 1992. Mobilization of aluminium in the weathering profiles of the African surface in Malawi. *Earth Surface Pro-*

cesses and Landforms, **17**, 789–805.

—— & HEYDEMAN, M. T. 1984. Some aspects of kaolinite dissolution by a laterite-indigenous micro-organism. *Geo-Eco-Trop.*, **8**, 73–91.

MADSEN, E. L., SINCLAIR, J. L. & GHIORSE, W. C. 1991. In situ biodegradation: microbiological patterns in a contaminated aquifer. *Science*, **252**, 830–833.

MADINI. 1992. *Stratigraphy and Geochronology of Dodoma, Tanzania*, Unpublished report.

MAJI. 1988. *Hydrogeology of the Makutuapora aquifer: Results of deep drilling.* Government printers, Dares Salaam.

MANLEY, S. L. & DASTOOR, M. N. 1988. Methyl iodide (CH_3I) production by kelp and associated microbes. *Marine Biology*, **98**, 477–482.

MANN, A. W. 1983. Hydrogeochemistry and weathering on the Yilgarn block, Western Australia-ferrolysis and heavy metals in continental brines. *Geochimica et Cosmochimica Acta*, **47**, 181–190.

NEIDHARDT, F. C., INGRAHAM, J. L. & SCHAECHTER, M. 1990. *Physiology of the Bacterial Cell. A Molecular Approach.* Sinauer Associates Inc., Sunderland, Massachusetts.

NOVARINO, G., WARREN, A., KINNER, N. E. & HARVEY, R. W. 1994. Protists from a sewage-contaminated aquifer on Cape Cod, Massachusetts, U.S.A. *Geomicrobiology Journal*, **12**, 23–36.

OGBUKAGU, I. K. 1984. Hydrology of groundwater resources of the Aguta area, SE Nigeria. *Journal of African Earth Science*, **2**, 109–117.

PAUL, J. H. & MEYERS, J. 1982. Fluorimetric determination of DNA in aquatic microorganisms by use of Hoechst 33258. *Applied and Environmental Microbiology*, **43**, 1393–1399.

SHOTYK, W. 1984. Metal-organic species in natural waters. *In:* FLEET, M. E. (ed.), *MAC Short Course Handbook, Environmental Geochemistry.* 45–66

SINCLAIR, J. L. & GHIORSE, W. C. 1987. Distribution of protozoa in subsurface sediments of a pristine groundwater study site in Oklahoma. *Applied and Environmental Microbiology*, **53**, 1157–1163.

—— & —— 1989. Distribution of aerobic bacteria, protozoa, algae, and fungi in deep subsurface sediments. *Geomicrobiology Journal*, **7**, 15–31.

SMITH, B. J., BREWARD, N., CRAWFORD, M. B., GALIMAKA, D., MUSHIRI, S. M. & REEDER, S. 1996. The environmental geochemistry of aluminium in tropical terrains and its implications to health. *This volume.*

SMITH, O. L. 1982. *Soil Microbiology: a Model of Decomposition and Nutrient Cycling*, CRC, Boca Raton.

TRESCASES, J-J. 1992. Chemical Weathering. *In:* BUTT, C. & ZEEGERS, H. (eds), *Regolith Exploration Geochemistry in Tropical and Subtropical terrains. Handbook of Geochemistry*, vol. IV, 25–40.

TSUNGAI, S. & SASE, T. 1969. Formation of iodide-iodine in the ocean. *Deep Sea Research*, **16**, 484–487.

VELBEL, M. A. 1989. Mechanisms of saprolization, isovolumetric weathering, and replacement during weathering – A review. Geochemistry of the earth's surface *Chemical Geology*, **84**, 17–18.

—— 1992. Constancy of silicate-mineral weathering-rate ratios between natural and experimental weathering implications for hydrologic control of differences in absolute rates. *Chemical Geology*, **105**, 89–99.

WHITE, D. C., BOBBIE, R. J., HERRON, J. S., KING, J. & MORRISON, S. 1979. Biochemical measurements of microbial biomass and activity from environmental samples. *In: Native Aquatic Bacteria: Enumeration, Activity and Ecology.* ASTM, Special Technical Publication, **695**, American Society for Testing and Materials, Philadelphia, 69-81.

WHO 1971 *International Standards for Drinking Water.* Geneva.

WILLIAMS, C. T., SYMES, R. F. & DIN, V. K. 1993. Mobility and fixation of a variety of elements, in particular B, during the metasomatic development of adinoles at Dinas Head, Cornwall. *Bulletin of the Natural History Museum (Geology Series)*, **49**, 81–98.

WRIGHT, E. P. 1992. The hydrogeology of basement complex regions of Africa with particular reference to southern Africa. *In:* WRIGHT, E. P. & BURGESS, W. P. (eds) *Hydrogeology of Crystalline Basement Aquifers in Africa*, Geological Society, London, Special Publication, **66**, 21–58.

Water quality and dental health in the Dry Zone of Sri Lanka

C. B. DISSANAYAKE

Institute of Fundamental Studies, Hantana Road, Kandy, Sri Lanka
(Present address: Department of Geology, University of Peradeniya, Peradeniya, Sri Lanka)

Abstract: The Dry Zone of Sri Lanka has mostly poor rural folk, who live in very close association with the immediate physical environment, depending on it for their living. Many of these people have lived in such a confined environment throughout their lives and it is reasonable to assume that at least some aspects of their health could be correlated with the geochemistry of their immediate environment.

Hydrogeochemical investigations of surface well water and deep well water in parts of the Dry Zone notably in the North Central Province, have shown that the fluoride concentrations often reach anomalously high levels, of the order of $10\,\text{mg}^{-1}$ in some instances. As a result of this, dental fluorosis is increasing in its prevalence among school children. With over 13 000 tube wells with hand pumps, mostly in the Dry Zone, the problem of dental fluorosis will assume serious proportions in the future unless methods of defluoridating household water supplies are developed.

An attempt has been made to implement a programme of defluoridating water supplies with a defluoridator using charcoal and charred bone meal. However, a major awareness programme is needed to educate the villagers of the dangers of excessive fluoride in drinking water and the need for community participation in maintaining household defluoridators.

The close relationship between human health and the physical and geochemical environment is clearly seen in Sri Lanka with its varied topography, climate, soils and geology. The effect of soil and water chemistry on the dental health of the Sri Lankan population, most of whom live in rural areas, can be seen most directly in the fluoride content of their water supplies since fluorides enter the body primarily from drinking water supplies, which in Sri Lanka come mainly from groundwater.

Fluoride is considered to be an essential element (Wood 1974), although dental health problems may arise from an excess of fluoride. Many water supply schemes, particularly in developing countries where dug wells and tube wells form the major water sources, contain excess fluoride and as such are harmful to dental health (Dissanayake 1991).

As shown by Dharmagunawardhane & Dissanayake (1993), although rural water supply projects based on tube wells have provided safe drinking water to the rural community, in the areas with fluoride-rich groundwater these projects have been of limited benefit to the people. This is because defluoridation at the village level has still not been successful in Sri Lanka and, therefore, these waters can cause dental fluorosis.

The majority of people living in the Dry Zone of Sri Lanka (Figs 1, 2), particularly the children, are thus prone to dental diseases caused by an excess of fluoride in drinking water. In order to take appropriate remedial action, the water supply authorities must be provided with information on the distribution of high and low-fluoride zones in an area. This requires geochemical mapping and incorporating the information into plans for shallow and deep wells, settlement schemes and water projects.

This paper examines the general water quality of the Dry Zone of Sri Lanka with particular emphasis on the fluoride geochemistry in relation to the dental health of the people.

Geographical and geological setting

Figure 1 illustrates the climatic, physiographic and major geological divisions of Sri Lanka. The dry Zone of Sri Lanka with an annual average rainfall of 1500 mm lies on the periphery of the highlands and covers the major part of the country. Geomorphologically, the Dry Zone is generally characterized by lowlands with an undulating ground surface and scattered discontinuous ridges. Over most parts of the year, in spite of an average 1500 mm of rainfall, which compared to non-tropical countries appears high, the groundwater tends to be deep and due to evaporation and evapotranspiration salts tend to build up on the surface causing problems of salinity. Such areas are often observed in the Anuradhapura, Polonnaruwa and Hambantota districts (Fig. 1).

Geologically, Sri Lanka consists of over 90%

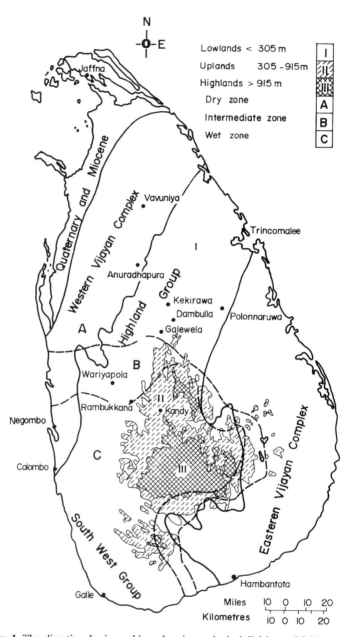

Fig. 1. The climatic, physiographic and major geological divisions of Sri Lanka.

metamorphic rocks of Precambrian age and these form three major units, the Highland Group, the east and west Vijayan Complex and the southwest Group. Recently some authors (Kröner *et al.* 1991) have re-named the east Vijayan Complex as the Wanni Complex. The Dry Zone of Sri Lanka as shown in Fig. 1 consists of rocks belonging to the Highland Group and the two Vijayan Complexes.

A suite of metasedimentary and metaigneous rocks formed under granulite facies conditions comprises the Highland Group. Among the metasedimentary rocks, quartzites marbles, quartzo-feldspathic gneisses and metapelites form the major constituents.

The Wester Vijayan rocks consist of leuco-

Fig. 2. The geochemical classification of groundwater in Sri Lanka (modified after Dissanayake & Weerasooriya 1986).

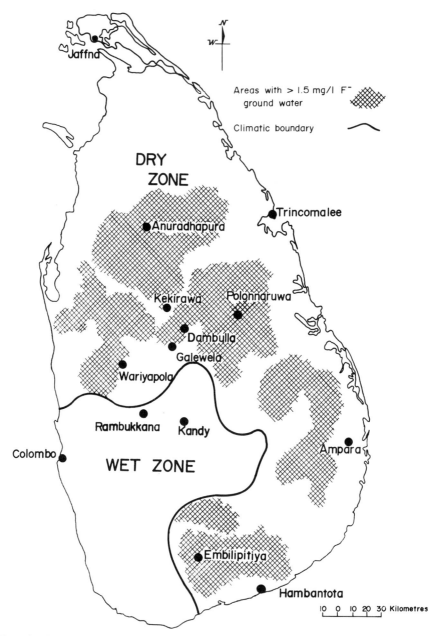

Fig. 3. The main high fluoride-bearing groundwater areas in Sri Lanka. This map has been compiled using data obtained from hydrogeochemical surveys carried out by several water supply organizations.

cratic biotite gneisses, migmatites, pink granitic gneisses and granitoids. The Eastern Vijayan is composed of biotite/hornblende gneisses, granitic gneisses and scattered bands of metasediments and charnockitic gneisses. Present in some of these rocks are fluoride-bearing minerals such as micas, hornblende and apatite. Additionally, fluorite, tourmaline, sphene and topaz are present and these contribute to the general geochemical cycle of fluorine.

Water quality of the Dry Zone

Figure 2 illustrates the geochemistry of the

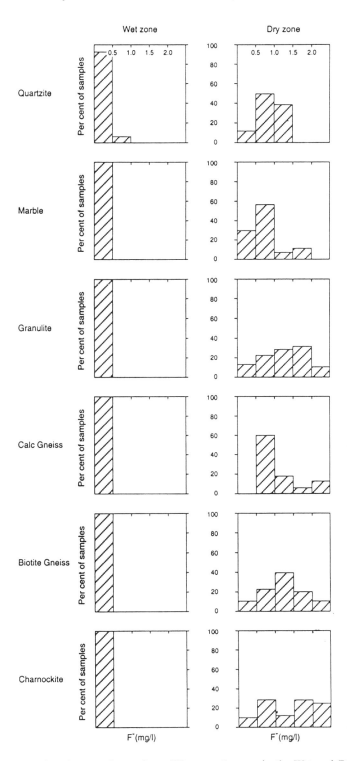

Fig. 4. Fluoride concentrations in groundwater from different rock types in the Wet and Dry Zones (after Dharmagunawardhane & Dissanayake 1993).

groundwater of Sri Lanka. It is seen that in the Dry Zone, sodium/potassium ions and calcium ions are dominant. Much of it is dominated by Na+K–Cl with salinity approaching undesirable levels in certain agroecological zones.

Figure 3 shows a generalized map indicating the distribution of fluoride in groundwater above $1.5\,\mathrm{mg\,l^{-1}}$, as obtained from data collected by water supply organizations in Sri Lanka. The effect of the climate on the geochemistry of fluoride in water is clearly apparent by the fact that the wet zone in spite of having similar lithology has negligibly small areas containing fluoride greater than $1.5\,\mathrm{mg\,l^{-1}}$ in the groundwater. It is likely that in the Wet Zone where the annual rainfall exceeds 5000 mm, fluoride is easily leached out from primary and secondary minerals. In the Dry Zone on the other hand, evaporation tends to concentrate the soluble ions upwards due to capillary action in soils.

Groundwater in the crystalline rock terrain of Sri Lanka mainly occurs within (1) the weathered overburden of the basement rocks; and (2) basement rocks, the secondary porosity being due to fractures, joints, faults, fissures and solution cavities. The water supply boreholes tap groundwater from depths where fractured or jointed zones are present. At present there are over 13 000 such tube wells with hand pumps mostly in the Dry Zone of Sri Lanka. Some wells reach a depth of 80 m while the average depth is in the region of 50 m. The upper parts of these boreholes are cased in the weathered overburden into hardrock with well casings.

Figure 4 compares the fluoride content in groundwater from different rock types under dry and wet conditions. Irrespective of the rock type, the groundwater remains low in fluoride in the Wet Zone while in the Dry Zone fluoride reaches levels as high as $10\,\mathrm{mg\,l^{-1}}$. It is particularly relevant to note that unlike drinking water with high levels of dissolved iron, which has both a colour and an objectionable taste, water containing excess fluoride is colourless and tasteless, chemical analyses being required to detect its presence.

Aspects of dental health

The prevalence of dental fluorosis in the Dry Zone of Sri Lanka has been highlighted by several workers (Senewiratne et al. 1974; Dissanayake & Senaratne 1982; Dissanayake 1991; Dharmagunawardhane & Dissanayake 1993). With the large number of shallow wells and deep boreholes now being constructed, the problem of excessive ingestion of fluorides through drinking water with resulting dental fluorosis is causing serious concern among community health workers (Fig. 5a, b). Warnakulasuriya et al. (1990) in a study on the prevalence of dental fluorosis from different areas in Sri Lanka, showed that in the endemic zone the prevalence of dental fluorosis was 51–78% while in a non-endemic area it was 5.4%. With some tube well water containing fluoride levels as high as $10\,\mathrm{mg\,l^{-1}}$ and bearing in mind that the WHO recommended danger level is $1.5\,\mathrm{mg\,l^{-1}}$ F^- for drinking water, one could even anticipate the prevalence of skeletal fluorosis. A man from the Anuradhapura area who appeared to show signs of a bone disease, when examined at the Faculty of Medicine, University of Peradeniya, was found to be a case of skeletal fluorosis (Prof. C. Ratnatunga, pers. comm.).

Warnakulasuriya et al. (1990) examined a total of 380 children aged 14. Table 1 shows the results for different areas (see Fig. 3 for locations). This work revealed that Kekirawa is a highly endemic area and nearly a third of the children examined had extensive dental fluorosis. Galewela and Wariyapola were moderately endemic while Rambukkana in the Sabaragamuwa Province was considered as a low endemic area.

Table 1. *Prevalence (% incidence) of dental fluorosis in 4 selected areas in Sri Lanka*

Area	Boys	Girls	Total
Galewela	5.2	53.5	52.2
Wariyapola	55.6	47.3	51.0
Kekirawa	73.7	78.2	78.4
Rambukkana	5.4	5.5	5.4
Total	44.6	49.7	46.4

After Warnakulasuriya et al. (1990). See Fig. 1 for locations.

One of the key issues that arose out of investigations on the prevalence of dental fluorosis in Sri Lanka was the determination of the optimal levels of fluoride in drinking water for hot and dry climates as exemplified by Sri Lanka. The WHO guidelines for the upper limit of fluoride in drinking water have long been felt unsuitable for developing countries with a hot, dry climate.

Dean (1945) originally recommended $1.0\,\mathrm{mg\,l^{-1}}$ as the optimal level of fluoride in drinking water basing his observations on the relation between caries inhibition and the severity of dental fluorosis. Myers (1978) showed that in temperate

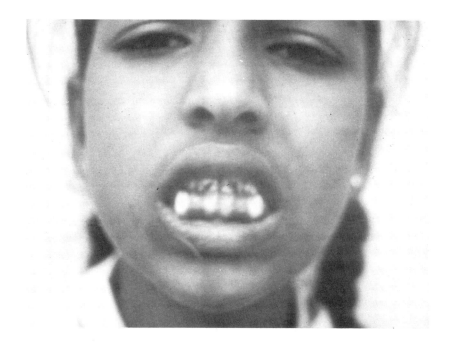

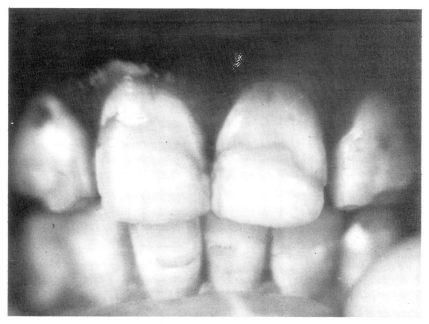

Fig. 5. School children in the village of Polpitigama in the Kurunegala District showing marked development of dental fluorosis.

climates levels up to $1.5\,\text{mg}\,\text{l}^{-1}$ in drinking water produce only questionable and mild fluorosis of no public health significance. This therefore is the desirable upper limit recommended by the World Health Organisation (WHO). Several workers (Galagan 1953; Richards et al. 1967; Moller 1982) had commented that the upper limit of fluoride concentration that causes dental fluorosis may change from country to country and that for tropical countries the recommended levels should be lower than those for temperature countries. Manji et al. (1968a, b) in a case study from Kenya showed that with fluoride levels of 0.1–$1.0\,\text{mg}\,\text{l}^{-1}$ a very high prevalence and severity of dental fluorosis were observed.

Apart from the geology, the varying levels of fluorides that cause dental fluorosis in different countries may also reflect the relative socio-economic levels. It is very likely that low protein diet populations are more prone to fluorosis and therefore it is possible that a combination of factors causes greater fluorosis in developing countries.

The recent work of Warnakalasuriya et al. (1992) adds further evidence to the fact that in hot dry climates where consumption of water is far higher, there could be dental fluorosis even in areas containing groundwater fluorides lower than $0.3\,\text{mg}\,\text{l}^{-1}$, as in the case of Sri Lanka. Their studies showed that among those consuming drinking water with $< 1.0\,\text{mg}\,\text{l}^{-1}$ fluoride, 32% of the children had mild forms and 9% severe forms of dental fluorosis. This work provided further reasons for a need to change the WHO guidelines for the upper limit of fluoride in drinking water. It was recommended that this level be $0.8\,\text{mg}\,\text{l}^{-1}$ for those living in hot, dry climates.

In a comparative study Rugg-Gunn et al. (1993) investigated the urinary fluoride excretion in 4 year old children in Sri Lanka and England. Urine had been collected over 24 hours from 53 children from Dambulla, Sri Lanka (see Fig. 1 for location) and 44 from Newcastle, England, both localities with drinking water containing 0.8–$1.1\,\text{mg}\,\text{l}^{-1}\,\text{F}$. The mean fluoride concentrations were $1.19\,\text{mg}\,\text{l}^{-1}$ in Sri Lanka and $1.02\,\text{mg}\,\text{l}^{-1}$ in England. A number of possible explanations were put forward by the authors for this small difference in fluoride urinary excretion.

1. Because of the higher air temperature at Dambulla (27°C) compared to Newcastle (12°C), fluoride intake from greater amounts of water or from other dietary sources may have been higher in Sri Lankan children.

2. The proportion of ingested fluoride excreted in urine could have been different.

The author suggested that the amount of water ingested may be of prime importance in determining the amount of fluoride excreted in England, but a more complicated relationship existed in Sri Lanka.

In a similar study Nunn et al. (1993) investigated the distribution of developmental defects of enamel on ten tooth surfaces in children aged 12 years living in areas with different water fluoride levels in Sri Lanka and England, data being presented for 168 and 379 subjects respectively. While the maxillary central incisors were affected most often in England and in the $0.1\,\text{mg}\,\text{l}^{-1}\,\text{F}$ area in Sri Lanka, they found that this was not the case in the 0.5 and $1.0\,\text{mg}\,\text{l}^{-1}\,\text{F}$ areas in Sri Lanka where prevalence was highest in premolar and canine teeth, with demarcated and diffuse opacities predominating in the $1.0\,\text{mg}\,\text{l}^{-1}\,\text{F}$ areas in both countries. Hypoplastic lesions were also prevalent in Sri Lanka in the 0.1 and $0.5\,\text{mg}\,\text{l}^{-1}\,\text{F}$ areas especially in maxillary incisor teeth, with nearly half the lesions extending to more than one-third of the tooth surface in the $1.0\,\text{mg}\,\text{l}^{-1}\,\text{F}$ areas. This study of Nunn et al. (1993) indicated that maxillary canine and premolar teeth are affected much more in high fluoride areas in Sri Lanka and it was suggested that this may be due to their later development relative to incisors and first molars. On account of the great aesthetic importance of the maxillary incisor teeth, the onset of fluorosis is at present causing great social problems among school children living in the high fluoride areas of the Dry Zone of Sri Lanka (Fig. 5a, b).

Defluoridation of water supplies and community participation

The construction of tube wells with hand pumps in rural community water supply programmes in Sri Lanka has increased drastically since the beginning of the last decade. At present there are over 13 000 tube wells with hand pumps in the island, mostly in the Dry Zone. With the increasing number of tube wells, the problems of water quality pertaining to fluoride have also become significant. Many of these tube wells have fluoride levels exceeding $1.5\,\text{mg}\,\text{l}^{-1}$ and their impact on the dental health of the population is apparent by the fact that 40–75% of the children using tube well water are affected by dental fluorosis to varying degrees. Interestingly, the old villagers who did not have tube wells, but used lake and stream water for their domestic requirements, did not show any symptoms of dental fluorosis.

Due to the fact that by far the largest proportion of people who rely on drinking

The ICOH defluoridator

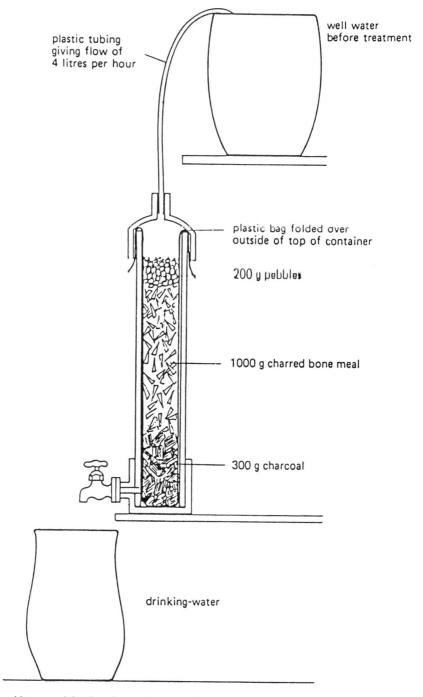

Fig. 6. The defluoridator used in the pilot project at Polpitigama, Kurunegala District, Sri Lanka. This defluoridator (after Phantumvanit et al. 1988) was developed by the Intercountry Centre for Oral Health, Chiang Mai, the Dental Faculty of Chulalongkorn University Bangkok and the WHO.

water from high fluoride tube wells in the Dry Zone live in remote villages, the only method of defluoridating their drinking water supplies is by the use of household defluoridators. The mechanism of defluoridation of household drinking water in remote villages in a developing country such as Sri Lanka, presents several problems. (a) The capital investment and the maintenance and treatment costs of defluoridation of water should be small. (b) The device should be simple in design. (c) The village community should be able to prepare and change the substances used in the household defluoridator with ease. (d) It should have the capacity to reduce the fluoride content from c $5\,\text{mg}\,l^{-1}$ to less than $0.5\,\text{mg}\,l^{-1}$. (e) It should improve the water quality in general, rather than make it harmful by the addition of chemicals. (f) The substances used in the defluoridator should maintain their activity for an acceptable period of time.

Such a defluoridator using charcoal and charred bore meal (Fig. 6) (Phantumvanit et al. 1988) was introduced into a village in the endemic fluorosis zone of Sri lanka. The initial response of the villagers was very good and the attempts made to eradicate their dental problems were gratefully acknowledged. However, a major awareness programme is needed to educate the village folk of the dangers of excess fluorides in their drinking water supplies and the need for community participation in maintaining household defluoridators.

An international collaborative programme involving dental epidemiologists specializing in community medicine and hydrogeochemists would be of immense national importance to Sri Lanka.

Thanks are due to Mrs Sandra Paragahawewa, Mrs Harshini Aluwihare and Ms Anne George for their assistance.

References

DEAN, H. T. 1945. On the epidemiology of fluorine and dental caries. *In:* GIES, W. J. (ed.) *Fluorine in Dental Public Health*. Academy of Medicine, New York, 19–30.

DHARMAGUNAWARDHANE, H. A. & DISSANAYAKE, C. B. 1993. Fluoride problems in Sri Lanka. *Environmental Management and Health*, **4**, 9–16.

DISSANAYAKE, C. B. 1991. The fluoride in the groundwater of Sri Lanka – Environmental Management and Health. *International Journal of Environmental Studies*, **38**, 137–156.

—— & SENARATNE, A. 1982. Geochemical provinces and geographic distribution of some diseases in Sri Lanka. *Water, Air and Soil Pollution*, **17**, 17–28.

—— & WEERASOORIYA, S. V. R. 1986. *The Hydrogeochemical Atlas of Sri Lanka*. A publication of the Natural Resources Energy and Science Authority of Sri Lanka, Colombo.

GALAGAN, D. 1953. Climate and controlled fluoridation. *Journal of the American Dental Association*, **47**, 159–170.

KRÖNER, A., COORAY, P. G. & VITANAGE, P. W. 1991. Lithotectonic subdivision of the Precambrian basement in Sri Lanka. *In: Geological Survey Department, Sri Lanka, Professional Paper*, **5**, 5–21.

MANJI, F., BAELUM, V. & FEJERSKOV, O. 1986a Fluoride, altitude and dental fluorosis. *Caries Research*, **20**, 473–480.

——, ——, —— & GEMERT, W. 1986b. Enamel changes in two low fluoride areas of Kenya. *Caries Research*, **20**, 371–380.

MOLLER, I. J. 1982. Fluoride and dental fluorosis. *International Dental Journal*, **32**, 135–147.

MYERS, H. M. 1978. *Fluorides and dental fluorosis*. Monograph on Oral Science, Basel, Karger, 7.

NUNN, J. H., EKANAYAKE, L., RUGG-GUNN, J. & SAPARAMADU, K. D. G. 1993. Distribution of development defects of enamel on ten tooth surfaces in children aged 12 years living in areas receiving different water fluoride levels in Sri Lanka and England. *Community Dental Health*, **10**, 259–268.

PHANTUMVANIT, P., SONGPAISAN, Y. & MOLLER, I. J. 1988. A defluoridator for individual households. *World Health Forum*, **9**, 555–558.

RICHARDS, L. F., WESTMORELAND, W. W., TASHIRO, M., MAKAY, C. H. & MORRISON, J. T. 1967. Determining optimum fluoride levels for community water supplies in relation to temperature. *Journal of American Dental Association*, **74**, 389–397.

RUGG-GUNN, A. J., NUNN, J. H., EKANAYAKE, L., SAPARAMADU, K. D. G. & WRIGHT, W. G. 1993. Urinary fluoride excretion in 4-year old children in Sri Lanka and England. *Caries Research*, **7**, 478–483.

SENEWIRATNE, B., THAMBIPILLAI, S., HETTIARACHCHI, J. & SENEWIRATNE, K. 1974. Endemic fluorosis in Ceylon. *Transactions of the Royal Society of Tropical Medicine and Hygiene*, **68**, 105–113.

WARNAKULASURIYA, K. A. A. S., BALASURIYA, S. & PERERA, P. A. J. 1990. Prevalence of dental fluorosis in four selected schools from different areas in Sri lanka. *Ceylon Medical Journal*, **35**, 125–128.

——, ——, —— & PEIRIS, L. C. L. 1992. Determining optimal levels of fluoride in drinking water for hot, dry climates – a case study in Sri Lanka. *Community Dental and Oral Epidemiology*, **20**, 367–367.

WOOD, J. M. 1974. Biological cycles for toxic elements in the environment. *Science*, **183**, 1049–1052.

The environmental geochemistry of aluminium in tropical terrains and its implications to health

B. SMITH,[1] N. BREWARD,[1] M. B. CRAWFORD,[1] D. GALIMAKA,[2] S. M. MUSHIRI,[3] & S. REEDER[1]

[1] *British Geological Survey, Keyworth, Nottingham NG12 5GG, UK*
[2] *Geological Survey and Mines Department, PO Box 9, Entebbe, Uganda*
[3] *Chemistry and Soil Research Institute, Ministry of Lands, Agriculture and Water Development, PO Box 8100, Causeway, Harare, Zimbabwe*

Abstract: The link between environmental exposure to aluminium, mammalian toxicity and agricultural productivity has focused attention on the relationship between the environmental geochemistry of Al and its potential toxicity. Recent work has questioned the accuracy of previously reported baseline levels of aluminium in tropical well waters. In this work we investigate the validity of these observations and their influence on criteria for the determination of water quality.

Data are presented for two contrasting study areas in Uganda in which c. 50 water samples were collected from a variety of water sources and subsequently analysed to identify concentrations of dissolved Al and of ligands that potentially control its environmental mobility and toxicty. Determinands included: pH, temperature, redox potential, conductivity, alkalinity, major and trace cations and anions and TOC.

Determination of Al by ICP-AES before and after sonication confirmed that sonication of samples filtered to 0.40 μm during sampling resulted in increased Al concentrations in solution. This effect is considered to be consistent with the aggregation of a non-labile colloidal Al on the addition of acid and confirms observations made by other researchers in Malawi. However, on-site analysis of samples using pyrocatechol violet indicated negligible free or labile Al present in the sampled waters and suggests that the majority of Al undetected by routine ICP-AES analysis is present in a non-labile form of relatively low potential toxicity.

Concern over the possible link between environmental exposure to aluminium, its mammalian toxicity (Martin 1986; Martyn *et al.* 1989; Harrington *et al.* 1994) and agricultural productivity (Baker & Schofield 1982; Fagerial *et al.* 1988) has focused attention on the relationship between the environmental geochemistry of Al and its potential toxicity.

Whilst the majority of studies to date have been focused on temperate regions of the northern hemisphere affected by acid rain (Schindler 1988; Sposito 1989), recent work questioning the accuracy of previously reported baseline levels of aluminium in tropical well waters (McFarlane 1991; McFarlane & Bowden 1992) suggests potential cause for concern in the developing world. This concern arises from the relatively rapid expansion in the use of groundwater as a source of potable water within developing countries and epidemiological and aetiological links between elevated Al concentrations in drinking water and Alzheimer's disease (Martyn *et al.* 1989; Crapper-McLachlan *et al.* 1991).

The ability to quantify accurately and precisely the concentration of aqueous Al is an obvious prerequisite to investigating its distribution within the environment and consequential toxicity. The analytical determination of Al in natural waters is well studied (Bloom & Erich 1989) although considerable difficulties still exist in the determination of Al because of its low concentration in many waters, the large number of forms or species of dissolved Al, and because of its ability to form stable natural colloids with dissolved organic carbon, fluoride and sulphate (Fig. 1).

Observations by McFarlane & Bowden (1992) indicating that filtration of water samples obtained from tropical wells reduces the Al content is consistent with the relatively high abundance of clay mineral particulates in such highly weathered unsaturated zones. Similar observations are documented in the literature (Kennedy *et al.* 1974; Wagemann & Brunskill 1975; Hem 1985; Goenaga *et al.* 1987). The observation that the addition of acid (1% HNO_3) after filtration further reduces the

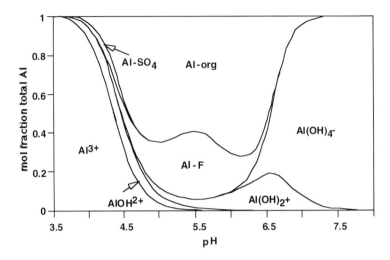

Fig. 1. Phase diagram illustrating the complex solution chemistry of Al for a typical uncontaminated stream water with a moderate dissolved organic carbon content (after Driscoll & Schecher 1990).

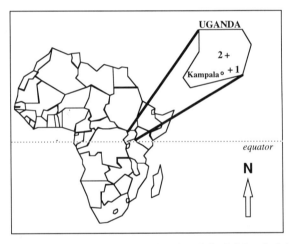

Fig. 2. Geographical position of site 1 (Mukono) and site 2 (Nawaikoke) in Uganda.

observed concentration of Al is, however, inconsistent with standard preservation procedures for trace metals (British Standards Institution 1986) and clearly warrants further more detailed investigation in order to define the speciation of Al at the time of sampling, its potential toxicity and the mechanism of removal.

In their previous work in Malawi, McFarlane & Bowden (1992) determined Al by graphite furnace atomic absorption spectrometry following filtration, acidification and sonication. Unfortunately they were unable to determine any other parameters, with the exception of electrical conductivity, making it impossible to test their observations further, either through the use of computer modelling (e.g. Tipping et al. 1991) or by the preparation of synthetic water samples on which to determine Al speciation experimentally (Driscoll 1984).

In this work we focus upon exposure to Al in shallow groundwaters and review results of experiments performed to investigate the observations of McFarlane & Bowden (1992) with regard to the analytical determination of aqueous Al. Data collected during two field exercises

Table 1. *Summary of sample preservation and analysis regime*

Sample treatment	Code	Analysis
Untreated	n/a	pH, temperature, conductivity, redox potential, total alkalinity
Untreated	n/a	labile Al by PCV
Untreated	UF/UA	no specific test carried out
Unfiltered, 1% HNO$_3$	UF/A	total aluminium (sonicated to include < 0.40 μm and > 0.4 μm particulates)
0.4 μm filtered	F/UA	major anions, fluoride, TOC, TIC, Si
0.4 μm filtered, 1% HNO$_3$, unsonicated	F/A	major cations, Fe, Si, total dissolved aluminium
0.4 μm filtered, 1% HNO$_3$, sonicated	F/A	total dissolved and < 0.40 μm particulate aluminium

UF, unfiltered; UF/A, unfiltered acidified; UF/UA, unfiltered unacidified; F/A, filtered acidified.

in central Africa are presented along with experiments performed on samples collected in the field to measure spatial variations in and factors controlling environmental concentrations of Al in the near surface environment.

Methodology

Description of study areas

Two representative study areas (Fig. 2) were selected in Uganda on the basis of soil type (wet, high productivity red clay loams and quartzitic sands of the Mukono region, very dry red clays and lateritic soils of the Nawaikoke region), climate, geomorphology and the availability of suitable sampling sites (i.e. the relative abundance of shallow and deep groundwater reserves).

Mukono (32.6°E, 0.5°N) *c.* 30 km northeast of Kampala (Department of Lands and Surveys 1992) is representative of a relatively high rainfall zone with high agricultural productivity characterized by banana plantations, remnants of tropical forest and coffee/tea plantations. The area is supplied with a relatively well developed system of shallow hand-dug wells and protected springs developed in highly weathered, Fe rich, kaolinitic soils occasionally associated with laterite and underlying undifferentiated gneiss (Geological Survey of Uganda 1957).

Nawaikoke (33.2°E, 1.1°N) *c.* 80 km north of Kampala (Department of Lands and Surveys 1963) is representative of a relatively low lying, low rainfall zone with limited agricultural productivity. The area has been equipped with a large number of recently constructed, deep boreholes (60–100 m) extending into a fractured crystalline basement of undifferentiated gneiss and cased off to below the lateritic regolith which typically extends to 20 m below ground level (Department of Lands and Surveys 1963).

Sample collection, preservation and field analysis

Water samples were collected from a variety of sources including surface waters (streams, pools and rivers), hand-dug wells and boreholes. Where appropriate, water was flushed through the sampling system before collection into containers previously rinsed with the sample.

Temperature, pH, redox potential, conductivity and total alkalinity of the samples were determined on collection. Aliquots of the water sample were then filtered through 0.4 μm Nucleopore™ membrane filters and preserved according to the specifications given in Table 1. During the second field exercise in 1994, aliquots of untreated sample were also complexed with pyrocatechol violet (PCV), according to the method described by Driscoll (1984) and refined by Trick *et al.* (1994), in order to determine the amount of labile aluminium.

Duplicate and blank samples were collected in order to check on the reproducibility of sampling and preservation, the precision of the analytical determinations and to identify any problems associated with contamination.

Analytical methodology

Samples were later analysed for major ions, Fe, Si and Al by inductively coupled plasma-atomic emission spectrometry (Perkin Elmer Plasma II sequential scanner) which is capable of measuring colloidal and particulate species provided that they are suspended at the time of sample introduction. Filtered samples, preserved by the addition of Aristar™ nitric acid (1% with respect to the concentrated acid), were analysed for aluminium by ICP-AES, both with and without sonication of the sample (to disperse precipitated particulate/sorbed material) for one hour before immediate introduction into the

Table 2a. *Summary of analytical results for Uganda giving average concentrations and the range of observed concentrations of each species determined for the study sites at Mukono and Nawaikoke*

Element	Units	Mukono District			Nawaikoke District		
		Average conc.	Range max.	Range min.	Average conc.	Range max.	Range min.
Temp.	°C	24.2	28.0	21.9	26.4	27.6	25.5
Eh (uncorrected)	mV	117	208	< 91.0	3.90	173	< 80.0
pH	–	6.02	7.40	5.23	6.56	6.89	6.14
Conductivity	μS/cm	116	370	47.7	684	2220	213
Ca	mg l^{-1}	9.60	41.6	2.05	67.3	238	11.2
Mg	mg l^{-1}	1.80	4.98	0.29	31.5	84.8	5.90
Na	mg l^{-1}	6.22	12.4	3.07	57.5	284	21.6
K	mg l^{-1}	4.60	14.3	1.76	4.77	10.5	2.25
HCO$_3$	mg l^{-1}	54.4	229	11.4	277	576	117
Cl	mg l^{-1}	< 5.00	7.59	< 5.00	42.4	124	< 5.00
SO$_4$	mg l^{-1}	3.52	23.2	< 5.00	110	803	< 5.00
TON (as NO$_3$)	mg l^{-1}	4.34	18.6	< 2.00	14.5	83.0	< 2.22
F	mg l^{-1}	0.16	0.35	0.06	0.56	1.11	0.05
TOC	mg l^{-1}	2.59	19.0	< 0.10	0.19	0.84	< 0.10
TIC	mg l^{-1}	14.8	53.7	5.10	60.3	118	32.0
Si (F/A)	mg l^{-1}	19.4	32.0	12.0	30.7	39.6	20.8
Si (UF/A)	mg l^{-1}	20.6	31.9	12.0	36.7	49.9	21.3
Fe	mg l^{-1}	1.93	25.9	0.03	2.43	7.00	0.26
Al Unsoni. (F/A)	mg l^{-1}	0.03	0.25	< 0.02	< 0.02	0.09	< 0.02
Al Soni. (F/A)	mg l^{-1}	0.17	0.68	< 0.02	0.02	0.08	< 0.02
Al Soni. (UF/A)	mg l^{-1}	0.66	4.09	< 0.02	0.17	1.60	< 0.02
Al PCV (1994)	mg l^{-1}	< 0.02	0.11	< 0.02	na	na	na

na, not analysed.

plasma. These analyses represent the concentrations of total Al less than 0.4 μm (i.e. including colloidal Al) and dissolved Al respectively. Unfiltered, acidified samples from the Mukono district of Uganda were also analysed after sonication in order to obtain a 'total' aluminium content.

Additionally, major anions were determined by ion chromatography (Dionex 2000i ion chromatograph). Total organic and inorganic carbon contents were determined using a Shimadzu TOC-5000 analyser and fluoride was determined by ion selective electrode. A full description of the methods employed for these analyses, including relevant QA protocols, is given in Reeder et al. (1994). Results for these analyses are summarized in Table 2a and presented in full in Table 2b.

Discussion

Analytical artefacts and determination of Al

In these studies, Al was analysed in the laboratory by ICP-AES. The accuracy and precision of these determinations were checked by: (i) analysis of Aquacheck^{TM} samples (distributed by Water Research Centre, Medmenham, UK) and (ii) replicate analysis of 4 representative samples spiked with a certified standard solution of Al at levels of 1, 2, 5 and 10 mg l^{-1} Al (NIST 1993).

Analysis of Aquacheck^{TM} samples over distribution numbers 33 through 62, which included analysis by ICP-AES up to and including the analysis of samples performed in these studies, gave an average difference, compared to consensus values of 10%, with a standard deviation of 10% at two sigma. During this period the consensus value for Al in distributed samples ranged from 0.020–0.700 mg l^{-1}.

In the spiked analysis, average recoveries > 97% were obtained for the four samples; inter-sample precision (4 replicates) was 1.5% at two sigma (Fig. 3). No difference was observed between sonicated and unsonicated levels of Al in the spiked solutions.

To compare our results with those of McFarlane & Bowden (1992), Al determinations were made before and after sonication (one hour) by ICP-AES on samples filtered to 0.40 μm and preserved with HNO$_3$ (F/A) and on unfiltered acidified samples (UF/A). Results of these analyses confirmed that sonication increased measured Al concentrations in the F/A samples

Table 2b. Analytical results from Mukono and Nawaikoke field sites

Sample Code	Class	Temp °C	Eh mV	pH	Cond. μS cm⁻¹	Ca ppm	Mg ppm	Na ppm	K ppm	HCO₃ ppm	Cl ppm	SO₄ ppm	NO₃ ppm	F ppm	TOC ppm	Si ppm	Total Fe ppm	Al (us) ppb	Al(s) ppb	Al (s > 0.4 μm) ppb
UG93/1	4SBH	25.8	184.0	6.46	230	22	3.0	6.6	4.2	79	<5.0	13	4.8	0.16	19	13	0.05	42	150	470
UG93/2	4SBH	28.0	-91.0	6.54	370	42	5.0	12	14	230	7.6	5.6	5.3	0.28	18	29	3.4	110	500	440
UG93/3	2HDW	24.8	208.0	5.75	57	5.0	1.1	3.7	3.8	22	<5.0	<5.0	2.2	0.12	1.1	15	0.30	<20	31	920
UG93/4	2HDW	25.1	-20.0	6.30	20	10	4.2	6.5	3.2	120	<5.0	<5.0	<2.2	0.22	0.10	20	26	64	270	460
UG93/5	2HDW	26.2	108.0	6.21	130	8.9	2.1	6.6	3.8	72	<5.0	5.67	<2.2	0.19	1.0	25	0.25	<20	20	76
UG93/6	2HDW	25.5	104.0	6.10	102	17	1.1	5.1	4.1	40	<5.0	<5.0	<2.2	0.18	1.0	21	2.3	<20	33	92
UG93/7	4SBH	25.5	74.0	5.97	80	7.8	0.79	5.8	3.3	41	<5.0	<5.0	3.4	0.12	0.68	19	0.63	<20	173	1900
UG93/8	2HDW	26.0	126.0	5.89	67	4.7	0.93	4.5	4.7	26	<5.0	6.4	5.3	0.10	3.5	15	0.07	100	400	1300
UG93/9	5DBH	25.7	183.0	6.28	146	7.9	2.6	7.6	4.5	75	<5.0	5.4	<2.2	0.23	0.65	25	1.1	<20	<20	<20
UG93/10	2HDW	23.3	92.0	5.83	95	8.7	1.1	4.5	5.0	30	<5.0	10	7.1	0.10	1.1	16	0.26	<20	47	2400
UG93/11	3SP	23.5	137.0	5.52	68	3.3	1.3	6.0	3.7	23	<5.0	<5.0	8.0	0.12	0.79	19	0.04	<20	73	88
UG93/12	3SP	24.0	140.0	5.48	68	3.1	1.4	5.9	3.5	25	<5.0	<5.0	8.1	0.11	1.4	19	0.03	<20	49	100
UG93/13	4SBH	24.1	116.0	6.10	155	14	4.8	6.1	6.3	53	<5.0	23	5.2	0.35	0.79	21	1.9	28	<20	21
UG93/14	2HDW	23.3	127.0	5.23	48	2.0	0.29	3.1	4.0	11	<5.0	5.8	7.4	0.06	0.73	14	0.20	250	670	1500
UG93/15	2HDW	24.7	170.0	5.54	63	4.3	0.44	4.2	1.8	16	<5.0	6.4	13	0.06	4.2	14	0.03	<20	71	56
UG93/16	2HDW	23.2	167.0	5.72	91	4.4	1.4	11	3.5	40	<5.0	6.7	<2.2	0.20	0.65	32	0.93	<20	70	110
UG93/17	2HDW	24.4	144.0	5.42	66	3.5	0.5	6.8	3.1	21	<5.0	<5.0	19	0.11	0.69	17	0.42	220	540	770
UG93/18	2HDW	24.3	123.0	6.09	83	2.6	1.5	12	3.1	34	<5.0	<5.0	9.9	0.30	1.0	27	0.01	<20	51	150
UG93/19	na	na	na	na	na	<0.05	<0.10	<0.02	<0.50	<5.0	<5.0	<5.0	<2.2	0.01	1.6	<0.05	0.73	<20	<20	<20
UG93/20	2HDW	23.2	140.0	5.80	83	4.5	1.7	5.9	4.7	32	<5.0	5.5	7.0	0.18	0.30	22	0.26	<20	30	400
UG93/21	2HDW	26.1	94.0	7.14	225	38	2.4	7.5	6.7	150	<5.0	5.7	<2.2	0.31	5.7	24	0.04	<20	85	21
UG93/22	2HDW	21.9	173.0	6.40	238	8.7	3.3	7.4	5.4	150	<5.0	6.5	<2.2	0.25	<0.10	24	2.0	240	680	4100
UG93/23	3SP	22.4	91.0	7.40	78	3.8	2.2	4.8	3.6	34	<5.0	6.9	<2.2	0.10	1.2	19	1.9	<20	<20	970
UG93/24	5DBH	22.2	94.5	5.91	112	6.9	2.4	7.2	8.3	45	<5.0	9.1	4.2	0.20	0.77	24	0.09	<20	29	22
UG93/25	2HDW	23.1	30.0	6.75	160	23	1.6	6.7	6.1	86	<5.0	11	7.1	0.14	6.5	15	0.75	33	<20	31
UG93/26	2HDW	23.3	154.0	5.45	68	3.0	1.0	5.2	4.7	17	<5.0	8.1	9.6	0.13	0.95	19	0.24	<20	26	180
UG93/27	2HDW	23.4	107.0	5.87	70	5.5	1.4	4.2	3.0	25	<5.0	<5.0	5.3	0.10	0.62	17	1.4	<20	380	510
UG93/28	2HDW	23.6	126.0	5.94	90	8.4	1.2	4.6	3.6	41	<5.0	5.5	5.4	0.12	0.80	15	0.04	<20	28	990
UG93/29	3SP	22.9	140.0	5.53	52	2.7	0.71	3.9	3.4	17	<5.0	6.0	3.9	0.08	1.3	14	0.20	<20	24	na
UG93/30	3SP	23.8	163.0	5.78	57	2.4	0.74	5.1	3.6	18	<5.0	<5.0	7.4	0.15	1.1	16	0.83	83	560	1600
UG93/31	5DBH	26.3	68.0	6.73	1010	102	39	64	6.6	300	97	88	8.3	0.65	<0.10	27	2.2	<20	39	160
UG93/32	4SBH	26.8	14.7	6.30	400	21	12	60	2.3	200	5.9	<5.0	13	0.58	<0.10	36	6.3	<20	66	650
UG93/33	5DBH	26.5	-41.0	6.57	660	71	30	38	5.5	310	69	24	14	0.57	<0.10	30	0.80	<20	<20	110
UG93/34	5DBH	26.8	-30.0	6.75	340	20	27	22	4.2	200	10	10	16	0.61	<0.10	21	4.3	<20	<20	100
UG93/35	?	26.3	-33.0	6.58	703	55	28	60	4.9	310	67	27	10	0.72	<0.10	31	ns	<20	<20	164
UG93/36	5DBH	26.0	-80.0	6.86	627	ns	ns	ns	ns	ns	ns	ns	ns	0.50	0.82	40	5.6	ns	ns	ns
UG93/37	5DBH	25.8	-20.0	6.14	213	11	5.9	26	3.2	120	5.9	5.7	8.1	0.67	<0.10	32	2.6	<20	<20	140
UG93/38	4SBH	25.5	-35.0	6.35	264	23	10	22	3.8	150	<5.00	13	7.3	0.70	<0.10	34	1.2	<20	<20	140
UG93/39	5DBH	26.5	-5.0	6.47	365	25	12	43	2.9	220	12	<5.0	5.6	0.46	<0.10	28	0.26	<20	<20	110
UG93/40	5DBH	26.7	80.0	6.66	462	34	32	31	4.1	310	6.1	8.5	2.3	0.59	0.21	29	1.6	<20	<20	120
UG93/41	4SBH	25.9	173.0	6.32	356	32	13	28	4.0	190	10	16	6.6	0.60	0.55	26	0.42	62	42	90
UG93/42	5DBH	26.0	57.0	6.64	1600	240	82	69	7.4	360	120	520	13	1.1	<0.10	21	ns	51	80	<20
UG93/43	5DBH	25.8	-4.0	6.89	1950	240	85	180	11	470	120	700	12	0.50	0.82	30	5.6	95	43	140
UG93/44	4SBH	25.5	-60.0	6.82	2220	220	82	280	7.9	580	120	800	31	0.40	<0.10	33	1.8	<20	<20	110
UG93/45	5DBH	25.9	-50.0	6.52	673	63	39	38	3.8	330	44	60	10	0.84	0.79	39	7.0	<20	<20	120
UG93/46	5DBH	26.8	-40.0	6.52	820	78	31	56	5.1	280	110	49	25	0.40	0.84	33	3.4	<20	30	90
UG93/47	4SBH	27.1	13.3	6.56	310	19	16	28	7.4	230	9.3	<5.0	11	0.39	<0.10	33	1.1	<20	49	160
UG93/48	4SBH	26.3	12.7	6.52	475	35	25	36	4.5	230	33	6.7	26	0.52	0.47	33	0.30	<20	<20	<20
UG93/49	5DBH	26.4	46.0	6.34	252	24	9.4	29	3.4	170	<5.0	6	2.7	0.37	0.46	39	1.8	<20	43	150
UG93/50	5DBH	26.2	-15.0	6.57	510	45	29	41	5.0	320	33	<5.0	2.5	0.56	<0.10	39	1.1	<20	49	<20
UG93/51	4SBH	26.8	14.0	6.67	320	23	23	22	4.5	250	<5.0	6.7	<2.2	0.05	0.44	33	2.6	<20	<20	150
UG93/52	?	27.6	20.0	6.51	520	39	31	31	3.7	270	41	<5.0	4.1	0.56	0.66	31	3.8	<20	<20	45

UG93/1 to 30 are from Mukono; UG93/31 to UG93/52 are from Nawaikoke; ?, uncertain data; Eh, uncorrected Eh; ns, not sampled ; na, not analysed; s, sonicated; us, unsonicated; 1SW, surface water; 2HDW, hand dug well; 3SP, spring water; 4SBH, shallow borehole; 5DBH, deep borehole

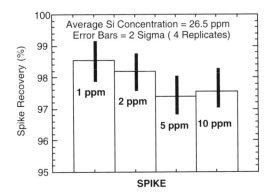

Fig. 3. Recovery of spiked Al from four representative samples selected from the Mukono region that contained negligible concentrations of Al (UG93/5, UG93/13, UG93/16 and UG93/24).

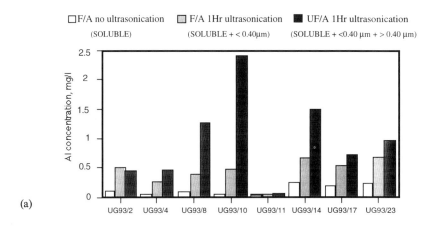

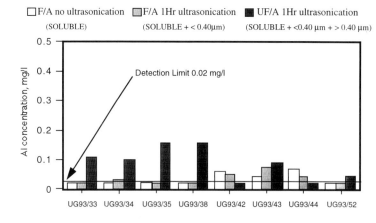

Fig. 4 (a). Results from a subset of Mukono samples showing effect of sonication on samples. **(b).** Results from a subset of Nawaikoke samples showing effect of sonication on samples.

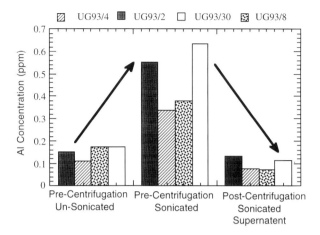

Fig. 5. The effect of centrifugation on the concentration of Al measured by ICP-OES before and immediately after sonication.

from Mukono by up to an order of magnitude (Fig. 4a), whilst it appeared to have a negligible effect on samples from Nawaikoke (Fig. 4b). Concentrations of Al in samples from Mukono ranged from < 0.02 to $0.250\,\mathrm{mg\,l^{-1}}$ in unsonicated F/A samples and < 0.02 to $0.680\,\mathrm{mg\,l^{-1}}$ in sonicated F/A samples. Whilst these represent the high end of the world values for surface water (Driscoll 1989), they are not unusually high for many shallow groundwaters (Hem 1985).

Of the Nawaikoke samples, only UG93/42, UG93/43 and UG93/44 had significant concentrations of Al in the < 0.40 μm fraction (Al(ns) and Al(s); Table 2b). Analysis of the emission spectra of these samples, however, showed spectral interference due to the high TDS content of these particular samples and this could account for a significant proportion of the apparent observed Al.

Sonicated UF/A samples gave results, up to $c.\,4\,\mathrm{mg\,l^{-1}}$, which were in some cases significantly higher than corresponding filtered sonicated values, indicating that some samples, such as UG93/10, contained significant particulate Al > 0.40 μm. In these cases, the measured difference between Si in F/A and UF/A samples showed a strong positive correlation (R = 0.94) with the corresponding difference in Al and indicated an Al:Si ratio in these particulates of close to 1. Increased Si contents in the UF/A fraction in samples with negligible Al indicate the presence of Si rich particulates. No increase in Al concentration with sonication was observed in the < 0.40 μm samples collected from Nawaikoke.

These results generally confirm the observations made by McFarlane & Bowden (1992) and suggest that Al is either being sorbed onto the walls of the sample containers (Nalgene™ HDPE) or precipitating/aggregating subsequent to filtration under acid conditions. Similar effects are not observed for either Si or Fe. However, in the case of Si it is likely that any observed differences due to the presence of Al–Si particulates is masked by the small absolute differences in Al concentration compared to the high concentrations of Si determined in the same samples. The good recoveries obtained during spiking experiments suggest that the former is unlikely, and the latter was tested by: (i) observing the effect of centrifugation (5000 rpm, 10 minutes) on the observed Al concentration in sonicated samples and (ii) filtering sonicated samples (2 ml) through ultra filtration membranes (Centricon™ 30 (30 000 MWt membranes), Amicon Ltd, UK, equivalent to $c.\,4\,\mathrm{nm}$) and visualizing the collected particulates by Scanning Electron Microscopy, SEM (Cambridge Instruments, Stereo-scan S250 fitted with a Link 860 X-ray micro-analyser).

The effect of centrifugation on sonicated samples is illustrated in Fig. 5 which shows that the increase in Al concentration observed after sonication is reversed by centrifugation. This is consistent with the observed increase on sonication arising from the uptake of particulate matter into the ICP.

Table 3. *Summary of results from SEM analysis of filter membranes (of c. 4 nm pore size) from ultra filtration of sonicated samples filtered to 0.40 μm and acidified during sample collection (F/A)*

Sample	Amount of material	Nature of material	Chemistry
Blank	*	Isolated particles > 1 μm	none/organic
UG93/2	*	Isolated particles > 1 μm, minor sub-micron nebulous particles	too small for analysis
UG93/4	**	Isolated particles > 1 μm, rare sub-micron nebulous particles	too small for analysis
UG93/8	***	Mainly isolated particles, high proportion are < 0.5 μm. Minor agglomeration	Al + Si +/− Fe, also silica
UG93/10	***	Minor sub-micron particles	Al + Si +/− Fe also silica

*, minimal amount of material similar to blank; **, significant amount of material but noticeably less than *** which showed a considerable quantity of deposited material.

SEM studies of 4 nm filters collected after centrifugation of acidified, previously 0.40 μm filtered samples and of precipitates collected by centrifugation of UF/UA samples, confirm the presence of Al–Si micro-particulates (4 μm to < 0.5 μm) in samples previously filtered to 0.40 μm during collection (Table 3).

The presence of such particulates must therefore be due to aggregation after filtration. Possible mechanisms that might promote such aggregation/precipitation include the destabilization and aggregation of small colloids < 0.40 μm by the addition of acid (in which the particulate matter appears to be relatively insoluble) as a result of the increased ionic strength, or, alternatively, the precipitation of relatively insoluble species at the reduced temperatures of sample storage (4°C).

The above tests and those performed by McFarlane & Bowden (1992) are limited to the determination of Al in samples transported back to the analytical laboratory, and it is impossible to define the exact lability and form of Al at the time of sampling. To study this, a spectrophotometric field methodology, used extensively in studying the abundance and speciation of Al in natural waters (Mcavoy et al. 1992) was further developed to allow the rapid determination of Al in waters relatively rich in dissolved Fe, TOC and F (Trick et al. 1994). Subsequent use of this methodology in Uganda during the second field visit in 1994 shows (Table 2a, b) that the majority of Al determined by ICP-AES in the Mukono region, following sonication, is present in a relatively non-labile, 'PCV unreactive' form, as defined by Driscoll (1984), even at the time of sampling. This does not necessarily mean that Al exists in this form in pore-waters, in which a higher pCO_2 and abundance of organic ligands, may promote dissolution (McFarlane et al. 1992; Warren et al. 1993), but reflects the concentration and speciation in water at the point of use. This is perhaps a better indication of the potential toxicity of Al.

Controls on Al abundance and speciation in sampled waters

Analytical results described previously give an indication of the total concentration of Al and its speciation in terms of relative lability under natural and acidic (1% HNO_3) conditions. To better define the relative importance of hydrochemical factors controlling Al concentration and speciation in these waters, calculations were performed using the PHREEQEV code (Crawford, 1993), which combines the inorganic speciation code PHREEQE (Parkhurst et al. 1980) with Model V (Tipping & Hurley 1992) to model organic complexation. The database used in these calculations was based on HATCHES (Cross & Ewart 1989) with data for Al reviewed and amended where appropriate.

Prior to sonication, Al was detected only in samples from Mukono. Since the extra Al detected after ultrasound treatment appears to be both relatively non-labile and associated with particulates or colloids, its availability to complex with other solutes is limited and its inclusion in equilibrium speciation calculations is inappropriate. Therefore, all samples have been speciated using the unsonicated Al levels and, where Al was not detected, half the detection limit (10 mg l^{-1}) has been used as an initial estimate.

Major cation and anion concentrations, pH and pCO_2 all tend to increase with depth and therefore, presumably, with increased ground-

Table 4. *Summary of speciation results using PHREEQEV for representative samples from Mukono*

Sample	Al (ppb)	F (ppm)	TOC (ppm)	% Al^{3+}	% Al–OH	% Al–F	% Al–TOC
UG93/3	10*	0.12	1.1	0	0	5	95
UG93/4	64	0.22	<0.1	0	55	40	5
UG93/5	10*	0.19	1.1	0	2	2	96
UG93/7	10*	0.12	0.7	0	2	7	91
UG93/8	102	0.10	3.6	0	1	4	95
UG93/10	10*	0.10	1.1	0	0	3	97
UG93/14	248	0.06	0.7	12	35	29	24
UG93/15	10*	0.06	4.2	0	0	0	100
UG93/16	10*	0.20	<0.1	0	3	88	9
UG93/17	215	0.11	0.7	0	28	47	25

* assumed concentration of these samples was taken as equivalent to half the detection limit for the unsonicated sample (20 ppb).

water residence time. The Mukono springs are most dilute and the waters generally evolve from Mukono hand-dug wells through Mukono boreholes and Nawaikoke shallow boreholes to the most concentrated Nawaikoke deep borehole samples. Major cation abundances are compatible with evolution through soil–water interaction and alteration of the soil mineral assemblage. The dominant overall reaction appears to be conversion of feldspar to kaolinite and this is corroborated by mineralogical analysis of soils, where general X-ray diffraction and thermogravimetric analysis yield kaolin contents of between 19 and 51%. Mixing between deeper and shallower waters might explain increasing Na and K abundances in waters from the upper soil zones where feldspar has all dissolved and kaolinite is now dissolving.

Given that pH values are above 5.0, the dominant Al species at Mukono are hydroxyl complexes with organic and fluoro-complexes becoming important in waters containing high concentrations of these ligands (Table 4). This means that concentrations of the free cation are extremely small and generally less than 1% of the measured total Al concentration. However, all the samples have pCO_2 values significantly higher than that of the atmosphere (> 72.5 compared to atmospheric values of 73.5), suggesting that degassing of CO_2 could have been significant during sampling and this would have raised the pH and led to changes in speciation. The measured field pH values must therefore be regarded as maxima; lower *in situ* pH values would result in the free Al cation being more favourable.

Si concentrations are high and reflect solution activities of H_4SiO_4 between 72.8 and 73.3. These levels are compatible with those that would be observed given control by average soil silica (−3.1; Lindsay 1979) and lie between those for control by quartz (−4.0) and amorphous silica (−2.7) as shown in Fig. 6. However, the scatter could hide a trend indicative of control by an aluminosilicate such as kaolinite. The high Si concentrations and neutral pH values in the Nawaikoke waters mean that Al concentrations of only 0.0001 mg l^{-1} would be required for kaolinite saturation. However, the Al levels in Mukono samples with detectable Al are unaccountably higher than those that would be expected for control by kaolinite or gibbsite even allowing for microbially mediated dissolution. In these cases it may be that the dissolution of feldspar yields a metastable Al–Si-based solid in solution prior to the formation of kaolinite (e.g. Paces 1978), but it is not feasible to include such a process in an equilibrium model.

Health considerations

Whilst the consumption of large doses of Al and its salts through medicinal preparations (for example antacids) have serious indirect effects through the complexation of phosphate (Deichman & Gerade 1969), the chronic effects of low doses of Al are the subject of continued debate (Harrington *et al.* 1994). The presence of dissolved Al in water supplies has been considered to be implicated in the aetiology of neurological disorders by some authors (Martyn *et al.* 1989; Crapper-McLachlan *et al.* 1991). This is despite the fact that Al present in drinking water typically provides less than 4% of the normal total daily intake (88 mg per day; WHO 1984), although this figure is subject to a high degree of uncertainty as typical daily intakes may be as low as 5 mg per day (Delves *et al.* 1993).

Whilst uptake of Al from food or potable

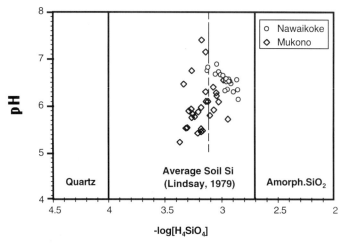

Fig. 6. Variation in H_4SiO_4 activity (approximated to concentration in these dilute, moderate pH waters) with pH for waters from Mukono and Nawaikoke Uganda. Superimposed are stability ranges for amorphous silica, quartz and average soil silica (Lindsay 1979).

water into the body ultimately depends upon its speciation in the gastrointestinal lumen (Powell & Thompson 1993), its speciation directly prior to ingestion and the lability of ingested species to gastrointestinal conditions also constrain its uptake. In the case of waters sampled during this work, solution speciation of Al at the time of sampling is predicted to be dominated by hydroxy, fluoro and organically-bound species. However, the high activities of dissolved H_4SiO_4 also observed in these waters are consistent with the formation of semi-colloidal Al–Si complexes which have been shown to influence both the uptake of Al by fish (Birchall et al. 1989) and man (Edwardson et al. 1993; Birchall 1993). Whilst the predictive modelling of such species is beyond the scope of thermodynamic equilibrium models as used in this work, experimental results indicating the predominance of particulate species during analysis are consistent with the majority of the total measured Al being in this physio-chemical form.

Data presented in Table 2a and b show that: (i) levels of Al in the $< 0.40\,\mu m$ unsonicated fraction are always below the WHO guideline value of $0.2\,mg\,l^{-1}$ (WHO 1993) and (ii) the levels of Al (as total Al or as dissolved Al) are less in the Nawaikoke region in which water is abstracted from an aquifer hosted within the crystalline basement. This would therefore result in a lower risk of Al toxicity.

The maximum amount of Al (particulate Al $> 0.40\,\mu m$ + colloidal Al $< \mu m$ + soluble Al species) received by a member of the public from ingestion of Al from drinking water sampled in the course of this work is $c.\,8\,mg$ per day (based on the consumption of 2 l of water a day). This equates to about 9% of the normal daily intake as estimated by WHO (1984) or up to about the normal daily intake as estimated by others (Delves et al. 1993). However, on the basis of this work, the total proportion of this Al assimilated into the body is likely to be between one and two orders of magnitude lower.

Conclusions

Sampling and analytical artifacts in the determination of Al have been confirmed and reemphasize the importance of using carefully considered sampling protocols, especially where an element's solution chemistry is likely to be controlled by the presence of colloidal species. In assessing water quality it is recommended that total Al is determined immediately after sonication of an unfiltered, unacidified sample, prior to an assessment of its availability by use of a spectrophotometric method such as PCV.

From a site specific view, data presented in this work are consistent with a negligible risk to the population from Al in drinking water, even if limits were lowered by WHO in response to heightened concerns within the developed world. The presence of high concentrations of dissolved silica would limit toxicity even if total Al concentrations were to be used as a basis for estimating risk. The relative insolubility of Al-rich particulates observed in this study, in samples containing 1% HNO_3, is consistent with a high degree of stability; together with relatively high concentrations of dissolved Si this points to a low biological uptake. There is no evidence from either consideration of speciation, or bulk chemistry, that any greater risk is attributable to the drinking of water from deeper wells that abstract directly from aquifers hosted in rocks of the crystalline basement than from shallow hand-dug wells.

On a regional basis, similar trends in major ion chemistry within water sources representative of both shallow and deeper groundwaters would be expected to yield similarly low risks of Al toxicity, except where natural pH is significantly lowered by the presence of oxidizable sulphides, or where recharge is affected by acid rain or industrial activities.

Experimental work performed in the course of these investigations illustrates the complexity of the collection and interpretation of analytical data on which to base epidemiological studies of Al and trace element excess/deficiency, and the requirement for a holistic approach to sampling, analysis and interpretation in the environmental sciences.

The authors acknowledge support through the Overseas Development Agency (Engineering Division TDR contract 92/10), Geological Survey and Mines Department of Uganda and the Ministry of Lands, Agriculture and Water Development, Zimbabwe without which this work would not have been possible.

This paper is published by permission of the director of the British Geological Survey (NERC).

References

BAKER, J. P. & SCHOFIELD, C. L. 1982. Aluminium Toxicity to Fish in Acidic Waters. *Water, Air, and Soil Pollution*, **18**, 289–309.

BIRCHALL, J. D. 1993. Dissolved silica and bioavailabilty of aluminium. *The Lancet*, **342**, 299.

―――, EXLEY, C., CHAPPELL, J. S. & PHILLIPS, M. J. 1989. Acute toxicity of aluminium to fish eliminated in silicon-rich acid waters. *Nature*, **9**, 146–148.

BLOOM, P. R. & ERICH, M. S. 1989. The quantitation of aqueous aluminium. *In:* SPOSITO, G. (ed.) *Environmental Chemistry of Aluminium*, CRC, Boca Raton, 1–27.

BRITISH STANDARDS INSTITUTION. 1986. *Guidance on the preservation and handling of samples*. BS 6068: Section 6.3: ISO 5667/3-1985.

CRAPPER-MCLACHLAN, D. R., KRUC, T. P., LUKIN, W. J. & KRISMAN, S. S. 1991. Would decreased aluminium ingestion reduce the incidence of Alzheimers disease? *Canadian Medical Association Journal*, **145**, 793–804.

CRAWFORD, M. B. 1993. *PHREEQEV; the incorporation of Model V into the inorganic speciation code PHREEQE to model organic complexation in dilute solutions*. British Geological Survey, Report, WE/93/19.

CROSS, J. E. & EWART, F. T. 1989. HATCHES – A thermodynamic database management system. *Radiochimica Acta*, **52/53**, 421–422.

DEICHMAN, W. B. & GERADE, H. W. 1969. *Toxicology of drugs and chemicals*. Academic, New York, 88.

DELVES, H. T., SIENIAWASKA, C. E. & SUCHAK, B. 1993. Total and bioavailable aluminium in food and beverages. *Analytical Proceedings*, **30**, 358–360.

DEPARTMENT OF LANDS AND SURVEYS. 1963. *Nawaikoke*. East Africa Series, Sheet 52/4. 1:50,000 (Uganda). Department of Lands and Surveys, Entebbe, Uganda.

―――― 1992. *Lugazi*. East Africa Series, Sheet 71/2, 1:50,000 (Uganda). Department of Lands and Surveys, Entebbe, Uganda.

DRISCOLL, C. T. 1984. A procedure for the fractionation of aqueous aluminium in dilute acidic waters. *International Journal of Environmental Analytical Chemistry*, **16**, 267–283.

―――― 1989. The chemistry of aluminium in surface waters. *In:* SPOSITO, G. (ed.) *Environmental Chemistry of Aluminium*, CRC, Boca Raton, 1–27.

―――― & SCHECHER, W. D. 1990. The chemistry of aluminium in the environment. *Environmental Geochemistry and Health*, **12**, 28–50.

EDWARDSON, J. A., MOORE, P. B., FERRIER, I. N., LILLEY, J. S., NEWTON, G. W. A., BARKER, J., TEMPLAR, J. & DAY, J. P. 1993. Effect of Silicon on gastrointestinal absorption of aluminium. *The Lancet*, **342**, 211–212.

FAGERIAL, N. K., BALIGAR, V. C. & WRIGHT, R. J. 1988. Aluminum toxicity in crop plants. *Journal of Plant Nutrition*, **11**, 303–319.

GEOLOGICAL SURVEY OF UGANDA. 1957. *Kampala Sheet North A36/U-IV*. Solid and Drift, 1:100,000. Geological Survey Department, Entebbe, Uganda.

GOENAGA, X., BRYANT, R. & WILLIAMS, D. J. A. 1987. Influence of sorption processes on aluminium determinations in acidic waters. *Analytical Chemistry*, **59**, 2673–2678.

HARRINGTON, C. R., WISCHIK, C. M., MCARTHUR, F. K., TAYLOR, G. A., EDWARDSON, J. A. & CANDY, J. M. 1994. Alzheimer's-disease-like changes in tau protein processing: association with aluminium accumulation in brains of renal dialysis patients. *The Lancet*, **343**, 993–997.

HEM, J. D. 1985. *Study and interpretation of the chemical characteristics of natural water*. United States Geological Survey, Water-Supply Paper, Report, **2254**.

KENNEDY, V. C., ZELLWEGER, G. W. & JONES, B. F. 1974. Filter pore-size effects on the analysis of Al, Fe, Mn, and Ti in Water. *Water Resources Research*, **10**, 785–790.

LINDSAY, W. L. 1979. *Chemical Equilibria in Soils*. Wiley, New York.

MCAVOY, D. C., SANTORE, R. C., SHOSA, J. D. & DRISCOLL, C. T. 1992. Comparison between pyrocatechol violet and 8-hydroxyquinoline procedures for determining aluminium fractions. *Soil*

Science Society of America Journal, **56**, 449–455.

McFarlane, M. 1991. Aluminium menace in tropical wells. *New Scientist*, 38–40.

—— & Bowden, D. J. 1992. Mobilisation of aluminium in the weathering profiles of the African surface in Malawi. *Earth Surface Processes and Landforms*, **17**, 789–805.

——, —— & Giusti, L. 1992. Some aspects of microbially-mediated Al mobility in weathering profiles in Malawi – the implications for groundwater quality. *Proceedings, 6th International Symposium on Microbial Ecology (ISME), Barcelona, September 1992*.

Martin, R. B. 1986. The Chemistry of Aluminium as Related to Biology and Medicine. *Clinical Chemistry*, **32**, 1797–1806.

Martyn, C. N., Barker, D. J. P., Osmond, C., Harris, E. C., Edwardson, J. A. & Lacey, R. F. 1989. Geographical relation between Alzheimer's disease and Al in drinking water. *The Lancet*, **i**, 59–62.

NIST. 1993. *Al Standard Reference Material (SRM 3101a)*. National Institute of Standards and Technology, Gaithersburg, MD 20899, USA.

Paces, T. 1978. Reversible control of aqueous aluminium and silica during the irreversible evolution of natural waters. *Geochimica et Cosmochimica Acta*, **42**, 1487–1493.

Parkhurst, D. L., Thorstenson, D. C. & Plummer, L. N. 1980. *PHREEQE – a computer program for geochemical calculations*. United States Geological Survey, Water-Resources Investigations, Report **80-96**.

Powell, J. J. & Thompson, R. P. H. 1993. The chemistry of aluminium in the gastrointestinal lumen and its uptake and absorption. *Proceedings of the Nutrition Society*, **52**, 241–253.

Reeder, S., Cave, M. R., Green, K. A., Trick, J. K. & Blackwell, P. A. 1994. *Standard protocols for the chemical characterisation of aqueous samples*. BGS, Technical Report, **WI/93/17**.

Schindler, D. W. 1988. Effects of Acid Rain on Freshwater Ecosystems. *Science*, **239**, 149–157.

Sposito, G. 1989. *The Environmental Chemistry of Aluminium*. CRC, Boca Raton.

Tipping, E., Woof, C. & Hurley, M. A. 1991. Humic substances in acid waters; modelling aluminium binding, contribution to ionic charge balance, control of pH. *Water Research*, **25**, 425–435.

—— & Hurley, M. A. 1992. A unifying model of cation binding by humic substances. *Geochimica et Cosmochimica Acta*, **56**, 3627–3641.

Trick, J. K., Reeder, S. & Smith, B. 1994. *The development and validation of a method for the determination of specific forms of aluminium in natural waters*. BGS, Technical Report, **WI/93/6**.

Wagemann, R. & Brunskill, G. J. 1975. The effect of filter pore-size on analytical concentrations of some natural trace elements in filtrates of natural water. *International Journal of Environmental Analytical Chemistry*, **4**, 75–84.

Warren, A., McEldowney, S. & Bowell, R. J. 1993. Biogeochemical factors affecting groundwater quality in tropical environments. *In: Environmental Geochemistry and Health in Developing Countries*, Geological Society, London, October 1993.

WHO 1993. *Guidelines for drinking-water quality: Volume 1. Second Edition, Recommendations*, WHO, Geneva.

WHO 1984. *Guidelines for drinking-water quality: Volume 2. Health criteria and other supporting information*. WHO, Geneva.

Sources and pathways of arsenic in the geochemical environment: health implications

IAIN THORNTON

Environmental Geochemistry Research Group, Centre for Environmental Technology, Imperial College of Science, Technology and Medicine, London SW7 2BP, UK

Abstract: Arsenic is ubiquitous in the environment, being present in small amounts in all rock, soil, dust, water and air. It is associated with many types of mineral deposits and in particular those containing sulphide minerals. The most common arsenic mineral is arsenopyrite, $FeAsS_2$. Elevated concentrations are sometimes found in fine grained argillaceous sediments and phosphorites. Some marine sediments may contain as much as $3000\,mg\,kg^{-1}$. Arsenic is co-precipitated with iron hydroxides and sulphides in some sedimentary rocks, and is precipitated as ferric arsenate in soil horizons rich in iron.
This paper reviews current knowledge on the natural geochemical sources of arsenic in several countries where high concentrations in soils, dusts, surface and groundwaters may present a hazard to human health. The chemistry and behaviour of arsenic within the weathering zone are discussed in relation to pathways leading to human exposure.

Arsenic is a metalloid element, but is often inaccurately referred to as a metal. It is a well known poison and as little as 0.1 g of arsenic trioxide may be lethal to humans (Jarup 1992). Acute arsenic poisoning is now rare, though chronic poisoning is widely recognized as a result of occupational exposure (WHO 1981). For more than a century this element at high doses has been known to be a human carcinogen and it is well established that ingestion of inorganic arsenic may cause skin cancer, whereas inhalation may produce respiratory cancer (Jarup 1992).

More recently there has been increasing concern as to the possible adverse health effects from exposure to elevated concentrations of arsenic in the natural geochemical environment. Scientific interest was initially stimulated by the results of an epidemiological study undertaken in southwest Taiwan in the 1960s which clearly showed a relationship between high concentrations of arsenic in drinking water and skin cancer, keratosis and Blackfoot Disease (a type of gangrene) (Tseng *et al*. 1968; Tseng 1977). Further investigations in Taiwan have established relationships between high arsenic exposure and cancers of the bladder and other internal organs (Chen *et al*. 1988, 1992).

Non-carcinogenic effects of inorganic arsenic have been reviewed by Abernathy & Ohanian (1992) and include hyperkeratosis and skin lesions, vascular effects including gangrene of the extremities in Taiwan (Tseng 1977), and vasoconstriction and acrocyanosis in Chile (Borgono *et al*. 1977). Neurological effects including tingling, numbness and pheripheral neuropathy have also been noted (Heyman *et al*. 1956). Hepatic effects have been recorded in India (Mazumder *et al*. 1988) and there is also evidence linking arsenic contamination of well waters with severe skin lesions (Das *et al*. 1994). Other health problems from chronic arsenic toxicity have been noted in Mexico, China and the Argentine.

Arsenic contamination of the environment has arisen as a result of mining and smelting activities in several countries. An example, where there is present-day concern, is in southwest England, where it has been suggested that the distribution of arsenic may possibly account for the high incidence of melanoma of the skin (Clough 1980), although, as yet no adverse health effects have been established (Thornton 1995). Arsenic enrichment in agricultural soils and pasture has also been recorded in association with lead/zinc mineralization elsewhere in Britain (Li & Thornton 1993). In the United States there is concern about the possible implications to human health of arsenic exposure from (a) drinking water in more arid areas, (b) exposure to arsenic used extensively as a pesticide on agricultural soils over the past 100 years and (c) exposure of communities in the vicinity of metal smelting and processing plants, power stations, etc.

The weight of evidence from the above studies has led the US Environmental Protection Agency (EPA) to consider reducing the permissible level of arsenic in drinking water from the present limit of $50\,\mu\mathrm{g\,l^{-1}}$ to as low as 2–$8\,\mu\mathrm{g\,l^{-1}}$. In parallel the World Health Organisation (WHO) has revised their recommendation for arsenic in drinking water to a provisional guideline of $10\,\mu\mathrm{g\,l^{-1}}$ in the place of the previously accepted $50\,\mu\mathrm{g\,l^{-1}}$ (WHO 1994).

It is also important to realize that there is evidence to support the hypothesis that arsenic in small quantities is an essential nutrient. Studies by Neilsen et al. (1975) showed growth retardation in rats fed on an arsenic depleted diet, while Anke et al. (1976) showed reproductive effects, depressed growth and elevated mortality in arsenic-deprived goats and miniature pigs. There is, however, no evidence to support the concept of arsenic deficiency in the human, whose possible requirement has been calculated to be as little as $12\,\mu\mathrm{g}$ per day (Uthus 1992). This author draws attention to the reported arsenic contents from diets from various parts of the world, indicating that the average human intake of arsenic is in the range 12–$40\,\mu\mathrm{g}$ per day, thus exceeding the possible human requirement. For example, the most recent Food and Drug Administration (FDA) Total Diet Study showed total mean As intakes by US adults to range from 20–$35\,\mu\mathrm{g}$ per day; fish and seafood accounted for 90% of the total food arsenic exposure, with all other foods the remaining 10% (FDA 1993).

In response to increasing concern, a Task Force was set up in 1993 by the Society for Environmental Geochemistry and Health, with funding from EPA and other sources, to examine health related arsenic issues. Subsequently a major international conference on Exposure and Health Effects of Arsenic was held in New Orleans in July 1993. This paper examines some of the issues raised and identifies some of the principal areas in which research is now urgently required.

Geochemistry and chemical behaviour of arsenic

Arsenic is ubiquitous in the environment, being present usually in small amounts in all rock, soil, dust, water and air. It is the main constituent of more than 200 mineral species, of which c. 60% are arsenates, 20% sulphides and sulphosalts and the remaining 20% include arsenides, arsenites, oxides, silicates and elemental As (Onishi 1969). Arsenic is found associated with many types of mineral deposits and in particular those containing sulphide minerals. The element is common in iron pyrite, galena, chalcopyrite, and more rarely in sphalerite (Goldschmidt 1954). Arsenic is in fact used as an indicator or pathfinder for gold in geochemical surveys. The most common arsenic mineral is arsenopyrite, $FeAsS_2$. The average concentration of As in igneous and sedimentary rocks is approximately $2\,\mathrm{mg\,kg^{-1}}$, and common concentrations in most rocks range from 0.5–$2.5\,\mathrm{mg\,kg^{-1}}$ (Kabata-Pendias & Pendias 1984), though higher concentrations are found in finer grained argillaceous sediments and phosphorites. Arsenic is concentrated in some reducing marine sediments which may contain as much as $3000\,\mathrm{mg\,kg^{-1}}$. In some sedimentary rocks arsenic is co-precipitated with iron hydroxides and sulphides. Thus iron deposits and sedimentary iron ores are rich in arsenic, as are manganese nodules. The arsenic contents of metamorphic rocks reflect those of the igneous and sedimentary rocks from which they were formed.

Weathering of rocks may mobilize arsenic as salts of arsenous acid and arsenic acid (Irgolic et al. 1995). The average concentration of arsenic in soil of about 5–$6\,\mathrm{mg\,kg^{-1}}$ is higher than that of rocks (Peterson et al. 1981) but will vary from region to region. Non-mineralized, uncontaminated soils usually contain 1–$40\,\mathrm{mg\,kg^{-1}}$ As. Lowest concentrations of arsenic are found in sandy soils and those derived from granites, with higher levels in alluvial soils and those rich in organic matter (Kabata-Pendias & Pendias 1984). Soils close to or derived from sulphide ore deposits may contain up to $8000\,\mathrm{mg\,kg^{-1}}$ As (Levander 1977). High concentrations are also found in soils and groundwaters affected by geothermal activity (Reay 1972). The roasting of arsenic-containing (sulphide) ores and burning of arsenic-rich coal releases arsenic trioxide, which may react in air with basic oxides, such as alkaline earth oxides, to form arsenates (Irgolic et al. 1995). These inorganic arsenic compounds can then be deposited onto soils and may be leached into surface and ground waters.

Under oxidizing conditions, in aerobic environments, arsenates (containing pentavalent arsenic) are the thermodynamically stable species. Arsenic is precipitated as ferric arsenate in soil horizons rich in iron. Arsenic derived from the weathering of pyritic slates in Alberta, Canada, has been found to leach from surface soils and accumulate up to several hundred $\mathrm{mg\,kg^{-1}}$ in the subsoil by adsorption onto secondary iron oxides (Dudas 1984). Elevated concentrations of arsenic (8–$40\,\mathrm{mg\,kg^{-1}}$) in acid sulphate soils in Canada and New Zealand are associated with the presence of pyrite (Dudas 1987), which typically holds up to 0.5% As through lattice substitution for sulphur. Iron-rich bauxites have been recorded containing in excess of 500 mg kg^{-1} As_2O_3. Under reducing conditions (< 200 mV) arsenites (containing trivalent arsenic) should be the predominant arsenic compounds. Because the reduction of arsenate to arsenite is slow, systems may not be at equilibrium. Inorganic arsenic compounds can be converted to methylated arsenic species by microorganisms, by plants, by animals, and by man. The oxidative methylation reactions act on trivalent arsenic compounds and produce methylarsonic acid, dimethylarsinic acid and trimethylarsine oxide (Irgolic et al. 1995). Under reducing conditions these pentavalent arsenic compounds can be reduced to volatile and easily oxidized methylarsines.

The chemical behaviour of arsenic in soil has been reviewed by O'Neill (1990) and Yan-Chu (1994). The ranges of Eh and pH in soils can lead to the presence of either As(V) or As(III), with microbial activity causing methylation, demethylation and/or change in oxidation state. The behaviour of arsenate (AsO_4^{3-}) resembles that of phosphate and vanadate. Arsenates of Fe and Al are the dominant phases in acid soils and are less soluble than calcium arsenate which dominates in many calcareous soils (Woolson et al. 1973; Fordyce et al. 1995). The presence of S species may, if the redox potential is low enough, favour the formation of arsenic sulphide minerals. A further complicating factor may be the presence of clay minerals, Fe, Mn and Al oxides and organic matter which can influence sorption, solubility and rate of oxidation of As species. Arsenic solubility is also controlled by adsorption reactions and biological activity.

The concentration of As in unpolluted fresh waters typically ranges from 1–10 μg l^{-1}, rising to 100–5000 μg l^{-1} in areas of sulphide mineralization and mining (Fordyce et al. 1995). Studies in the Obuasi gold-mining area of Ghana showed arsenic concentrations in drinking water from streams, shallow wells and boreholes to range from < 2–175 μg l^{-1} (Smedley et al. 1996). Both these sets of authors discuss the hydrogeochemistry of arsenic. In summary, at moderate or high redox potentials, As can be stabilized as a series of pentavalent (arsenate) oxyanions, H_3AsO_4, H_2AsO_4, $HAsO_4^{2-}$ and AsO_4^{3-}. However, under most reducing (acid and mildly alkaline) conditions, the trivalent arsenite species (H_3AsO_3) predominates. It has been noted that the retention of As in solution is constrained by co-precipitation with elements such as Fe, Ba, Co, Ni, Pb and Zn (Fordyce et al. 1995). In Ghana the highest concentrations of arsenic were found in deeper more reducing waters where it was thought that the element had built up as a result of longer residence times of groundwater in a deeper part of the aquifer (Smedley et al. 1996).

In the marine environment, more complex organic arsenic compounds such as tetramethylarsonium salts, arsenocholine, arsenobetaine, dimethyl(ribosyl)arsine oxides, and arsenic-containing lipids have been identified (Irgolic et al. 1995). However, only a very minor fraction of the total arsenic in the oceans remains in solution in sea water, due to removal by suspended particulate material.

Geochemical mapping

Geochemical maps based on the systematic sampling and analysis of stream sediments, soils and/or waters have proved useful as a source of baseline data for arsenic distribution on a regional scale. At the same time, maps have delineated anomalous areas associated with natural arsenic enhancement in sulphide minerals and/or contamination with arsenic-rich mineral wastes or flue dusts. For example, a map showing the regional distribution of arsenic in stream sediments was published as part of the Wolfson Geochemical Atlas of England and Wales (Webb et al. 1978) and clearly shows extensive contamination of soils in southwest England, with in excess of 700 km^2 affected (Webb et al. 1978; Abrahams & Thornton 1987).

Geochemical surveys undertaken by the British Geological Survey have indicated anomalous levels of As in the southwest Highlands and the Grampian Highlands of Scotland, the English Lake District and North Wales, associated with specific geological units (Plant et al. 1989, 1991; Simpson et al. 1993). These identify all the known gold mineralization, and are also associated with Cu dominated multi-element mineralization and complex Pb–As mineralization, together with environments in which As has been complexed and absorbed in stream sediments by organic matter and hydrous oxides in areas of impeded drainage. It may be concluded that arsenic in stream sediments occurs mainly as As_2O_3 and As_2S_3, as heavy metal arsenates, and sorbed onto ferric hydroxides.

Collaborative geochemical and hydrogeochemical studies into mining-related arsenic contamination have been undertaken in the Ron Phibun District of Thailand by the British Geological Survey and the Government of Thailand Department of Mineral Resources (Fordyce et al. 1995). Here the sources of arsenic are high-grade arsenopyrite waste piles and alluvial mineral deposits. Alluvial soils contained up to 5000 mg g^{-1} As. It was found that the waste materials contained only a small component of the primary arsenopyrite, with most of the arsenic present as secondary arsenate minerals, in particular the relatively insoluble scorodite ($FeAsO_4 \cdot 2H_2O$). Concentrations of arsenic ranged up to in excess of 5000 $\mu g \, l^{-1}$ in shallow wells used for drinking water, while water from deeper boreholes was much less contaminated.

Geochemical mapping of Finland has clearly shown anomalies in the south of the country relating to sulphide mineralization and other geological features (Koljonen et al. 1992), and draws attention to possible sources of arsenic enrichment in well waters and potential increased human exposure (I. Niinisto pers. comm.).

Systematic sampling of soils and dusts in and around the ancient lead mining and smelting site at Lavrion, Greece, has indicated extensive contamination with arsenic as well as lead, and has instigated present-day studies into possible health implications to the local community (Stavrakis et al. 1994).

Sources of arsenic leading to human exposure

The presence of arsenic in sulphide mineral deposits has been noted above and associated mining and smelting activities have led to high concentrations ranging up to 1000 mg kg^{-1} or more of this element in both agricultural and garden soils and house dusts in old mining areas of Cornwall and Greece (Colbourne et al. 1975; Abrahams & Thornton 1987; Stavrakis et al. 1994). A further study of arsenic in surface waters in Cornwall has shown soluble arsenic in specific catchments to range from 10–50 $\mu g \, l^{-1}$ (Aston et al. 1975). However, abstraction of surface waters for processing and distribution avoids waters contaminated by past mining activities and water processing using aluminium hydroxide removes the majority of soluble arsenic. As a result arsenic in drinking water in Cornwall rarely exceeds 10 $\mu g \, l^{-1}$ (MAFF 1982). There are, however, some 20–30 000 private suppliers of water in south west England for which few data are available and on which further study would be advisable.

The sources of arsenic in tube wells in India and in well waters in Arizona and California are again geological, though the natures of the arsenic enriched deposits are as yet unknown. It is probable that these are sulphur- and/or iron-rich deposits of sedimentary origin or fine grained argillaceous marine deposits in which arsenic from biogenic sources has been concentrated. There is also a possibility that arsenic has been derived as a result of geothermal activity.

It has been estimated that agricultural soils in the USA contaminated with pesticide residues used over the past 100 or so years, with the use of wood preservatives, and with the current use of arsenic acid as a cotton defoliant may extend to as much as 100 000 to 1 million hectares exceeding 200 mg kg^{-1} As, with 10 million hectares with 20–30 mg kg^{-1} (Chaney pers. comm.). The extent of similar contamination in the UK and elsewhere in Europe, and indeed on a worldwide basis is difficult to estimate but is likely to be considerable.

Other anthropogenic sources of arsenic include metal processing plants, chemical works, coal combustion and geothermal power plants. In Poland sources of arsenic have been attributed to metal smelting, coal burning and use of As-rich phosphatic fertilizers (Kabata-Pendias 1994). The phytotoxicity of arsenic added to soil depends on soil type, with 90% growth reduction recorded at 1000 mg kg^{-1} on heavy soils and at 100 mg kg^{-1} on light soils (Woolson et al. 1973).

Exposure pathways

The main routes of environmental as opposed to industrial exposure to arsenic result from the ingestion of contaminated drinking waters, foodstuffs, soil and dust.

Taiwan

Exposure has been calculated only from the consumption of arsenic-rich drinking water. Sixty-five thousand artesian wells were sampled in the early 1970s in which concentrations of arsenic in waters ranged up to 600 $\mu g \, l^{-1}$ As. Unfortunately the majority of the wells were only sampled on one occasion and it is not known whether waters were filtered or not prior to analysis. The analytical method used, based on a mercury bromide stain, was probably not applicable for concentrations less than 100 $\mu g \, l^{-1}$ thus rendering many of the analytical results suspect (K. T. Irgolic, pers. comm.). Estimation of human exposure has been based on an assumption of 3–4.5 l consumed per day. However, it is realized that in a hot climate this could have been much greater, perhaps up to 8–10 l per day. Thus the dose may have been greatly underestimated. Exposure to arsenic in food, ingested soil/dust and by dermal contact was not taken into consideration in the overall estimate of human exposure.

Speciation of arsenic in the Taiwan well-waters showed an approximate ratio of As^{3+}: As^{5+} of 3:1, though this is known to change on storage and transport due to the oxidation of the former to the latter.

India: west Bengal

Drinking water from within an area of 35 000 km² along the River Ganga has been found to exceed 50 μg l^{-1} in several thousand tube-wells, ranging up to 200/400 μg l^{-1} (Das et al. 1994). Groundwater depletion as a result of irrigation requirements, brought about by the green revolution, has led to increasing arsenic concentrations in well waters. The reason for this has not yet been clearly established, though one possibility is that arsenic-rich pyritic beds have been exposed due to a lowering of the water table and oxidation has led to the leaching of arsenic from these beds into the well-water. There is also the possibility that the use of arsenic-rich irrigation water will have resulted in the accumulation of arsenic in surface soils used for the production of food crops. However, total arsenic exposure in local populations from both drinking water and food has not yet been determined.

Southwest England

Of 9000 km surveyed in Devon and Cornwall, 1.3% of surface soils have been described as highly contaminated with arsenic (exceeding the 90th percentile of 190 μg g^{-1} As), and 6.6% as moderately contaminated (exceeding the 70th percentile, ranging from 110–190 μg g^{-1} As; Abrahams & Thornton 1987). However, arsenic sometimes ranges as high as 0.1–1% or more in surface soils near old roasting ovens and smelter stacks.

Possible exposure of local populations through the consumption of dairy products and locally grown food crops has been examined. Arsenic uptake into pasture grass was found to be relatively small (it has been suggested by Kabata-Pendias & Pendias (1984) that plants take up arsenic passively with water flow), with a maximum concentration of 9.6 μg g^{-1} As dry matter in washed grass growing on soil containing 1000 μg g^{-1} As or more (Thoresby & Thornton 1979). However, it was subsequently shown that the majority of arsenic intake by grazing cattle resulted from the accidental ingestion of soil along with grass. Arsenic ingested in soil, determined in 11 herds of cattle, was found to range from c. 50–80% of the total arsenic intake (Thornton & Abrahams 1983). On moderately contaminated land, in which arsenic ranged from 160–250 μg g^{-1}, this resulted in a mean total daily intake of c. 50 mg arsenic.

Arsenic uptake into vegetable crops was studied in household gardens with soils ranging from around 150–900 μg g^{-1} As (Xu & Thornton 1985). Arsenic concentrations in each of the six crops tested increased with the arsenic content of the soil, though only in lettuce did this exceed 1 μg g^{-1} dry matter. Arsenic uptake was influenced by the iron and phosphorus content of the soil. All the vegetable samples fell below the UK statutory limit for arsenic in foods offered for sale of 1 mg kg^{-1} As freshweight (Arsenic in Food Regulations 1959), and it is not thought that exposure through locally grown foods is in any way hazardous.

It is now considered that accidental ingestion of contaminated dust and soil is the main route of exposure to arsenic in this situation, and that this will be particularly important for the young child of 3–36 months of age. A recent study in several old Cornish mining villages showed a significant relationship between arsenic levels in garden soils and in house dust (Elgali 1994). One study based on 70 households in the Camborne–Hayle area of Cornwall indicated that a young child could ingest as much as 42 μg of arsenic per day by this route (Johnson 1983). This was assuming that a 2 year-old child ingests 100 μg dust per day. A further study of 23 households in old mining villages, in which arsenic in soils ranged up to 770 μg g^{-1} As, and in housedusts up to 460 μg g^{-1}, showed amounts of arsenic on children's hands ranging up to 3.5 μg As. Total arsenic intake by hand-to-mouth activity was then estimated at between 35 and 46 μg arsenic per day, though it was accepted that this estimate was at the best semi-quantitative and that further study was required (Harding 1993). It is possible, however, that a young child in this heavily contaminated environment could have an arsenic intake exceeding the WHO Provisional Tolerable Daily Intake for inorganic arsenic of 2 μg kg^{-1} body weight (WHO 1983). A recent pilot study showed raised levels of arsenic in dust removed from children's hands by wet-wipes and in the hair of children sampled from old mining villages compared with a control location (Elgali 1994). However, a preliminary study of inorganic arsenic and its methylated metabolites in urine from adults and children from this Cornish mining area showed only slightly elevated urinary arsenic contents (Johnson & Farmer 1989).

Assessment of exposure and risk

To date, much of the concern to human exposure to arsenic has been confined to amounts of inorganic arsenic consumed in drinking water. In a recent review, Warner-North (1992) draws attention to some of the problems inherent in the current approach to risk assessment used by USEPA which has resulted in high risk estimates for skin cancer from drinking water containing arsenic even at the present Maximum Contaminant Level (MCL) of $50\,\mu g\,l^{-1}$, which is based on systemic effects and potential carcinogenicity. This approach is based on the 'default' methodology from the Guidelines for Carcinogen Risk Assessment (USEPA 1986) and assumes a linear dose–response relationship. There is, however, clear evidence for non-linearity for the association for arsenic exposure and cancer of internal organs, and the EPA approach does not reflect the increasing evidence indicating either a threshold or sub-linear dose–response relationship for low doses of arsenic (Marcus & Rispin 1988; Petito & Beck 1990). The detoxification pathway for arsenic in the human body is one of methylation in which ingested inorganic arsenic species which are potentially toxic are converted into organic non-toxic forms. It has been proposed that an explanation for the sub-linear dose–response relationship is due to the saturation of this methylation process (USEPA 1989; Carlson-Lynch et al. 1994). However, the methylation threshold hypothesis for the toxicity of inorganic arsenic is not accepted by all research workers (Hopenhayn-Rich et al. 1993), and was not even considered by Smith et al. (1992) in estimating potential risks from low-level arsenic exposures typical of the US population.

It is noted that EPA is currently addressing the potential status of arsenic regulation development and is considering a potential range of MCL options from $1\,\mu g\,l^{-1}$ to $20\,\mu g\,l^{-1}$ (Shank-Givens & Auerbach 1993).

Patel (1994) has listed relevant factors to risk assessment for arsenic as (a) speciation, particularly as inorganic As is considered to be more toxic than organic; (b) exposure, including non-water sources; (c) metabolism and efficiency of detoxification by methylation; (d) genotoxicity; (e) the results of carcinogenicity studies in animals which may not be predictive of arsenic toxicity in humans; and (f) the results of epidemiological studies in Taiwan, etc. Assuming either a threshold dose or a non-threshold dose for arsenic carcinogenicity, Patel (1994) lists risk estimates derived from several studies for concentrations of arsenic in drinking water ranging from $10-20\,\mu g\,l^{-1}$, and concludes that risk assessment for arsenic in drinking water involves a lot of uncertainty.

Sage (1994) has reviewed risk assessment procedures used by WHO and within the UK for As in drinking water, contaminated land and waters. UK government departments apply quantitative risk assessments by WHO and the International Agency for Research on Cancer, who have developed practical guidelines that form the foundation for EU directives and in the interpretation of UK law. Sage (1994) draws attention to a range of areas requiring further consideration:

(i) improving the sensitivity of epidemiological studies;
(ii) creating accurate measurement techniques for individual species of arsenic at low concentrations;
(iii) the adequate quantification of all exposure routes;
(iv) obtaining bioavailability data for arsenic to plants and humans;
(v) additional research into mechanisms by which arsenic acts in the body, especially synergistic and antagonistic effects;
(vi) constructing representative models for responses at low doses;
(vii) methods to express clearly remaining uncertainties in a form which can be systematically built into risk management.

None of the risk assessment procedures for arsenic applied to date has taken into account total exposure from ingestion of diet and water, ingestion of soil and dust and inhalation of atmospheric particles.

Research requirements

The above sections review current information on the sources, chemical behaviour, exposure pathways and risk assessment to human health of arsenic in the environment. There are clearly many gaps in present day knowledge that must be filled if we are to improve our understanding of the dangers that environmental arsenic poses to human health. Priority research requirements are as follows.

Geochemistry. While it is accepted that most sources of arsenic in waters are of geological origin, there is as yet little clear understanding of the geochemical nature of As-enriched strata nor of the mineral and chemical forms of arsenic present. In the course of chemical weathering in the surface environment, arsenic from sulphide minerals and from anthropogenic sources will, with time, interact with chemical, physical and biological components of the lithosphere. Dispersion by wind and water will in due course lead to accumulation of arsenic in river and estuarine sediments and in inshore environments. Again, little is known of the mineral and chemical forms of arsenic in the soil, sediment and marine environments. Mobility of arsenic from waste materials and contaminated soils is to be expected under oxidizing conditions and perhaps this will be accelerated as a result of acid precipitation and environmental change. Mobility will, however, be reduced by the formation of secondary minerals, such as scorodite, which have low solubilities. There is a need for research into the processes controlling solubilization and migration pathways through the soil–rock system and the implications to contamination of ground and surface waters. Drainage from disused metalliferous and coal mines is frequently a source of both soluble and particulate arsenic in surface river systems. There is a need for research into the chemical and physical factors controlling downstream dispersion in soluble, suspended particulate and sediment phases and interactions with iron, manganese and other chemical species, and processes leading to accumulation of arsenic in sediment 'sinks'.

Exposure pathways. Human exposure is thought to be mainly through the digestion route, via the diet (mainly seafood) and drinking water. In specific situations this exposure may be added to through the inhalation of atmospheric particulates derived from industrial emissions or from suspended arsenic-rich soils and waste materials. The majority of studies to date have focused on the most obvious exposure pathway, i.e. water in Taiwan and India. It is now necessary to reassess total arsenic intakes of exposed communities, taking into account both inorganic and organic forms of arsenic in the diet, soluble and particulate arsenic in drinking water (and its speciation), together with involuntary ingestion of arsenic-rich dusts and soils by hand-to-mouth activity in young children and through atmospheric contamination of food stuffs.

Risk assessment. Current risk assessment strategies have all been based on relationships established between epidemiological studies in Taiwan and arsenic concentrations in drinking water. It is now thought that some of the environmental measurements in this study are flawed due to sampling and analytical errors.

There is now an urgent need to establish a new multi-disciplinary study incorporating reliable environmental measurements, assessment of total human exposure from ingestion and inhalation routes, and carefully planned epidemiological investigations. Such a study would form the basis for the development of a risk assessment study in which uncertainty was minimized.

References

ABERNATHY, C. O. & OHANIAN, E. V. 1992. Non-carcinogenic effects of inorganic arsenic. *Environmental Geochemistry and Health*, **14**, 35–41.

ABRAHAMS, P. W. & THORNTON, I. 1987. Distribution and extent of land contaminated by arsenic and associated metals in mining regions of South-west England. *Transactions of the Institution of Mining and Metallurgy (Section B: Applied Earth Science)*, **96**, B1–B8.

ANKE, M., GRUN, M. & PARTSCHEFELD, M. 1976. The essentiality of arsenic for animals. *In:* HEMPHILL, D. D. (ed.) *Trace Substances in Environmental Health*, University of Missouri, Columbia, 403–409.

ARSENIC IN FOOD REGULATIONS. 1959. S.I. 831.

ASTON, S. R., THORNTON, I. & WEBB, J. S. 1975. Arsenic in stream sediments and waters of south west England. *The Science of the Total Environment*, **4**, 347–358.

BORGONO, J. M., VINCENT, P., VENTURINO, H. & INFANTE, A. 1977, Arsenic in the drinking water of the City of Antofagasta: epidemiological and clinical study before and after the installation of treatment plant. *Environmental Health Perspectives*, **19**, 103–105.

CARLSON-LYNCH, H., BECK, B. D. & BOARDMAN, P. D. 1994. Arsenic risk assessment. *Environmental Health Perspectives*, **102**, 4, 354.

CHEN, C-J., KUO, T-L. & WU, M-M. 1988. Arsenic and cancers. *The Lancet*, **1**, 414–415.

——, CHEN, C. W., WU, M-M. & KUO, T.-L. 1992. Cancer potential in liver, lung, bladder and kidney due to ingested inorganic arsenic in drinking water. *British Journal of Cancer*, **66**, 888–892.

CLOUGH, P. 1980. Incidence of melanoma of the skin in England and Wales. *British Medical Journal*, 12 January 1980.

COLBOURNE, P., ALLOWAY, B. J. & THORNTON, I. (1975). Arsenic and heavy metals in soils associated with regional geochemical anomalies in south west England. *The Science of the Total Environment*, **4**, 359–363.

DAS, D., CHATTERJEE, A., SAMANTA, G., MANDAL, B., CHOWDHURY, T. R., ET AL. 1994. Arsenic contamination in groundwater in six districts of West Bengal, India: the biggest arsenic calamity in the world. *Analyst*, **119**, 168–170.

DUDAS, M. J. 1984. Enriched levels of arsenic in postactive acid sulphate soils in Alberta. *Journal of the Soil Science Society of America*, **48**, 1451–1452.

—— 1987. Accumulation of native arsenic in acid

sulphate soils in Alberta. *Canadian Journal of Soil Science,* **67,** 317–331.

ELGALI, A. L. 1994. *Evaluation of the Exposure of Young Children to Arsenic Contamination in South-West England.* M.Sc Thesis, Imperial College of Science, Technology and Medicine, London.

FDA (Food and Drug Administration). 1993. *Guidance Document for Arsenic in Shellfish.* US Food and Drug Administration, Washington, D.C., 25–27.

FORDYCE, F. M., WILLIAMS, T. M., PAIJITPAPAPON, A. & CHAROENCHAISEI, P. 1995. *Hydrogeochemistry of Arsenic in an Area of Chronic Mining-related Arsenism, Ron Phibun District.* British Geological Survey.

GOLDSCHMIDT, V. M. 1954. *Geochemistry.* Clarendon, Oxford.

HARDING, E. R. 1993. *Arsenic contamination of garden soils and housedusts in West Cornwall.* BSc Dissertation, Imperial College of Science, Technology and Medicine, London.

HEYMAN, A., PFEIFFER, J. B., WILLETT, R. W. & TAYLOR, H. M. 1956. Peripheral neuropathy caused by arsenical intoxication. *New England Journal of Medicine,* **254,** 401–409.

HOPENHAYN-RICH, C., SMITH, A. H. & GOEDEN, H. M. 1993. Human studies do not support the methylation threshold hypothesis for the toxicity of inorganic arsenic. *Environmental Research,* **60,** 1616–1677.

IRGOLIC, K. T., GRESCHONIG, H. & HOWARD, A. G. 1995. Arsenic. *In: Analyst: the Encyclopedia of Analytical Science.* Academic, London.

JARUP, L. 1992. *Dose–response Relations for Occupational Exposure to Arsenic and Cadmium.* National Institute of Occupational Health, Sweden.

JOHNSON, L. R. 1983. *A study of arsenic in housedusts and garden soils with relation to geochemistry and health.* MSc Thesis, University of London.

—— & FARMER, J. G. 1989. Urinary arsenic concentrations and speciation in Cornwall residents, *Environmental Geochemistry and Health,* **11,** 39–44.

KABATA-PENDIAS, A. 1994. Biogeochemistry of arsenic and selenium. *In: Arsenic and Selenium in the Environment: Ecological and Analytical Problems,* Polska Akademia Nauk Warszawa, 9–16.

—— & PENDIAS, H. 1984. *Trace Elements in Soils and Plants,* CRC, Boca Raton.

KOLJONEN, T., GUSTAVSSON, N., NORAS, P. & TANSKANEN, H. 1992. *The Geochemical Atlas of Finland,* Geological Survey of Finland, Espoo.

LEVANDER, O. A. 1977. *Arsenic.* National Academy of Sciences, Washington, D.C.

LI, X. & THORNTON, I. 1993. Arsenic, antimony and bismuth in soil and pasture herbage in some old metalliferous mining areas in England. *Environmental Geochemistry and Health,* **15**(2/3), 135–144.

MARCUS, W. L. & RISPIN, A. S. 1988. Threshold carcinogenicity using arsenic as an example. *In:* COTHERN, C. R., MEHLMAN, M. A. & MARCUS, W. L. (eds) *Risk assessment and risk management of industrial and environmental chemicals.* Princeton Scientific Publishing Co., Princeton, NJ, 133–158.

MAZUMDER, D. N. G., CHAKRABORTY, A. K., GHOSE, A., JUPTA, J. D., CHAKRABORTY, D. P., DEY, S. B. & CHATTOPADHYAY, N. 1988, Chronic arsenic toxicity from drinking tubewell water in rural West Bengal. *Bulletin, WHO,* **66,** 499–506.

MAFF (Ministry of Agriculture, Fisheries and Food). 1982. *Survey of Arsenic in Food.* Food Surveillance Paper, **8,** HMSO, London.

NIELSEN, F. H., GIVAND, S. H. & MYRON, D. R. 1975. Evidence of a possible requirement for arsenic by the rat. *Proceedings of the Federation of American Societies of Experimental Biology,* **34,** 923.

O'NEIL, P. 1990. Arsenic. *In:* ALLOWAY, B. J. (ed.) *Heavy Metals in Soils,* Blackie, London, 83–99.

ONISHI, H. 1969. *In:* WEDEPOHL, K. H. (ed.) *Handbook of Geochemistry.* Springer, New York.

PATEL, A. 1994. The risk assessment of arsenic in drinking water. *In:* VARNAVAS, S. P. (ed.) *Proceedings of 6th International Conference on Environmental Contamination, Delphi, Greece.* CEP Consultants, Edinburgh, 17–19.

PETERSON, P. J., BENSON, L. M. & ZEIVE, R. 1981. Metalloids, Section 3 – Arsenic. *In:* LEPP, M. W. (ed.) *Effect of Heavy Metal Pollution on Plants.* Applied Science, London, Vol. 1, 299–322.

PETITO, C. T. & BECK, B. D. 1990, Evaluation of evidence of nonlinearities in the dose–response curve for arsenic carcinogenesis. *Trace Substances in Environmental Health,* **24,** 143–176.

PLANT, J. A., BREWARD, N., FORREST, M. D. & SMITH, R. T. 1989. The gold pathfinder elements As, Sb and Bi – their distribution and significance in the southwest Highlands of Scotland. *Transactions of the Institution of Mining and Metallurgy* (Section B: Applied Earth Science), **98,** B91.

——, COOPER, D. C., GREEN, P. M., REEDMAN, A. J. & SIMPSON, P. R. 1991. Regional distribution of As, Sb and Bi in the Grampian Highlands of Scotland and English Lake District: implications of gold metallogeny. *Transactions of the Institution of Mining and Metallurgy* (Section B: Applied Earth Science), B135.

REAY, P. F. 1972. The accumulation of arsenic from arsenic-rich natural waters by aquatic plants. *Journal of Applied Ecology,* **9,** 557–565.

SAGE, C. 1994. *Environmental Arsenic in England and Wales: the Assessment and Management of Risk.* Unpublished MSc. Thesis, Centre for Environmental Technology, Imperial College of Science, Technology and Medicine, London.

SHANK-GIVENS, H. L. & AUERBACH, J. 1993. *The regulatory development process and the status of arsenic regulation development.* Paper presented at American Water Works Association Water Quality Technology Conference, November 1993.

SIMPSON, P. R., EDMUNDS, W. M., BREWARD, N., COOK, J. M., FLIGHT, D., HALL, G. E. M. & LISTER, T. R. 1993. Geochemical mapping of stream water for environmental studies and mineral exploration in the UK. *Journal of Geochemical Exploration,* **49,** 63–88.

SMEDLEY, P. L., EDMUNDS, W. M. & PELIG-BA, K. B.

1996. Mobility of arsenic in groundwater in the Obuasi gold-mining area of Ghana: some implications for human health. *This volume.*

SMITH, A. H., HOPENHAYN-RICH, C., BATES, M. N., GOEDEN, H. M., HERTZ-PICCIOTTO, I., ET AL. 1992. Cancer risks from arsenic in drinking water. *Environmental Health Perspectives*, **97**, 259–267.

STAVRAKIS, P., DEMETRIADES, A., VERGOU-VICHOU, K., THORNTON, I., FOSSE, G., MAKROPOULOS, V. & VLACHOYIANNIS, N. 1994. A multidisciplinary study on the effects of environmental contamination on the human population of the Lavrion urban area, Hellas. *In:* VARNAVAS, S. P. (ed.) *Proceedings of 6th International Conference on Environmental Contamination, Delphi, Greece.* CEP Consultants, Edinburgh, 20–22.

THORESBY, P. & THORNTON, I. 1979. Heavy metals and arsenic in soil, pasture herbage and barley in some mineralised areas in Britain: Significance to animal and human health. *In:* HEMPHILL, D. D. (ed.) *Trace Substances in Environmental Health.* Vol. 13. University of Missouri, 93–103.

THORNTON, I. 1995. Sources and pathways of arsenic in south-west England: health implications. *In:* CHAPPELL, W. R., ABERNATHY, C. O. & COTHERN, C. R. (eds) *Exposure and Health Effects of Arsenic.* Science Reviews, Northwood, 61–70.

—— & ABRAHAMS, P. 1983. Soil ingestion – A major pathway of heavy metals into livestock grazing contaminated land. *The Science of the Total Environment*, **28**, 287.

TSENG, W. P. 1977. Effects of dose–response relationships on skin cancer and Blackfoot Disease with arsenic. *Environmental Health Perspectives*, **19**, 109–119.

——, CHU, H. M., HOW, S. W., FONG, J. M., LIN, C. S. & YEH, S. 1968. Prevalence of skin cancer in an endemic area of chronic arsenism in Taiwan. *JNCI*, **40**, 453–463.

USEPA. (United States Environmental Protection Agency). 1986. Guidelines for carcinogen risk assessment. *Federal Register*, **51**, 33992–34003.

USEPA (United States Environmental Protection Agency). 1989. *Science Advisory Board's review of the arsenic issues relating to the phase II proposed regulations from the Office of Drinking Water.* EPA-SAB-EHC-89-038, Memorandum to William K. Reilly, Washington, DC.

UTHUS, E. O. 1992. Evidence for arsenic essentiality. *Environmental Geochemistry and Health*, **14**, 55–58.

WARNER-NORTH, D. 1992. Risk assessment for ingested inorganic arsenic: a review and status report. *Environmental Geochemistry and Health*, **4**, 59–62.

WEBB, J. S., THORNTON, I., THOMPSON, M., HOWARTH, R. J. & LOWENSTEIN, P. L. 1978. *The Wolfson Geochemical Atlas of England and Wales.* Oxford University Press, Oxford.

WOOLSON, E. A., AXLEY, J. H. & KERNEY, P. C. 1973. The chemistry and phytotoxicity of arsenic in soils, II. Effects of time and phosphorus. *Proceedings of the Soil Science Society of America*, **37**, 254.

WHO. 1981. *Arsenic Environmental Health Criteria.* World Health Organisation, Geneva, **18**, 174.

WHO. 1983. *27th Report of the Joint FAO/WHO Expert Committee on Food Additives, 29.* Technical Report Series 696, World Health Organisation, Geneva.

WHO. 1994. *Guidelines for drinking-water quality. Volume 1: Recommendations.* Second edition, World Health Organisation, Geneva.

YAN-CHU, H. 1994. Arsenic distribution in soils. NRIAGU, J. O. (ed.) *In: Arsenic in the Environment. Part 1. Cycling and Characterization.* Wiley, New York, 17–49.

XU, J. & THORNTON, I. 1985. Arsenic in garden soils and vegetable crops in Cornwall, England: implications for human health. *Environmental Geochemistry and Health*, **7** (4), 131–133.

Mobility of arsenic in groundwater in the Obuasi gold-mining area of Ghana: some implications for human health

P. L. SMEDLEY,[1] W. M. EDMUNDS[1] & K. B. PELIG-BA[2,3]

[1] *British Geological Survey, Maclean Building, Wallingford OX10 8BB, UK*
[2] *Water Resources Research Institute, PO Box M32, Accra, Ghana*
[3] *Department of Biotechnology, University for Development Studies, PO Box 1350, Tamale, Ghana*

Abstract: Arsenic in drinking water from streams, shallow wells and boreholes in the Obuasi gold-mining area of Ghana range between < 2 and 175 μg l^{-1}. The main sources are mine pollution and natural oxidation of sulphide minerals, predominantly arsenopyrite (FeAsS). Streamwaters have been most affected by mining activity and contain some of the highest As concentrations observed. They are also of poor bacteriological quality. Some of the streams have a high As(III) content (As(III)/As$_{total}$ > 0.5), probably as a result of methylation and reduction reactions mediated by bacteria and algae. Concentrations of As in groundwaters reach up to 64 μg l^{-1}, being highest in deeper (40–70 m) and more reducing (220–250 mV) waters. The As is thought to build up as a result of the longer residence times undergone by groundwaters in the deeper parts of the aquifer. The proportion of As present as As(III) is also higher in the deeper groundwaters. Deep mine exploration boreholes (70–100 m) have relatively low As contents of 5–17 μg l^{-1}, possibly due to As sorption onto precipitating ferric oxyhydroxides or to localized low As concentrations of sulphide minerals.

Median concentrations of inorganic urinary As from sample populations in two villages, one a rural streamwater-drinking community and the other a suburb of Obuasi using groundwater for potable supply, were 42 μg l^{-1} and 18 μg l^{-1} respectively. The value for the community drinking groundwater is typical of background concentrations of urinary As. The slightly higher value for the streamwater-drinking community probably reflects different provenance of foodstuffs and higher As concentrations of water sources local to the village. The low value obtained for the inhabitants of the Obuasi suburb, living close to and abstracting groundwater from the area of major mining activity, suggests that groundwater can form a useful potable supply of good inorganic quality provided that deep, long residence time sources are avoided.

Arsenic has long been recognized as a toxin and carcinogen. A relatively high incidence of skin and other cancers has been noted in populations ingesting water with high As concentrations (WHO 1981) and other disorders, especially some dermatological (e.g. Bowen's disease, hyperkeratosis, hyperpigmentation), cardiovascular (Blackfoot disease, Raynaud's syndrome), neurological and respiratory diseases, are linked to arsenic ingestion and exposure (e.g. Gorby 1994). Recent epidemiological evidence has led the WHO to decrease its recommended maximum value for As in drinking water from 50 μg l^{-1} to 10 μg l^{-1} (WHO 1993) in recognition of the element's potential health risks, although many regulatory bodies such as the EC and USEPA have not yet followed suit. Arsenic is a relatively common trace element in the environment and this guideline maximum is frequently exceeded in drinking water supplies. Many cases of chronic and acute endemic As poisoning from drinking water have been documented, notably cases in Taiwan (Tseng et al. 1968; Chen et al. 1994), Argentina (Astolfi 1971), Chile (Zaldivar 1974), China (Wang & Huang 1994) and Mexico (Cebrián et al. 1994). Wang & Huang (1994) claimed that no morbidity cases were found where drinking water concentrations were less than 100 μg l^{-1}, but that morbidity increased exponentially as aqueous As increased. Mild As poisoning was observed between 100 and 200 μg l^{-1}.

The average concentration of As in the earth's crust is 1.5–2 mg kg^{-1} (National Academy of Sciences 1977). It occurs in many geological materials including clay minerals, phosphorites and iron and manganese ores, but is found in highest concentrations in sulphide minerals, especially arsenopyrite, as well as realgar (AsS), orpiment (As$_2$S$_3$) and in solid solution in pyrite (FeS$_2$). Terrains rich in sulphide-bearing minerals can therefore have high concentrations of dissolved As in rivers and groundwaters derived by sulphide oxidation. In southwest Taiwan for example, water from artesian wells in sulphide-rich black shales has concentrations of As as high as 1.8 mg l^{-1}. Here, a clear dose–response relationship has been found between As ingestion and occurrences of endemic skin cancer, hyperpigmentation, keratosis and Blackfoot disease (Tseng et al. 1968).

Mining of sulphide ore bodies can pose a particular problem of As pollution and many ore smeltering operations are well known to release large quantities of As into the environment (e.g. Díaz-Barriga et al. 1993; Lagerkvist & Zetterlund 1994). Concentrations of As up to 240 μg l^{-1} have been found in polluted river water draining a tin-mining area in southwest England (Hunt & Howard 1994). Wilson & Hawkins (1978) found concentrations of As ranging from 5 μg l^{-1} up to 1.2 mg l^{-1} in streamwaters draining the Fairbanks gold-mining district of Alaska. Acidic waters draining a gold-mining area in Zimbabwe were reported to contain up to 72 mg l^{-1} of As (Williams & Smith 1994) and high As contents have also been reported in lake sediments affected by former gold mining in the Waverley area of Nova Scotia (Mudroch & Clair 1986).

The town of Obuasi in the Ashanti Region of Ghana has been the centre of large-scale goldmining activity since the late 19th century. The main gold-bearing ore is arsenopyrite. Mining activity is known to have given rise to substantial airborne As pollution from the oreroasting chimney in the town as well as riverborne As pollution derived from nearby tailings dams. Some studies of As exposure of mine workers in Obuasi have been carried out (Amasa 1975) but little information is available about the concentrations of As and other potentially toxic metals in the drinking water of rural communities around the town, both as a result of pollution from the mining activity and from natural processes of water–rock interaction and sulphide oxidation. This paper investigates the concentrations of As in drinking water from streams, shallow dug wells and boreholes in a 40 × 40 km area around the town as well as the composition of deep groundwaters (70–100 m depth) from mine exploration boreholes and mining effluent. Arsenic has also been determined in urine samples from volunteers in two selected villages. The main processes of As mobilization in the natural groundwater environment of the Obuasi region and the impact of As on the health of local rural communities are assessed.

Hydrogeochemistry and biogeochemistry of arsenic

Arsenic can occur in the environment in several oxidation states (−3 to +5) but in natural waters is mostly found as an oxyanion in trivalent (arsenite) or pentavalent (arsenate) form. Equilibrium speciation of As has been described in detail by Ferguson & Gavis (1972) and Brookins (1988). In oxidizing environments, arsenate is the dominant form: under acidic conditions (pH less than $c.6.5$), $H_2AsO_4^-$ may be stabilized, whilst at higher pH, $HAsO_4^{2-}$ will be more stable (H_3AsO_4 and AsO_4^{3-} may be present in extremely acidic and alkaline conditions respectively). Under reducing conditions at pH less than $c.9$, the arsenite species $H_3AsO_3^0$ should predominate (Brookins 1988). Arsenic can therefore be stable in dissolved form over a wide range of Eh and pH conditions.

However, under reducing conditions in the presence of sulphur, As concentrations are limited by the low solubility of As sulphide minerals, such as realgar and orpiment. Little information is available about the rates of As reactions in natural waters and specific rate constants are largely unknown. At near-neutral pH, the rate of oxidation of As(III) to As(V) is known to be very slow, but may proceed faster (order of days) under more extreme pH conditions (Ferguson & Gavis 1972).

Water pH has an important impact on As release rates from minerals. Leaching experiments on river sediments contaminated by mine waste have shown that As release to solution is lowest at near-neutral pH; release is enhanced under both acidic and alkaline aqueous conditions (Mok & Wai 1989, 1990).

Most of the As in waters is present in inorganic form (e.g. Cebrián et al. 1994). However, organic arsenicals are also known to be stabilized by methylation reactions involving bacteria and algae (Mok & Wai 1994). Monomethylarsonic acid (MMAA) and dimethylarsinic acid (DMAA) have for example been observed, albeit in small quantities, in some river waters and porewaters by biotransformation of inorganic As compounds (e.g. Faust et al. 1987). Demethylation reactions are also known to occur in water.

Mobility of As in water is limited by sorption onto ferric oxyhydroxides, humic substances and clays (e.g. Mok & Wai 1994). Wauchope (1975) showed that sorption of As species onto sediments was in the order: arsenate > MMAA > arsenite > DMAA and Bowell (1994) found that sorption onto Fe oxyhydroxides and oxide minerals was in the order As(V) > DMAA = MMAA > As(III) at pH values less than 7, the degree of sorption being greater on amorphous oxyhydroxides (goethite) than on crystalline Fe oxides (haematite). Arsenic should therefore be less mobile in oxidizing environments, because of firstly the greater tendency for sorption of As(V) than As(III) and secondly, the lower solubility of ferric oxyhydroxides under such conditions (Peterson & Carpenter 1986). Dissolved arsenic is therefore more commonly found in reducing waters, often correlating well with dissolved Fe concentrations.

Arsenic toxicity is dependent on both its oxidation state and partitioning between organic and inorganic phases. Reduced forms tend to be more toxic than oxidized forms and inorganic species more toxic than organic, though toxicity of reduced organic forms such as trimethylarsine is relatively high. Toxicity decreases from arsine (−3) through arsenite (inorganic trivalent), arsenoxides (organic trivalent), arsenate (inorganic pentavalent), arsonium compounds to native arsenic (Morton & Dunette 1994). Arsenic(III) is known to be up to 60 times more toxic than As(V) due to its reactivity with, and inhibition of, sulphydryl-bearing enzymes in human metabolism (Squibb & Fowler 1983). However, some *in vivo* reduction of As(V) to As(III) may take place (Vahter & Envall 1983).

The most common sources of ingested As are drinking water and food (in the absence of direct pollution sources). Some foodstuffs are known to contain high concentrations of As, especially seafood (up to $40 \mu g\,g^{-1}$, National Academy of Sciences 1977), although this contains As almost entirely in organic form as arsenobetaine $((CH_3)_3As + CH_2CO_2^-)$ and to a lesser extent as arsenocholine $((CH_3)_3As^+ CH_2CH_2OH)$, both of which are of very low toxicity. Most other foodstuffs contain As (in either organic or inorganic form) in the range $0.1–1.0\,\mu g\,g^{-1}$. Since the most common species in water are inorganic, this represents potentially the most detrimental source of As in the human diet, especially if present in reduced form.

Arsenic tends not to accumulate in the body but is readily excreted via the kidneys (e.g. Vahter & Lind 1986). Urine is therefore a good indicator of environmental As exposure and many studies of populations exposed to airborne As from smelters have been reported. Since seafood is known to be a major source of As (in organic form), most studies report both inorganic and organic (seafood-derived) As content.

Background concentrations of inorganic As in urine are largely reported to lie in the range $10–23\,\mu g\,l^{-1}$ (Braman & Foreback 1973; Bencko & Symon 1977; Vahter & Lind 1986; Pan *et al.* 1993). Lin *et al.* (1985) quoted a somewhat higher background value of $63.4\,\mu g\,l^{-1}$ although the proportion of As present in inorganic form is not known.

Concentrations of As in urine from smelter workers exposed to airborne As are higher: Kodama *et al.* (1976) reported a value of $56\,\mu g\,l^{-1}$ and Lagerkvist & Zetterlund (1994) gave a mean value of $61\,\mu g\,l^{-1}$ in urine from smelter workers. Concentrations of As in urine of people having recently eaten seafood can reach up to $1000\,\mu g\,l^{-1}$ (Pinto *et al.* 1976), mostly as arsenobetaine.

Obuasi: local environment and geology

Ashanti region is in the tropical rainforest belt of southern Ghana. It receives a high annual rainfall (*c.* 1580 mm) with a pronounced wet season from March to November and a diurnal temperature range of 20–30°C. Evaporation is also high, about $1260\,mm\,a^{-1}$ although rainfall exceeds evaporation for about 8 months of the year (March to October). Much of the Ashanti region comprises natural forest, although many places have been cleared for cocoa and food-crop production. River flows in the region are mainly from northeast to southwest and are dominated by two rivers, the Oda and the Gymi rivers, separated by the Sansu-Moinsi hill range (maximum height *c.* 600 m).

The geology of the area is summarized in Fig. 1. Underlying bedrocks are of Birimian (Proterozoic) metasedimentary and metavolcanic rocks, the Lower Birimian comprising mainly schist, phyllite and metagreywacke as well as granite and the Upper Birimian comprising mainly metavolcanic rocks (Kesse 1985; Fig. 1). These were subjected to intensive folding and faulting and metamorphosed to greenschist grade during the Eburnian orogeny (1830–2030 Ma).

Gold-mining has been the major industry in the Ashanti region since the late 19th century. Ore deposits occur in a major shear zone with a proven lateral extent of 8 km and depth of 1600 m (Amanor and Gyapong 1988). Numerous shafts and surface operations occur along the shear zone in the vicinity of Obuasi (Fig. 1). The gold is present as both disseminated grains in quartz reefs and in association with sulphide minerals, particularly arsenopyrite (Bowell 1992).

Processing of the gold ore is carried out at the Pompora Treatment Plant (PTP) in Obuasi where the ore is crushed, roasted and extracted by a cyanide complexation process. Spent ore is collected in nearby tailings dams and although the liquid effluent from the dams is now recycled in the gold extraction process, at the time of field investigation it was discharged directly into a local stream, the Kwabrafo. This stream flows southerly on the east side of the town, joining the Pompo River 5 km southeast of Obuasi and thereafter flowing into the Gymi River. The

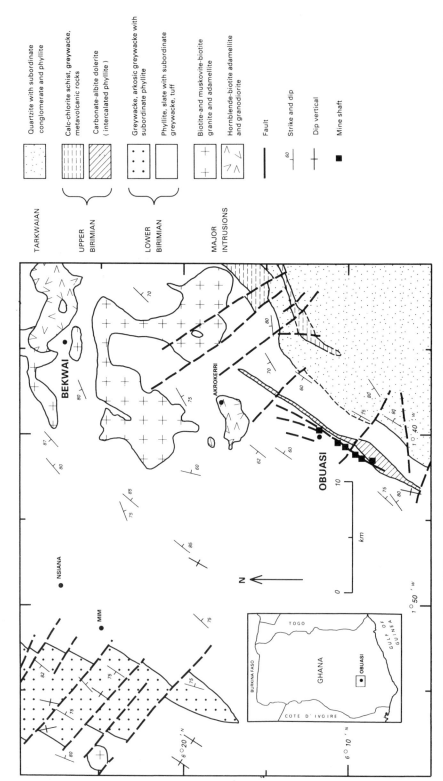

Fig. 1. Geological map of the Obuasi area of Ghana showing the mine shafts of the Ashanti mine complex (after Ghana Geological Survey maps, Obuasi sheets, 1:62 500; Bowell 1992). Location of the study area is indicated in the inset map.

effluent is known to contain large quantities of dissolved As, SO$_4$ and CN from the extraction and concentration process and as a result has caused serious pollution of the stream waters downstream of the input point. Additional discharges from the mine workings occur from tailings dam overflows, pumping of mine-water drainage into local streams and run-off from slime dams during periods of heavy rainfall.

Arsenic (as arsenic trioxide, As$_2$O$_3$) and sulphur (SO$_2$) were also emitted into the atmosphere via the PTP chimney until the recent installation of an As scrubber which reduced the levels of As emissions considerably (Ashanti Goldfields Corporation Environmental Laboratory, pers. comm. 1993). The prevailing wind direction is northerly and as a result of long-term emissions from the chimney, vegetation on the northerly slopes adjacent to the chimney has been killed or badly damaged. Defoliation as a result of both arsenic and sulphur poisoning has been observed up to 8 km north of Obuasi (Amasa 1975). This has resulted in severe erosion of bare slopes by high rainfall.

High As concentrations in hair samples of workers from the PTP attest to the former high incidence of airborne As pollution from the mining activity. Amasa (1975) found that hair samples from workers at the PTP ranged between $196 \pm 7\,\mu\mathrm{g\,g^{-1}}$ and $1940 \pm 62\,\mu\mathrm{g\,g^{-1}}$. Arsenic contents of soils and vegetation were also found to be high (up to $148\,\mu\mathrm{g\,g^{-1}}$ and $4700\,\mu\mathrm{g\,g^{-1}}$ respectively). Amasa (1975) noted that residents of villages 8–10 km north of the town claimed to have suffered eye inflammations as a result of mine emissions.

Soils in the Obuasi region are leached kaolinite–muscovite laterites with average thicknesses of 1–3 m, although the thickness depends on underlying saprolite lithology, topography and drainage (Bowell 1993). Bowell (1993) studied soil profiles 40 km east of Obuasi and noted that As was concentrated in the upper A horizon of the soils, in association with the higher content of organic matter in this layer. Concentrations of As were noted to be very low at 0.2–$0.3\,\mu\mathrm{g\,g^{-1}}$ in the A horizon of profiles not contaminated by mining activity, decreasing to $0.01\,\mu\mathrm{g\,g^{-1}}$ in the argillaceous (B) horizon below. Soil As in the Ashanti concession has much higher values of between 1 and $1530\,\mu\mathrm{g\,g^{-1}}$ (mean $41.2\,\mu\mathrm{g\,g^{-1}}$; N. Bailie, pers. comm. 1993). Saprolite thickness is variable but is mostly in the range 10–20 m (Gibb & Partners 1992). In much of the saprolite, arsenopyrite appears to have been replaced by secondary As- and Fe-bearing minerals, including scorodite (FeAsO$_4$.2H$_2$O), haematite, arsenolite, amorphous iron oxides and arsenates (Bowell 1992).

Groundwater flow in the Birimian aquifer is predominantly via fracture zones, mostly along quartz veins. Rest-water levels in the Obuasi area are mainly in the range 2.5–7.5 m below ground level with a seasonal variation of several metres (Gibb & Partners 1992), although the piezometric surface for the whole area studied is believed to reflect the surface topography closely. Groundwater flow direction is principally from northeast to southwest.

Health aspects

Many rural communities in the Obuasi area have either never had access to groundwater, or boreholes drilled over the last few years have fallen into disrepair through lack of maintenance. Use of surface waters for potable and domestic supply is therefore common. Education on the relationship between water quality and health is generally poor. As a result, water-borne diseases such as bilharzia and dysentery are common among streamwater drinkers. Guinea worm is rare though present in parts and occasional cholera outbreaks occur, especially during the wet season (J. Ansah, AGC Hospital, Obuasi, pers. comm. 1993). Such water-borne diseases usually have much lower prevalence among communities using groundwater.

Despite the occasional incidence of skin depigmentation and rashes, no diseases which could be unequivocally attributed to As ingestion were observed among the rural communities in the study area.

Sampling and analysis

Water samples from a 40×40 km area around Obuasi (Fig. 1) were collected in January 1993, towards the end of the dry season. It was intended to collect samples from boreholes as well as a few hand-dug wells in the area. However, it was found that many rural water-supply boreholes had been abandoned due to lack of ability or willingness to pay maintenance costs and many communities have therefore resorted to surface water sources which are generally highly coloured (humic substances) with high suspended solids and poor bacteriological quality. Since such water sources were the drinking water of many communities, these too were sampled. In all, 65 boreholes, 13 shallow wells and 26 streams used for drinking water were sampled in our study.

On-site analysis comprised temperature, specific electrical conductance (SEC, corrected to

25°C), HCO_3 (measured as total alkalinity), pH, Eh (temperature corrected) and dissolved oxygen, the last three of which were measured in an anaerobic flow-through cell attached in-line to borehole water outlets. Boreholes were pumped until stable readings for these parameters were obtained (usually 15–20 minutes). Streams and shallow wells were not pumped and so pH and dissolved oxygen were measured either at source or from a collected sample (on-site); no Eh readings were taken at these sites. Filtered (0.45 µm) samples were also collected in acid-washed polyethylene bottles for subsequent laboratory analysis. Samples for major-cation, SO_4 and trace element analysis were acidified to 1% HNO_3 (v/v, Aristar), unacidified samples were collected for anion analysis and separate HCl-acidified aliquots were collected for As(III) and total As analysis. Samples were stabilized at c. pH 4 for As(III) analysis and subsequently acidified to 2% HCl (v/v Aristar) for As_{total} analysis.

Major cations, SO_4, Si, Al, P, V, Fe_{total}, Mn, Sr and Ba were analysed by ICP-AES using an ARL 34000C optical emission spectrometer; Cl, N species, I and F by automated colorimetry, other trace elements by ICP-MS (Fison's PQ1 instrument) and As by hydride-generation ICP-AES (using an ARL 341 hydride generator). Total As was analysed by pre-reduction of all As(V) to As(III) in samples 24 hours before analysis using 5% KI (2.5 ml 20% KI in 7.5 ml sample solution), followed by on-line hydride generation using 1% w/v $NaBH_4$ in 0.1% w/v NaOH (Trafford 1986). Arsenic(III) was analysed using a modified version of the method given by Driehaus & Jekel (1992). Arsenic(III) was reduced to arsine by $NaBH_4$ at pH > 3.5 using 0.5 M acetic acid as buffer solution (to prevent As(V) also being reduced at lower pH; Driehaus & Jeckel 1992). The acetic-acid method has the limitation that it fails to separate As(III) from DMAA (Anderson et al. 1984) and therefore analyses quoted in this paper for As(III) are strictly for the two species. However, since methylated As species usually have very low concentrations in natural waters, the values listed are probably dominated by As(III). The As(III) method used is known to suffer from suppression of the As(III) signal by high concentrations (order of 1 mg l^{-1} and above) of Fe, Cu and Ni (e.g. Driehaus & Jeckel 1992). All samples except one of mine-tailings effluent had very low concentrations of Cu and Ni and suppression from these elements is thought to be insignificant. Suppression of the As(III) signal by Fe at the concentrations found in most samples should also be negligible except in some of the minewaters which have up to 20 mg l^{-1} Fe. The As(III) data for these, together with the single tailings effluent sample, may have suffered serious suppression of the As(III) signal and must therefore be taken as minimum values (Table 1).

Eighty per cent of samples analysed had ionic charge imbalances of less than 5%. Charge imbalances were poorer for some samples due to low TDS concentrations. The accuracy of ICP-AES and ICP-MS analyses was periodically checked using international reference standards. Mean analyses and 1σ standard deviations for the National Research Council of Canada standard SLRS1 analysed by ICP-AES just prior to the Obuasi samples included (certified values in parentheses; mg l^{-1}): Na: 9.7 ± 0.2 (10.4 ± 0.6), K: 1.30 ± 0.07 (1.30 ± 0.20), Ca: 29.0 ± 0.7 (25.1 ± 0.9), Mg: 5.8 ± 0.1 (5.99 ± 0.28), Ba: 0.020 ± 0.002 (0.022 ± 0.001), Fe: 0.036 ± 0.006 (0.031 ± 0.002) and Sr: 0.137 ± 0.003 (0.136 ± 0.003; n = 14). Precision of ICP-AES data is typically of the order of 2% RSD (Fe: 5%). Mean concentrations for the National Institute of Standards and Technology standard 1643C obtained by ICP-MS during the course of sample analysis were: Al: 119 ± 67 (certified value: 114 ± 5.1), Cr: 20.4 ± 1.3 (19.0 ± 0.6), Co: 25.8 ± 1.5 (23.5 ± 0.8), Ni: 63.0 ± 3.7 (60.6 ± 7.3), Cu: 22.6 ± 1.3 (22.3 ± 2.8), Zn: 70 ± 11 (73.9 ± 0.9), Sr: 269 ± 8 (263.6 ± 2.6), Mo: 111 ± 3 (104.3 ± 1.9), Cd: 11.7 ± 0.2 (12.2 ± 1.0), Ba: 50.9 ± 0.9 (49.6 ± 3.1), Pb: 35.5 ± 1.3 µg l^{-1} (35.3 ± 0.9; n = 3) and Rb: 12.1 ± 0.2 µg l^{-1} (11.4 ± 0.2; n = 4). The mean concentration for total As in 1643C by ICP-AES hydride during the course of sample analysis was 75.5 ± 6 µg l^{-1} (n = 4, certified value 82.1 ± 1.2 µg l^{-1}) and SPEX standard EP8 gave a value of 10.5 ± 0.4 µg l^{-1} (n = 4, certified value 10.0 µg l^{-1}). Detection limits are quoted as 3σ about the variation of the blank concentration although precision is poorer at these low levels. Detection limits are given as 6σ for As_{total} and As(III) analyses.

Urine samples were collected from sample populations in two villages (30 volunteers in each) in the Obuasi area. Samples were collected in 30 ml sterile containers and frozen before analysis. Each represented a single, rather than 24-hour bulked sample. Records were made of villagers' age, occupation, sex, general health and dietary intake over the previous 24 hours. Sample digestion for reducible As was achieved by adding 4 ml concentrated HCl to 2 ml of urine and refluxing in an air condenser for 2–3 hours. Condensers were then washed down and the wash added to the sample; 1 ml 20% KI was added and the sample made up to 20 ml using deionized water. The reducible fraction is strictly inorganic As(III), As(V) and the methylated metabolite MMAA but the method does not quantitatively reduce DMAA. Since MMAA is usually a minor constituent of urine (Naqvi et al. 1994), the reducible fraction can be taken mainly to represent the concentration of inorganic As species. Digestion for total As involved mixing 2 ml of urine with 1 ml of a concentrated HNO_3-$HClO_4$-H_2SO_4 solution in the ratio 20:10:1. Samples were heated from 150°C to 300°C in steps of 30°C every 15–20 minutes and the resultant salt residue then dissolved in 5 ml deionized water and 4 ml HCl. Finally, 1 ml of KI was added to each sample and made up to 20 ml using deionized water. All solutions were heated to 90°C for 20 minutes prior to analysis to ensure total reduction of all As to As(III). Both total As and reducible As were analysed by the National Poisons Unit, Guy's Hospital, London by hydride-generation atomic absorption spectrometry.

Results and discussion

Mine effluent discharge

Chemical data given by Gibb & Partners (1992) for tailings effluent from the Obuasi mine reveal it to be a highly toxic cocktail of chemicals including especially As and CN derived from the roasting and extraction process. These have historically been discharged directly to nearby streams with no prior treatment or control. One sample taken during this study of the effluent being pumped directly into the Kwabrafo stream on the east side of Obuasi (Table 1) showed this to be an alkaline solution (pH 8.3) incorporating a range of toxic substances including dissolved As of $1.8\,mg\,l^{-1}$, SO_4 of $1750\,mg\,l^{-1}$ and CN of $30\,mg\,l^{-1}$ (CN analysed for this study by Ashanti Goldfields Corporation laboratory). Concentrations of NH_4, P, B, Co, Ni, Cu, Zn, I, Mo, Sb, Ag, Au and W are also very high in this effluent (Table 1). The As(III) content of the sample is only $50\,\mu g\,l^{-1}$ and suggests that most of the As in the effluent is present as oxidized As(V). However, for reasons stated above, the low As(III) value may be an artefact of the analytical method and the ratio of As(III) to As total may be higher than that quoted. The high concentrations of dissolved solutes in the effluent are most likely due to the formation of CN complexes.

The concentration of As in the Kwabrafo stream just a few metres downstream of the effluent discharge point (sample collected January 1993) was $7.9\,mg\,l^{-1}$ and SO_4 $1130\,mg\,l^{-1}$. Concentrations of P, B, I, Ni, Co, Mo, Ag and Sb were $1.4\,mg\,l^{-1}$, $740\,\mu g\,l^{-1}$, $533\,\mu g\,l^{-1}$, $62\,\mu g\,l^{-1}$, $68\,\mu g\,l^{-1}$, $7.2\,\mu g\,l^{-1}$, $0.24\,\mu g\,l^{-1}$ and $10\,\mu g\,l^{-1}$ respectively. A sample of Kwabrafo streamwater taken at the same time from several hundred metres upstream of the effluent discharge point also revealed a high total As concentration (though much lower than the effluent; $0.35\,mg\,l^{-1}$), probably as a result of wash-in of pollutants to the stream from various points upstream in the mine area.

The quality of the Pompo and Gymi river systems downstream of the mine complex has not been investigated in this study although it is thought to have been seriously impaired over a distance of many kilometres. Concentrations of CN are expected to diminish downstream due to degradation in the aerobic environment. This may lead to decreased trace metal mobilization as CN complexes break down but will result in increased concentrations in stream sediments. Conversely, mobilization of many trace elements, notably As, is expected to be enhanced in the streams by binding to organic matter and colloids.

Regional water quality

Representative chemical data for streams, wells and boreholes are given in Table 1 and median values are summarized in Table 2. Waters are usually acidic (median values for streams, wells and boreholes are 6.4, 5.4 and 5.8 respectively) with low total dissolved solids contents (TDS; 70, 52 and $101\,mg\,l^{-1}$ respectively; Table 2). The low overall TDS concentrations suggest that the waters have had short contact times with host rock materials and that water interaction has been relatively small. Bicarbonate (HCO_3) concentrations are also usually low but reach higher values in the groundwaters (median value $67\,mg\,l^{-1}$; Table 2). This suggests that the deeper groundwaters have undergone a greater degree of water–rock interaction than the streams and shallow-well waters. All waters sampled are undersaturated with respect to calcite and reflect the general paucity of carbonate in the aquifer, although carbonate is present as a gangue mineral in the auriferous shear zones (Leube et al. 1990). The waters are saturated with respect to quartz, reflecting the dominance of silicate–water interaction processes.

The surface waters are almost universally brown, being rich in organic matter and colloidal Fe. Bacterial counts are usually high, with mean concentrations of aerobic hetero-

Table 1. Chemical analyses of representative water samples from streams, shallow dug wells and boreholes from the Obuasi area, Ghana.

Sample	Units	Odumase stream Akokoaso	Ankasa stream,	Dwete stream	Wumase well	Abedwum well	Mile Fourteen borehole	Meduma borehole	Manso Atwere borehole	Kofikurom borehole	Tweapease borehole	Obuasi mine, bore 1	Obuasi mine, bore 2	Obuasi, tailings outflow
Sample type		stream	stream	stream	well	well	borehole	borehole	borehole	borehole	borehole	exploration borehole	exploration borehole	tailings effluent
Longitude	(1°W)	39.23	38.20	35.67	41.91	39.63	49.61	35.64	51.22	40.92	44.33	39.57	39.56	39.71
Latitude	(6°N)	9.76	19.81	13.41	5.28	18.26	13.47	16.11	27.27	11.89	8.92	12.21	12.24	13.39
Water level	m				2.91	3.54	2.9	14.96		6.52	2.23	41.35	36.75	
Temperature	°C	24.2	22.7	25.6	25.0	25.0	25.5	26.3	25.5	26.4	26.1	26.6		32.3
Well depth	mbgl				3.76	5.85	64	53		38	43	70	70	
pH		6.53	6.65	5.99	5.57	5.58	6.20	5.87	6.18	5.44	5.70	5.94	6.17	8.45
Eh	mV										307			
DO	mg l⁻¹				1.1		<0.1	3.7	<0.1	1.6	<0.1			
SEC	µS cm⁻¹	169			82		332	419	296	399	152	279	258	
Ca	mg l⁻¹	4.6	4.4	1.4	7.29	5.4	213	110	275	113	10.12	956	631	2960
Mg	mg l⁻¹	6.9	55.9	50.3	0.81	67.4	22.14	5.92	43.6	6.81	4.01	87.3	58.7	435
Na	mg l⁻¹	4.18	2.51	1.05	6.5	6.6	6.03	3.34	4.17	3.83	16.8	77.1	42.5	9.76
K	mg l⁻¹	8.4	1.71	1.03	0.81	0.45	16.9	10.8	16.3	9.5	0.51	26.3	23.9	349
HCO₃	mg l⁻¹	0.54	6.8	3.5	28.6	3.1	0.71	3.33	0.4	0.92	82.4	0.76	0.74	33.8
SO₄	mg l⁻¹	54.3	0.4	<0.1	3.0	0.67	130	62.5	187	38.2	0.8	110	135	154
Cl	mg l⁻¹	1.9	25.5	8.1	8.2	25.2	0.3	0.2	<0.1	0.3	8	507	253	1746
NO₃-N	mg l⁻¹	6.2	0.7	2.2	<0.2	0.2	6.7	2.8	7	8.5	<5	11.2	10.2	184
NO₂-N	mg l⁻¹	<0.2	4.2	3.9	8	4	<0.2	<0.2	<0.2	3	<0.2	<0.2	<0.2	7.3
NH₄-N	mg l⁻¹	<5	<0.2	<0.2	<5	0.3	<5	<5	<5	<5	<5	14	6	408
Si	mg l⁻¹	0.01	<0.01	<0.01	<0.01	<0.01	<0.01	<0.01	<0.01	<0.01	<0.01	0.03	0.02	17.7
Al	mg l⁻¹	8.6	7.3	4.3	7.4	5.6	31.3	28.7	30.2	12.1	28.5	7.9	12.3	4.5
P_total	mg l⁻¹	0.04	0.09	0.04	0.17	<0.02	<0.02	<0.02	<0.02	0.05	<0.02	<0.02	<0.02	0.03
V	mg l⁻¹	<0.1	<0.1	<0.1	<0.1	<0.1	0.2	0.1	0.3	<0.1	0.2	<0.1	<0.1	1.4
	µg l⁻¹	<3	<3	<3	18	<3	<3	<3	<3	6	<3	4	4	<3
Fe_total	mg l⁻¹	1.08	0.68	0.14	0.21	0.067	0.032	<0.020	0.75	0.053	0.702	20.4	4.16	0.702
Mn	mg l⁻¹	0.048	0.027	0.004	0.053	0.032	0.201	0.001	0.56	0.029	0.066	8.79	5.01	0.079
As	µg l⁻¹	7	43	10	<2	<2	54	3	57	<2	10	8	5	1810
As(III)	µg l⁻¹	4	5	<3	<3	<3	12	<3	11	<3	4	.3	•.3	50
I	µg l⁻¹	12.3	6.5	6.0	6.0	20.0	9.6	4.0	5.2	8.0	12.6	16.0	21.2	775
FF	µg l⁻¹	80	30	30	30	40	310	90	130	30	200	350	170	490
Sr	µg l⁻¹	44.7	26.8	7.7	35.5	50.1	334	109	364	66.2	172	395	305	1496
Ba	µg l⁻¹	8.3	14.6	5.5	41.7	28.6	8.5	129	4.0	39.3	6.6	54.7	29	22.7
Li	µg l⁻¹	1.65	6.43	1.08	13.4	2.46	32.7	18.3	31.9	9.18	36.4	14.3	11.8	6.42
Be	µg l⁻¹	<0.3	<0.3	<0.3	0.3	<0.3	<0.3	<0.3	<0.3	<0.3	<0.3	<0.3	<0.3	<0.3
B	µg l⁻¹	<8	<8	<8	<8	<8	<8	<8	<8	<8	<8	<8	<8	343
Cr	µg l⁻¹	0.38	0.33	0.40	0.51	<0.30	<0.30	1.91	<0.30	<0.30	<0.30	<0.30	<0.30	0.62
Co	µg l⁻¹	0.13	0.48	0.15	2.18	1.60	4.60	0.05	3.48	1.93	2.52	65.9	22.41	261
Ni	µg l⁻¹	0.47	1.18	0.43	4.67	2.19	0.61	3.79	1.20	4.16	4.29	161	61.8	1471
Cu	µg l⁻¹	3.6	2.3	1.2	3.0	2.7	1.8	7.9	0.5	6.8	1.2	8.3	31	9219
Zn	µg l⁻¹	3.2	3.3	3.9	27.2	33.2	6.0	5.3	20.8	14.8	11.2	909	2502	631
Y	µg l⁻¹	0.024	0.36	0.11	0.28	0.47	<0.020	0.036	0.063	0.11	<0.020	35.3	1.42	0.062
Mo	µg l⁻¹	<0.6	<0.6	<0.6	<0.6	<0.6	<0.6	<0.6	<0.6	<0.6	<0.6	<0.6	0.7	173
Ag	µg l⁻¹	<0.02	<0.02	<0.02	<0.02	<0.02	<0.02	<0.02	<0.1	<0.02	<0.1	<0.1	<0.1	19.6
Cd	µg l⁻¹	<0.2	<0.2	<0.2	<0.2	<0.2	<0.2	<0.2	<0.2	<0.2	<0.1	1.2	0.6	0.2
Pb	µg l⁻¹	0.12	0.15	0.19	0.11	0.27	<0.07	0.37	<0.07	0.10	<0.07	0.23	<0.07	<0.07
Rb	µg l⁻¹	0.76	0.68	0.12	1.25	1.58	2.04	1.98	0.63	1.33	1.24	2.23	1.68	34
Cs	µg l⁻¹	<0.2	<0.2	<0.2	<0.2	<0.2	0.63	<0.2	0.41	0.32	<0.2	2.9	1.7	0.63
Sb	µg l⁻¹	0.27	<0.04	0.06	0.14	0.07	0.11	<0.04	<0.04	<0.04	<0.04	1.63	1.65	93
Tl	µg l⁻¹	<0.020	<0.020	<0.020	<0.020	<0.020	0.024	<0.020	<0.020	<0.020	<0.020	2.30	0.030	<0.020
U	µg l⁻¹	0.014	0.011	<0.005	0.076	0.028	0.098	0.060	0.13	0.008	<0.005	0.17	0.11	0.13
Au	µg l⁻¹													120
W	µg l⁻¹													911

All sources listed are used for drinking water supply apart from the Obuasi mine exploration boreholes and the Obuasi tailings effluent. DO, dissolved oxygen; SEC, specific electrical conductance (corrected to 25°C); bgl, below ground level; *represents a minimum value. Charge imbalances are all <3% except tailings outflow (−5.6%).

Table 2. *Median values for major elements and parameters of health significance in Obuasi drinking waters (streams, wells and boreholes)*

		Streams	Wells	Boreholes	WHO maximum
n		26	13	65	
Temperature	°C	24.3	25.2	26.3	
Well depth	m		3.5	48	
pH		6.42	5.37	5.84	
Ca	$mg\,l^{-1}$	4.8	3.9	7.0	
Mg	$mg\,l^{-1}$	3.1	0.82	3.9	
Na	$mg\,l^{-1}$	8.5	6.6	13.8	(200)
K	$mg\,l^{-1}$	1.4	1.1	0.55	
HCO_3	$mg\,l^{-1}$	41	21	67	
SO_4	$mg\,l^{-1}$	1.5	1.1	0.44	500
Cl	$mg\,l^{-1}$	6.2	6.2	6.0	(250)
NO_3-N	$mg\,l^{-1}$	< 0.2	0.3	< 0.2	10
NO_2-N	$\mu g\,l^{-1}$	< 5	< 5	< 5	910
NH_4-N	$mg\,l^{-1}$	< 0.01	< 0.01	< 0.01	(1000)
TDS	$mg\,l^{-1}$	70	52	101	
Si	$mg\,l^{-1}$	8.7	8.0	21.3	
Al	$mg\,l^{-1}$	0.03	<0.02	<0.02	0.2
Fe_{total}	$mg\,l^{-1}$	0.5	0.07	0.2	0.3
Mn	$mg\,l^{-1}$	0.03	0.03	0.08	0.5
Sr	$\mu g\,l^{-1}$	44	38	94	
Ba	$\mu g\,l^{-1}$	19	29	16	700
As(III)	$\mu g\,l^{-1}$	< 3	< 3	< 3	
As_{total}	$\mu g\,l^{-1}$	7	< 2	< 2	10
Cd	$\mu g\,l^{-1}$	< 0.2	< 0.2	< 0.2	3
Cu	$\mu g\,l^{-1}$	1.3	1.5	3.7	
Ni	$\mu g\,l^{-1}$	1.0	2.2	3.4	
Co	$\mu g\,l^{-1}$	0.34	1.55	1.33	
Zn	$\mu g\,l^{-1}$	3.9	27	18	
Sb	$\mu g\,l^{-1}$	< 0.04	< 0.04	< 0.04	5
Ag	$\mu g\,l^{-1}$	< 0.02	< 0.02	< 0.02	
Pb	$\mu g\,l^{-1}$	0.09	0.11	0.06	10
F	$\mu g\,l^{-1}$	65	50	70	1500
I	$\mu g\,l^{-1}$	9.1	10.1	6.7	

WHO (1993) maximum guideline values are given for comparison (values in parentheses are for elements for which no recommended guideline maximum is given, but the value represents the limit above which taste problems might occur). TDS, total dissolved solids.

trophic bacteria of 5700 colony-forming units ml^{-1}. Many of the shallow dug wells also have high bacterial populations (mean aerobic heterotrophs 4700 CFU ml^{-1}), but organic water quality is usually much better in the hand-pumped groundwaters (mean 860 CFU ml^{-1}; West *et al.* 1995).

Most of the groundwaters are oxidizing with Eh values between ≥ 300 and 500 mV but some deeper waters have lower redox potentials of around 220–250 mV. These are relatively high values for reducing waters but reflect the low pHs of the sources investigated.

Dissolved SO_4 concentrations are usually very low in the Obuasi waters, median values for streams, wells and boreholes being 1.5, 1.1 and 0.4 $mg\,l^{-1}$ respectively (Table 2). All drinking water sources are undersaturated with respect to barite. Iron contents are variable, the median values for streams being 0.5 $mg\,l^{-1}$, wells 0.07 $mg\,l^{-1}$ and the boreholes 0.2 $mg\,l^{-1}$.

Three groundwater samples collected from exploration boreholes penetrating to 70–100 m depth at the mine complex have different compositions with much higher TDS values of 500–800 $mg\,l^{-1}$. They are moderately reducing with Eh values of 260–280 mV. Dissolved oxygen was only measured at one site but gave a low value of 0.1 $mg\,l^{-1}$. These groundwaters have high dissolved SO_4 (200–500 $mg\,l^{-1}$) and Fe (4–20 $mg\,l^{-1}$) concentrations as well as other trace elements such as Mn, Co, Ni, Cu and Zn (e.g. Table 1). Iron was observed to oxidize and precipitate over the course of several minutes from water abstracted during sampling. These waters are saturated with respect to barite. The

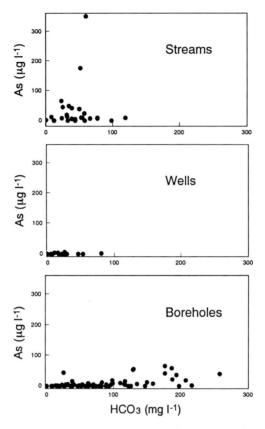

Fig. 2. Variation of As$_{total}$ with bicarbonate (HCO$_3$) in water from streams, shallow wells and boreholes in the Obuasi area.

high concentrations are most likely the result of oxidation of Fe and related sulphides. One sample of mine drainage being pumped from shafts into a local stream has a particularly high TDS of 1322 mg l^{-1}, high SO$_4$ of 926 mg l^{-1} and Fe of 7.6 mg l^{-1}. The sample is saturated with respect to goethite and ferrihydrite. This appears to have undergone extensive sulphide oxidation, probably as a result of introduction of atmospheric oxygen via the open shaft workings.

Distribution of arsenic. Total arsenic concentrations in the Obuasi waters vary between < 2 (detection limit) and 350 μg l^{-1}, although the highest value observed in drinking-water supplies was 175 μg l^{-1}. Twenty per cent of supplies studied exceed the WHO guideline maximum of 10 μg l^{-1}, although only 6% exceed the former WHO and current EC guideline maximum of 50 μg l^{-1}. Concentrations are plotted for streams, wells and boreholes against HCO$_3$ content as a measure of degree of water-rock (mainly silicate) interaction in Fig. 2. The streams have the highest values observed. The highest concentra-

tion (350 μg l^{-1}) is from the Kwabrafo upstream of the effluent discharge point and is manifestly affected by mine pollution. Unlike all other samples given in Fig. 2, this is not a drinking water supply source.

Shallow wells have universally low As$_{total}$ contents as well as HCO$_3$ as indicated in Table 2. The HCO$_3$ contents of the groundwaters range up to higher values (258 mg l^{-1} maximum) as would be expected for groundwaters having infiltrated to deeper levels in the aquifer and having had longer residence times for chemical reaction. The increase in HCO$_3$ also corresponds with increasing borehole depth (not shown).

Figure 2 shows that the total As concentration of the groundwaters notably increases with increasing HCO$_3$ concentration and suggests that As mobilization is achieved through greater aquifer residence time. This is corroborated by the observed correlation between As$_{total}$ and borehole depth in at least some of the groundwaters given in Fig. 3; deeper groundwaters should have had longer periods in contact with aquifer material and hence have undergone

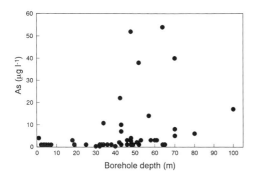

Fig. 3. Variation of As with total borehole depth (depth below ground level) in groundwaters (from wells and boreholes) in the Obuasi area.

greater water–rock reaction with As-bearing mineral phases, particularly sulphides. These deeper groundwaters also have lower redox potentials.

Investigation of the distributions of As_{total} in waters from the Obuasi area (Fig. 4) shows that the highest values in streams are mainly found to the north of Obuasi town. Groundwaters in this tract of land do not show correspondingly high As concentrations and so it is unlikely that the streamwater As is derived from the bedrock. Since the prevailing wind direction is towards the north, it is suggested that the high As streamwater values result from airborne pollution derived from the PTP chimney stack in Obuasi. Although an As scrubber has now been fitted to the plant, it is considered that the observed As represents relict pollution which has not yet been flushed from the natural surface water system because of sorption and precipitation processes involving the soil. Groundwater As_{total} concentrations reach the highest values in the western part (Fig. 4). This corresponds with deeper (and more reducing) groundwaters in this area.

Concentrations of As(III) (strictly As(III) plus DMAA) are presented in Tables 1 and 2 and ratios of As(III) to As_{total} are plotted in an Eh–pH diagram in Fig. 5. The smallest symbols shown in Fig. 5 represent those samples for which Eh and pH values were determined but either As(III) or As_{total} or both were below detection limits. Figure 5 shows that the more reducing groundwaters (Eh values $\leq 300\,mV$) not only have higher As_{total} contents, but also largely have higher $As(III)/As_{total}$ ratios. The higher As_{total} contents of deeper groundwaters may therefore also be due to the lower tendency for sorption of As(III) to Fe species than As(V), together with the lower potential for Fe(III) precipitation under reducing conditions. Since As(III) is much more toxic than As(V), these deeper groundwaters are more a cause for concern with respect to As than the oxidized shallower groundwaters.

Histograms of the distribution of As(III)/

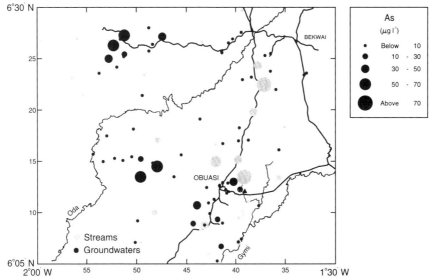

Fig. 4. Regional distribution of As_{total} in waters from the Obuasi area. Groundwaters are denoted by solid symbols and surface waters by stipple. As_{total} concentration is proportional to symbol size. Location of the Gymi and Oda rivers (and tributaries of the Gymi River) are indicated along with major roads. ▲, location of the Pompora Treatment Plant (PTP).

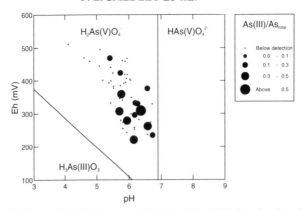

Fig. 5. Part of the Eh–pH diagram for the As system (after Brookins 1988) showing the principal stability fields of arsenate (As(V)) and arsenite (As(III)) species together with measured As(III)/As$_{total}$ ratios in water samples from the Obuasi area. The smallest symbols represent those samples for which Eh and pH determinations were made but As(III) or As$_{total}$ values were less than detection limits.

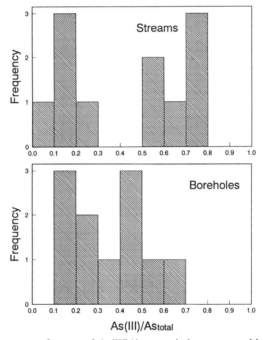

Fig. 6. Histograms of measured As(III)/As$_{total}$ ratio in streams and borehole waters.

As$_{total}$ in streams and borehole waters are given in Fig. 6. Whilst most of the groundwaters have ratios lower than 0.5 (the higher values corresponding to the deeper reducing waters), the streams have an apparent bi-modal distribution. Some have values less than 0.3 as would be expected for surface waters in contact with atmospheric oxygen. However, many have As(III)/As$_{total}$ ratios greater than 0.5. Remembering that the streams in the area have high humic and colloidal Fe contents and high bacterial counts, it is possible that oxidized As(V) has been reduced to As(III) by bacterial activity and that biomethylation reactions have generated DMAA in some of the surface waters.

Arsenic mobilization. Given the regional distribution of As-bearing minerals in the Obuasi area, the most likely source of the As in the waters is arsenopyrite (FeAsS) although minor amounts could also be derived by desorption from clays or secondary minerals (scorodite, arsenolite) formed after arsenopyrite oxidation. Oxidation of pyrite and arsenopyrite by oxygen may be described respectively by the equations:

$$4FeS_2 + 14O_2 + 4H_2O = 4Fe^{2+} + 8SO_4^{2-} + 8H^+ \quad (1)$$

and

$$4FeAsS + 13O_2 + 6H_2O = 4Fe^{2+} + 4AsO_4^{3-} + 4SO_4^{2-} + 12H^+ \quad (2)$$

The equations show that the oxidation process releases Fe and SO_4 (and arsenopyrite releases As) into solution. As noted above, with the exception of the minewaters, SO_4 contents of Obuasi groundwaters are generally quite low. However, the amount of arsenopyrite oxidation required to generate the observed dissolved As concentrations is also small. Assuming a congruent and complete reaction, oxidation of 0.85 mmol of arsenopyrite would be required to produce the highest observed As concentration in the drinking water boreholes ($64\,\mu g\,l^{-1}$). This would produce a corresponding SO_4 concentration of only $82\,\mu g\,l^{-1}$ and Fe of $48\,\mu g\,l^{-1}$. Iron is present in some of the waters in reducing acidic or organic-rich conditions, but molar Fe/S ratios are usually much lower than the value of 0.5 expected for stoichiometric oxidation of pyrite (and indeed the value of 1.0 expected for oxidation of arsenopyrite). This is likely due to Fe precipitation as ferric oxyhydroxide in aerobic waters. Even the reducing waters from the mine exploration boreholes have Fe/S ratios significantly less than 0.5 and suggest that some loss of Fe(III) has occurred by precipitation of minerals such as goethite and ferrihydrite, although the possibility of non-stoichiometric release of As from sulphide minerals under reducing conditions cannot be ruled out.

Arsenic concentrations in the mine-waters are relatively low compared to the highest values observed in the study area: the three boreholes sampled had groundwater concentrations of 5, 8 and $17\,\mu g\,l^{-1}$ As. This may be due to increased sorption of As by ferric oxyhydroxide in waters particularly enriched in dissolved iron, or due to oxidation of larger amounts of As-poor sulphide, particularly pyrite.

Despite the clear potential for As problems in the Obuasi groundwaters, only some of the deeper more reducing waters have concentrations above the WHO recommended limit and are considered potentially hazardous to health. Although streamwaters are also largely below the limit, concentrations are higher than in the shallow groundwaters, probably as a direct result of mining pollution. This concentration of As, together with their poor bacteriological quality renders them undesirable as sources of drinking water and their use should be discouraged.

Other elements. Although the high concentrations of solutes in the mine-tailings effluent from the Obuasi mine complex are the result of CN complexing rather than natural processes, the composition gives an indication of the range of other elements in local drinking waters which may be present and potentially problematic as a result of both pollution and natural sulphide oxidation. Equations (1) and (2) demonstrate that in the natural system, the oxidation process generates a large amount of acid (H^+), arsenopyrite even more so than pyrite. Consequent lowering of water pH gives greater potential for mineral dissolution and desorption reactions and under such conditions Fe, Mn and Al are more stable in solution. Also, other trace elements may be present in the sulphide ores which are released as oxidation proceeds, for example, Ni, Co, Cu, Zn, Mo and Ag.

Other elements of potential health concern have been examined in the Obuasi waters. Most are, however, not problematic: the waters largely have low TDS concentrations and so major elements are of low concentration. Chloride is mostly $<20\,mg\,l^{-1}$, SO_4 $<4\,mg\,l^{-1}$, Na $<30\,mg\,l^{-1}$, Ca in the range $0.2-54\,mg\,l^{-1}$ and Mg in the range $0.1-10\,mg\,l^{-1}$. Concentrations of constituents commonly attributed to agricultural pollution (NO_3, K and P) are also usually low: only one sample investigated has a concentration of NO_3-N higher than the WHO recommended limit and most are $<5\,mg\,l^{-1}$, K values are mostly $<3\,mg\,l^{-1}$ and P concentrations are $<0.3\,mg\,l^{-1}$.

Aluminium concentrations are quite high in some samples: around 10% of the drinking water supplies sampled had concentrations above the WHO guideline maximum of $200\,\mu g\,l^{-1}$, especially some of the streams and shallow wells, the Al largely occurring in colloidal form. The high concentrations reflect the acidic character of the waters (Table 1).

Iron, although not detrimental to health, is often mentioned together with Mn in a health context because of aesthetic and acceptability problems. Iron is often quite high in Obuasi waters, notably the more reducing sources and surface sources with heavy particulate matter and humic loads. The WHO guideline maximum value for Fe in drinking water is $0.3\,mg\,l^{-1}$ but an upper limit of $1\,mg\,l^{-1}$ should suffice for most purposes (WHO 1993). About 27% of drinking waters sampled had Fe concentrations greater than $1\,mg\,l^{-1}$. Manganese is less problematic, with 3% of sources having concentrations

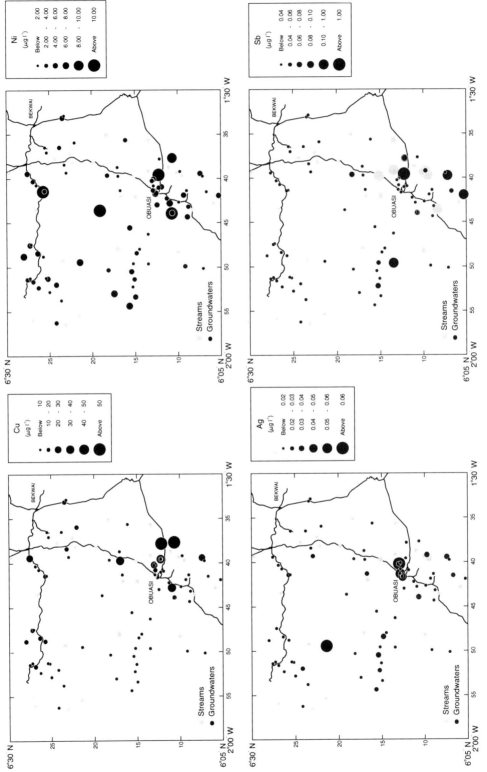

Fig. 7. Regional distributions of Cu, Ni, Ag and Sb in waters (streams, wells and boreholes) in the Obuasi area.

Table 3. *Median concentrations and ranges of As in urine from sample populations of two villages in the Obuasi area*

Village	Water source	No. villagers studied	Age range	Median total As ($\mu g\,l^{-1}$)	Range ($\mu g l^{-1}$)	Median inorganic As* ($\mu g\,l^{-1}$)	Range ($\mu g l^{-1}$)
Wumase	Stream	30	10–62	297	55–532	42	4–112
Kofikurom	Borehole	30	15–60	224	58–679	18	4–68

*Strictly includes the methylated metabolite MMAA.

greater than the provisional WHO guideline maximum of $0.5\,mg\,l^{-1}$.

Concentrations of halogens in drinking waters, notably I and F, have frequently been linked with health problems, principally goitre (I deficiency) and dental problems respectively. Median I concentrations in Obuasi waters are 7–10 $\mu g\,l^{-1}$ (Table 2) and are not thought to promote a health problem locally; indeed goitre has not been identified as an endemic problem in the Obuasi area. Fluoride concentrations are generally low (median values 50–70 $\mu g\,l^{-1}$, Table 2) and much lower than the WHO recommended maximum value of $1.5\,mg\,l^{-1}$. Dental caries (commonly associated with dietary F deficiency) was identified as a common problem in the study area, although high sugar intake and poor dental hygiene are additional likely causes.

Figure 7 shows the regional distributions of Cu, Ni, Ag and Sb in Obuasi waters. Antimony has been included because of its known association with As in mine wastes and high concentrations in mining areas (e.g. Mok & Wai 1990). Although the elements given in Fig. 7 are all well below levels considered detrimental to health, their concentrations are higher in the vicinity of Obuasi. This may largely reflect natural water–rock interaction processes in the main sulphide vein-bearing zone (Fig. 1). Concentrations of most of these elements are higher in groundwaters than streams but some streamwaters around Obuasi also have high Sb concentrations, probably due to binding with colloidal-Fe and organic materials. It is also possible that the trace element anomalies around Obuasi are caused by recharge of polluted water from the Pompo and Gymi river systems to the Birimian aquifer, rather than natural water–rock interaction processes. This would also give rise to elevated concentrations of the elements shown in Fig. 7 in the groundwaters as a result of mixing. Further investigation of the distributions of elements concentrated in the tailings effluent such as Cl, SO_4, CN, Sr, B and Rb would help to determine whether this localized area of trace element anomaly is pollution-derived or natural, although CN should not have long-term stability in oxidizing conditions in river systems. Further work on the hydrogeology of the Obuasi area would also help to determine whether the groundwater quality is severely affected by mine waste recharge.

Urine samples

Statistical results for urine samples from the two villages studied in the Obuasi area are given in Table 3. One village, Kofikurom, situated about 2 km west of the Obuasi mine, is a suburb of Obuasi where the inhabitants have used groundwater from borehole supplies for many years. The chemical analysis for borehole water from this village is given in Table 1: As concentration was low at $<2\,\mu g\,l^{-1}$. The other village, Wumase, is located 14 km south of Obuasi, where the water supply has traditionally been from a stream source (although a few shallow dug wells are available). Total arsenic concentration from one of the wells was $<2\,\mu g\,l^{-1}$ (Table 1) but higher in the local stream (the main source) at $5\,\mu g\,l^{-1}$. The As_{total} concentration of unfiltered streamwater from this site was found to be $8\,\mu g\,l^{-1}$, the excess over dissolved concentration probably bound to suspended solids or organic acids. People from both villages eat a diet of rice, cassava, plantain, yam, maize, occasional meat and fish (most volunteers having eaten fish, mostly local freshwater types, within the 24-hour period preceding the sampling). Villagers from Wumase obtain vegetables mainly from local farms, whilst those from Kofikurom obtain vegetables from Obuasi market and a farm at Anyenim (Obuasi). Ages of the volunteers were 10–62 and 15–60 in Wumase and Kofikurom respectively (Table 3). Occupations of volunteers in Wumase were mainly farmers, housewives and schoolchildren, whilst those in Kofikurom had a more varied range of occupations and included miners, farmers, traders, security staff and housewives.

Within each community, urinary As concentrations did not vary significantly with age, sex or occupation. Median values of total As concentration were 297 $\mu g\,l^{-1}$ for Wumase and

224 $\mu g\,l^{-1}$ for Kofikurom. These values are very high compared to concentrations quoted elsewhere in the literature, but reflect the content of organic As (arsenobetaine, arsenocholine) derived from recent ingestion of fish in both populations. Inorganic (reducible) As contents were much lower at 42 and 18 $\mu g\,l^{-1}$ respectively (Table 3). The median concentration for Kofikurom falls within the range of background urinary As values given by other studies. The value for Wumase is somewhat higher and could be a result of ingestion of drinking water with more As (albeit at a concentration below the WHO recommended maximum) or due to differences in provenance of foodstuffs such as vegetable crops: those purchased at Obuasi market by the Kofikurom inhabitants are likely to be exotic rather than grown locally in potentially As-rich soils. There is also the possibility that Wumase residents use water from the polluted Gymi river since this flows very close to the village. This is, however, unconfirmed.

Since Kofikurom is a suburb of Obuasi, situated close to the mine complex and using groundwater abstracted from the vicinity of the auriferous vein systems, the low concentration of inorganic As in Kofikurom urine samples is particularly encouraging and provides further evidence that local shallow groundwater is of good quality with respect to As.

Conclusions

The distribution of As and other trace elements has been investigated in streams, shallow wells and boreholes providing drinking water to rural communities in the Obuasi area. Concentrations of total As range between < 2 $\mu g\,l^{-1}$ and 175 $\mu g\,l^{-1}$, 20% of sources exceeding the WHO guideline maximum value for As of 10 $\mu g\,l^{-1}$ (although only 6% exceed the current EC and USEPA maximum of 50 $\mu g\,l^{-1}$). The high As loads of some local streams and groundwaters result from both mine-derived pollution and natural water–rock interaction processes. Pollution results from the discharging of tailings effluent and mine drainage into streams around Obuasi and historic atmospheric emissions of As trioxide and sulphur dioxide from roasting operations at Obuasi mine. High concentrations of As are also derived from natural oxidation of As-rich sulphide ores, notably arsenopyrite, as well as derivation from secondary As-bearing minerals.

Some of the streamwaters contain the highest concentrations of As, particularly those in the path of former mine emissions, to the north of Obuasi. Many of the As-rich streamwaters have high colloidal-Fe concentrations and are highly coloured, suggesting a high organic-acid content. The high As concentrations are likely to be facilitated by strong binding to these compounds. Arsenic(III)/As$_{total}$ ratios have a bimodal distribution, some with values less than 0.5 dominated by As(V) and some dominated by As(III) and possibly DMAA (values greater than 0.5) which are likely to have resulted from biomethylation and reduction reactions involving bacteria and algae.

Arsenic concentrations are universally low in shallow wells (median value < 2 $\mu g\,l^{-1}$) but reach higher levels (up to 64 $\mu g\,l^{-1}$) in groundwater from some of the deeper, more reducing boreholes (40–70 m depth). Deep mine-waters (70–100 m) rich in Fe and SO_4 indicative of Fe-sulphide oxidation have As concentrations between 5 and 17 mg l^{-1}. This range is lower than observed local maxima and probably results from As sorption by precipitating ferric oxyhydroxide although oxidation of As-poor sulphide minerals such as pyrite may also be responsible for the relatively low concentrations.

Positive correlations between As, borehole depth and bicarbonate content in groundwaters suggest that high As is controlled by depth of circulation and results from greater degrees of water–rock interaction with sulphide minerals, afforded by longer residence times in the aquifer. The higher concentrations in the deeper boreholes are generally accompanied by higher proportions of As(III) which is the stable species under more reducing conditions.

As(III) is known to be more toxic than As(V). The deeper borehole waters and some streams are therefore a greater health concern, both because of higher As$_{total}$ contents and higher proportions of As(III). However, possible risks from As(V) should not be ignored because of the potential for *in vivo* reduction to As(III).

The quality of Obuasi drinking waters with respect to other inorganic constituents is generally good. Concentrations of the major elements are low with Cl mostly < 20 mg l^{-1}, SO_4 < 4 mg l^{-1} and Na < 30 mg l^{-1}. There is also little evidence of significant agricultural pollution since concentrations of NO_3, K and P are usually low. Trace metals, particularly those associated with sulphide oxidation and gold-mining activity (e.g. Co, Ni, Cu, Zn, Pb, Ag, Sb) are also all below WHO guideline maxima in drinking waters analysed in this study. However, higher values are observed to the east and south of Obuasi, either as a result of recharge of polluted river water to the aquifer or from natural sulphide oxidation in the area where the greatest concentration of ore veins occurs.

Microbiological quality of some Obuasi drinking waters is less good, particularly streamwaters and some shallow wells which have high bacterial counts and are often associated with water-borne diseases (West et al. 1995). This, together with potentially high As concentrations, renders the streamwaters largely unsuitable for potable supply.

Use of borehole water should therefore be encouraged in the area, with the exception of deeper groundwater sources with potentially high As concentrations. The local geological and geochemical environment in Obuasi is typical of the Ashanti gold belt in general and it is therefore likely that the As mobilization processes outlined above are representative of the region as a whole.

Urinary total As concentrations in volunteers from two villages, one a rural community 14 km south of Obuasi and the other a suburb of the town, are very high with median values at 297 and 224 μg l^{-1} respectively and reflect the high intake of fish in the diet. Inorganic As contents are much lower at 42 and 18 μg l^{-1} respectively. Differences between the two probably reflect differences in As content of water supplies and provenance of vegetable foods. In particular, the low median value for the village using groundwater, despite its proximity to mining activity and to potential sources of As, attests to the overall good quality of shallow groundwater with respect to As in the Obuasi area.

Concentrations of As in river sediments downstream of the pollutant discharge points were not investigated in this study. This would be an interesting topic for further research since large amounts of aqueous As are known to sorb onto sediments. Since As release from sediments is thought to be achieved more readily under acidic conditions, the river sediments can have a large impact on potential for As dispersion downstream of the pollution injection point and on recharge to the aquifer. This has large implications for groundwater resource protection in the area to the south and east of Obuasi.

The assistance of staff of the Water Resources Research Institute (WRRI), Accra, especially N. B. Ayibotele and A. Paintsil is gratefully acknowledged, as is the great help provided by members of Ashanti Goldfields Corporation staff in Obuasi (R. Jenson, K. Asamoah, N. Laffoley, N. Bailie, C. Laidler) and the US Peace Corps (M. Fisher and K. M. Koporc). Considerable assistance and advice was also given by J. West (British Geological Survey). Chemical analysis was mostly carried out by J. M. Trafford, K. L. Smith and several workers at WRRI. Thanks are extended to G. N. Volans and I. M. House of the National Poisons Unit, Guy's Hospital for determination of As in urine samples and to J. D. Appleton, M. Williams and an anonymous reviewer for their constructive comments. We are especially indebted to the villagers in all communities visited, most notably Wumase and Kofikurom. The research has been funded by the British Overseas Development Administration as part of the Technology Development and Research Programme (Project R5552). This paper is published with the permission of the Director of the British Geological Survey.

References

AMANOR, J. A. & GYAPONG, W. A. 1988. *The Geology of Ashanti Goldfields*. Report of the Ashanti Goldfields Corporation, Obuasi, Ghana.

AMASA, S. K. 1975. Arsenic pollution at Obuasi goldmine, town and surrounding countryside. *Environmental Health Perspectives*, **12**, 131–135.

ANDERSON, R. K., THOMPSON, M. & CULBARD, E. 1984. Selective reduction of arsenic species by continuous hydride generation. Part I. Reaction media. *Analyst*, **111**, 1143–1152.

ASTOLFI, E. 1971. Estudio de arsenicismo en agua de consumo. *Prensa Médica Argentina*, **58**, 1342–1343.

BENCKO, V. & SYMON, K. 1977. Health aspects of burning coal with a high arsenic content. *Environmental Research*, **13**, 131–135.

BOWELL, R. J. 1992. Supergene gold mineralogy at Ashanti, Ghana: implications for the supergene behaviour of gold. *Mineralogical Magazine*, **56**, 545–560.

—— 1993. Mineralogy and geochemistry of tropical rain forest soils: Ashanti, Ghana. *Chemical Geology*, **106**, 345–358.

—— 1994. Sorption of arsenic by iron oxides and oxyhydroxides in soils. *Applied Geochemistry*, **9**, 279–286.

BRAMAN, R. S. & FOREBACK, C. C. 1973. Methylated forms of arsenic in the environment. *Science*, **182**, 1247–1249.

BROOKINS, D. G. 1988. *Eh-pH Diagrams for Geochemistry*. Springer, New York.

CEBRIÁN, M. E., ALBORES, A., GARCÍA-VARGAS., G., DEL RAZO, L. M. & OSTROSKY-WEGMAN, P. 1994. Chronic arsenic poisoning in humans: the case of Mexico. *In*: NRIAGU, J. O. (ed.) *Arsenic in the Environment, Part II: Human Health and Ecosystem Effects*. Wiley, New York, 93–107.

CHEN, S-L., DZENG, S. R., YANG, M-H., CHIU, K-H., SHIEH, G-M. & WAI, C. M. 1994. Arsenic species in groundwaters of the Blackfoot Disease area, Taiwan. *Environmental Science and Technology*, **28**, 877–881.

DÍAZ-BARRIGA, F., SANTOS, M. A., MEJÍA, J., BATRES, L., YÁÑEZ, L., ET AL. 1993. Arsenic and cadmium exposure in children living near a smelter complex in San Luis Potosi, Mexico. *Environmental Research*, **62**, 242–250.

DRIEHAUS, W. & JEKEL, M. 1992. Determination of As(III) and total inorganic arsenic by on-line pretreatment in hydride generation atomic absorption spectrometry. *Fresenius Journal of Analytical Chemistry*, **343**, 352–356.

FAUST, S. D., WINKA, A. J. & BELTON, T. 1987. An assessment of chemical and biological significance of arsenical species in the Maurice River Drainage Basin (N. J.) Part 1: Distribution in water and river and lake sediments. *Journal of Environmental Science and Health*, **A22**, 209–237.

FERGUSON, J. F. & GAVIS, J. 1972 A review of the arsenic cycle in natural waters. *Water Research*, **6**, 1259–1274.

GIBB, SIR A. & PARTNERS. 1992. *Water Sector Rehabilitation Project; Water Resources Report.* Unpublished Report, Sir Alexander Gibb and Partners, Kumasi, Ghana.

GORBY, M. S. 1994. Arsenic in human medicine. *In:* NRIAGU, J. O. (ed.) *Arsenic in the Environment, Part II: Human Health and Ecosystem Effects.* Wiley, New York, 1–16.

HUNT, L. E. & HOWARD, A. G. 1994. Arsenic speciation and distribution in the Carnon estuary following the acute discharge of contaminated water from a disused mine. *Marine Pollution Bulletin*, **28**, 33–38.

KESSE, G. O. 1985. *The Mineral and Rock Resources of Ghana.* Balkema, Rotterdam.

KODAMA, Y., ISHINISHI, N., KUNITAKE, E., INAMASU, T. & NOBUTONO, K. 1976. Subclinical signs of the exposure to arsenic in a copper refinery. *In:* NORDBERG, G. F. (ed.) *Effects and Dose-Response Relationships of Toxic Metals*, Elsevier, Amsterdam, 464–470.

LAGERKVIST, B. J. & ZETTERLUND, B. 1994. Assessment of exposure to arsenic among smelter workers: a five-year follow-up. *American Journal of Industrial Medicine*, **25**, 477–488.

LEUBE, A., HIRDES, W., MAUER, R. & KESSE, G. O. 1990. The early Proterozoic Birimian Supergroup of Ghana and some aspects of its associated gold mineralization. *Precambrian Research*, **46**, 139–165.

LIN, S. M., CHIANG, C. H. & YANG, M. H. 1985. Arsenic concentration in the urine and blood of patients with Blackfoot disease and Bowen's disease. *Biology and Trace-Element Research*, **8**, 11–19.

MOK, W. M. & WAI, C. M. 1989. Distribution and mobilization of arsenic species in the creeks around the Blackbird mining district, Idaho. *Water Research*, **23**, 7–13.

—— & —— 1990. Distribution and mobilization of arsenic species in the Coeur D'Alene River, Idaho. *Environmental Science And Technology*, **24**, 102–108.

—— & —— 1994. Mobilization of arsenic in contaminated river waters. *In:* NRIAGU, J. O. (ed.) *Arsenic in the Environment, Part I: Cycling and Characterization.* Wiley, New York, 99–118.

MORTON, W. E. & DUNETTE, D. A. 1994. Health effects of environmental arsenic. *In:* NRIAGU, J. O. (ed.) *Arsenic in the Environment, Part II: Human Health and Ecosystem Effects.* Wiley, New York, 17–34.

MUDROCH, A. & CLAIR, T. A. 1986. Transport of arsenic and mercury from gold mining activities through an aquatic system. *The Science of the Total Environment*, **57**, 205–216.

NAQVI, S. M., VAISHNAVI, C. & SINGH, H. 1994. Toxicity and metabolism of arsenic in vertebrates. *In:* NRIAGU, J. O. (ed.) *Arsenic in the Environment, Part II: Human Health and Ecosystem Effects.* Wiley, New York, 55–91.

NATIONAL ACADEMY OF SCIENCES. 1977. *Medical and Biological Effects of Environmental Pollutants: Arsenic.* National Academy of Sciences, Washington DC.

PAN, T-C., HUANG, K-C., LIN, T-H. & HUANG, C-W. 1993. Determination of arsenic and mercury concentrations in urine of patients with Blackfoot disease. *Japanese Journal of Toxicological and Environmental Health*, **39**, 148–154.

PETERSON, M. L. & CARPENTER, R. 1986. Arsenic distributions in porewaters and sediments of Puget Sound, Lake Washington, the Washington coast and Saanich Inlet, B. C. *Geochimica et Cosmochimica Acta*, **50**, 353–369.

PINTO, S. S., VARNER, M. O., NELSON, K. W., LABBER, A. L. & WHITE, L. D. 1976. Arsenic trioxide absorption and excretion in industry. *Journal of Occupational Medicine*, **18**, 677–680.

SQUIBB, K. S. & FOWLER, B. A. 1983. The toxicity of arsenic and its compounds. *In:* FOWLER, B. A. (ed.) *Biological and Environmental Effects of Arsenic*, Elsevier, New York, 233–269.

TRAFFORD, J. M. 1986. *Investigations into the storage of groundwater containing trace levels of arsenic, prior to analysis by hydride generation combined with ICP-AES.* British Geological Survey, Technical Report, **86/8**.

TSENG, W. P., CHU, H. M., HOW, S. W., FONG, J. M. LIN, C. S. & YEH, S. 1968. Prevalence of skin cancer in an endemic area of chronic arsenicism in Taiwan. *Journal of the National Cancer Institute*, **40**, 453–463.

VAHTER, M. & ENVALL, J. 1983. In vivo reduction of arsenate in mice and rabbits. *Environmental Research*, **32**, 14–24.

—— & LIND, B. 1986. Concentrations of arsenic in urine of the general population in Sweden. *The Science of the Total Environment*, **54**, 1–12.

WANG, L. & HUANG, J. 1994. Chronic arsenism from drinking water in some areas of Xinjiang, China. *In:* NRIAGU, J. O. (ed.) *Arsenic in the Environment, Part II: Human Health and Ecosystem Effects.* Wiley, New York, 159–172.

WAUCHOPE, R. D. 1975. Fixation of arsenical herbicides, phosphate and arsenate in alluvial soils. *Journal of Environmental Quality*, **4**, 355–358.

WEST, J. M., GARDNER, S. J., PAINTSIL, A., SMEDLEY, P. L., PELIG-BA, K. B. & DARTY, G. 1995. *Vulnerability of shallow groundwater due to natural geochemical environment – Obuasi, Ashanti Region and Bolgatanga, Upper East Region, Ghana. Results of microbiological and health studies.* British Geological Survey, Technical Report, **WE/93/6R**.

WHO. 1981. *Environmental Health Criteria 18: Ar-*

senic. World Health Organisation, Geneva.

WHO. 1993. *Guidelines for drinking-water quality. Volume 1: Recommendations*. Second edition, World Health Organisation, Geneva.

WILLIAMS, M. W. & SMITH, B. 1994. *Empirical and modelled hydrochemistry of acid mine drainage, Iron Duke mine, Mazowe, Zimbabwe*. British Geological Survey Technical Report **WC/94/78/R**.

WILSON, F. H. & HAWKINS, D. B. 1978. Arsenic in streams, stream sediments and ground water, Fairbanks area, Alaska. *Environmental Geology*, **2**, 195–202.

ZALDIVAR, R. 1974. Arsenic contamination of drinking water and foodstuffs causing endemic chronic poisoning. *Beitrage zur Pathologie*, **151**, 384–400.

Environmental impact of mining and smelting industries in Poland

E. HELIOS RYBICKA

University of Mining and Metallurgy, 30-059 Cracow, Al. Mickiewicza 30, Poland

Abstract: Mining and smelting industries cause tremendous devastation of both terrestrial and aquatic environments on the local and regional scale. Mines and smelters produce large quantities of waste, which must be deposited on land or into aquatic systems. The major effects in terms of contamination by heavy metals are in the pollution of air, soil, river water and groundwater systems. In Poland the most endangered surface water systems are the upper courses of both the Vistula and Odra Rivers, where the increase of contaminants, especially heavy metals and chloride ions, must be regarded as alarming. About 50% of surface water flows do not even meet the standards for quality class III. In 1990 the volume of wastes produced by mining and processing industries was more than 660 million tonnes of spoils and 490 million tonnes of tailings. It is estimated that in the period 1984–2000 about 900 million m^3 of spoils will be dumped in the area of the Upper Silesian Coal Basin alone. Taking into account the volume of sewage which should be treated, the emission of dusts and gases, and the volume of dumped wastes per km^2, 27 ecologically endangered regions have been distinguished. Almost half of these correspond to mining and smelting districts.

Mining and smelting activity imposes adverse, usually irreversible, effects on the terrestrial and aquatic environment. The most serious are: (1) changes in hydrogeological systems; (2) hydrological transformations of soils and surface water flows; (3) contamination of soils and surface and ground water resources; and (4) pollution of the atmosphere. In most cases the impact of mining on the environment is both local and regional. Very often the metallurgical industries are situated near the metal mines, and thermal power and heating plants are located in the vicinity of coal mines, so the geochemical characteristics of the mining areas are very complex.

Out of about 80 mineral commodities exploited recently in Poland, the most important are hard and brown coals; copper, zinc and lead ores; and native sulphur and rock-salt (Fig. 1).

The metallurgical industries in Poland are among some of the largest in Europe and the huge plants, located in heavily industrialized areas, have caused large-scale deterioration of the environment.

Upper Silesian district

Hard coals have recently been mined in Poland in three districts: the Upper Silesian Coal Basin (USCB: Katowice, Rybnik areas), the Lower Silesian Coal Basin (LSCB: Walbrzych and Nowa Ruda areas) and the Lublin Coal Basin (LCB: area east from Lublin). Adverse effects of mining activity are especially marked in the Upper Silesian Coal Basin where coal has been intensively exploited for about 150 years. In the 1980s, 62 coal mines in the Upper Silesia region extracted 190–200 million tonnes of coal per year (Nowicki 1993).

A large part of the coal extracted is burned in several power stations and in the heating-power plants of the Silesia conurbations, which have over three million residents. It is also converted into coke. In 1989 Katowice province extracted 98% of the total output of coal in Poland and 100% of zinc ores, generated 50% of the total steel production, and produced 34% of the coke (Przybylski 1991).

The hydrological system of the Upper Silesian Coal Basin (USCB) is influenced mainly by naturally mineralized and polluted waters which originate from dewatering of mines and are discharged into surface water drainage systems. In 1989 the operating coal mines in the USCB discharged about 720 000 m^3 of water per day from 83 discharge points. These waters are rich in Cl^- and SO_4^{2-} and contain high concentrations of heavy metals (Guziel 1988; Wilk et al. 1990). The total amount of chloride and sulphate ions released to the surface water systems reached 8000 tonnes per day in 1990, of which 5000 tonnes were finally transported into the Vistula River and 3000 tonnes into the Odra River. Concentrations of chloride and sulphate ions of more than 300 g/m^3 (reaching a maximum of 1600 g Cl^-/m^3) have been recorded over a distance of about 200 km in the Odra and Vistula rivers.

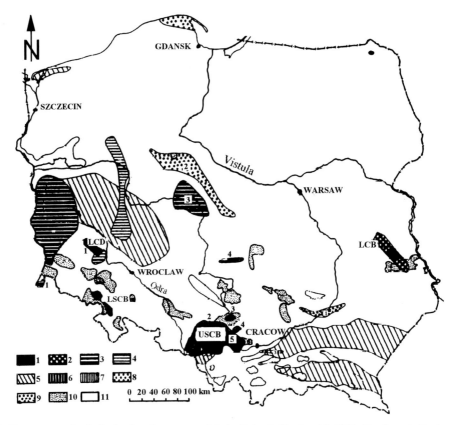

Fig. 1. Occurrence of principal mineral raw materials in Poland (Kozlowski 1983). Hard coal districts: 1, operating; 2, designed. Brown coal districts: 3, operating; 4, designed (1, Turoszów, 3, Konin;, 4, Beltchatów); 5, petroleum and gas districts. Metallic ore districts: 6, operating; 7, designed (1-copper, 2, 3, 4, 5, Pb–Zn); Chemical row material districts: 8, operating; 9, designed (1-sulphur, 2, 3, rock-salt), Industrial stones:. 10, operating; 11, designed.

Mineralized waters in the USCB coal mines have high concentrations of natural radioactive isotopes (Ney 1988; Wilk et al. 1990). This is caused by the reducing environment which predominates within the coal formations. In such conditions some elements (e.g. radium) are selectively leached from the wall rocks which results in a large excess of Ra over U and Th in the mine waters. The concentration of ^{226}Ra in waters discharged to both the Vistula and Odra rivers varies from $< 0.1–28.1\,\text{kBq/m}^3$.

In the Upper Silesian Zn–Pb mining district, the principal environmental problem is the formation of depression cones caused by the dewatering of mines. This affects an area of about 1000 km². In the Bytom subdistrict, where ores have been worked since the twelfth century, the mine waters have been almost completely drained off. Although the Bytom mines have been inactive since 1989, the water is still pumped and discharged into the surface drainage system (mainly to the Brynica stream). This discharge supplies 100 tonnes of TDS per day (90% sulphates) and up to $2\,\text{g/m}^3$ of base metals (mostly Zn and Pb). The mine waters also commonly contain phenols ($1–2\,\text{g/m}^3$) which result from the infiltration of surface contamination.

Industrial effluents are commonly discharged to surface water drainage systems after clarification in tailings ponds. Such waters may contain up to $300\,\text{g/m}^3$ sulphate ions and up to $3\,\text{g/m}^3$ Zn and Pb ions. An example of such strongly degraded surface water is the Luszówka stream (a tributary of the Vistula River) which receives 2.2 tonnes of sulphates and 22 kg of Zn and Pb per day.

A major problem is the pollution of both the bottom and flood-plain sediments of the two main rivers, the Vistula and the Odra, with heavy metals (Figs 2, 3). These are not only derived from mine waters but also are released

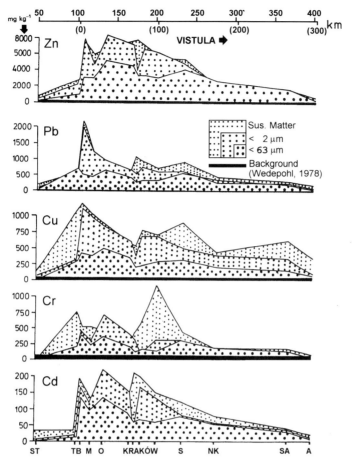

Fig. 2. Distribution of heavy metals among the suspended matter and bottom sediment grain size fractions of Upper Vistula River.

by Zn, Pb and Cu ore processing and smelting plants. Recent investigations (Macklin & Klimek 1992) have revealed very high concentrations of Zn (up to 11 000 mg kg^{-1}), Pb (over 1700 mg kg^{-1}) and Cd (up to 150 mg kg^{-1}) in the overbank alluvial sediments of Upper Vistula and Przemsza rivers.

Industrial waters discharged from Zn–Pb ore processing plants, as well as mine waters and meteoric waters infiltrating the waste dumps, are the principal pollutants of the Przemsza River, a tributary of the Vistula River. The concentrations of heavy metals in the bottom sediments of these two rivers are very high, and can reach maxima of 7000 mg kg^{-1} Zn, more than 900 mg kg^{-1} Pb and 200 mg kg^{-1} Cd in the Przemsza River, and up to 6000 mg kg^{-1} Zn, 800 mg kg^{-1} Pb and about 140 mg kg^{-1} Cd in the Vistula River sediments (Helios Rybicka 1992, 1993). The maximum concentration of cadmium is particularly high and approaches some of the highest levels recorded in river sediments anywhere in Europe (c. 200 mg kg^{-1}). The concentrations of heavy metals in the Odra River sediments are also high with concentrations up to 6700 mg kg^{-1} Zn, 4000 mg kg^{-1} Pb, 1800 mg kg^{-1} Cu, 350 mg kg^{-1} Ni and 12 mg kg^{-1} Cd (Fig. 3).

Soils in the neighbourhood of metal mines and smelters exhibit metal concentrations which are up to 100 times the natural background (Table 1, sites: Bukowno, Bytom, Katowice, Olkusz, Tarnowskie Góry; Kucharski et al. 1992). One important environmental problem in Poland is the contamination of soils with cadmium and lead. The ranges of metal concentrations in soils in the metal-processing area are as follows: 1665–13 800 mg kg^{-1} Zn, 72–2480 mg kg^{-1} Pb and 6–270 mg kg^{-1} Cd (Kabata-Pendias & Pendias 1992). Soils in the vicinity of

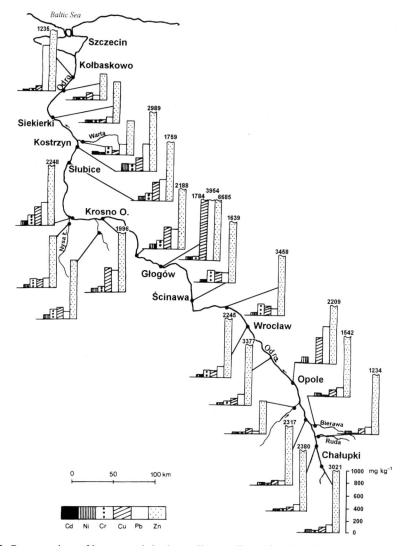

Fig. 3. Concentrations of heavy metals in the < 63 μm sediment fraction of the Odra River.

a Zn–Pb mining and smelting complex in Bukowno contain high concentrations of Zn, Pb and Cd (234–12 400, 42–3570 and 25–133 mg kg^{-1} respectively; Verner *et al.* 1994).

The metal industry also releases noxious gases including sulphur and nitrogen oxides. In the neighbourhood of the Miasteczko Śląskie Zn–Pb smelter, forest and other vegetation has been almost completely destroyed as a result of acid air pollution. Surface subsidence affects the mining fields of 43 USCB mines and has resulted in the formation of 322 ponds (data for 1987). The total submerged area is about 8 km^2 and this land is completely excluded from any new land use.

Mining activity yields several tens of millions of tonnes of coal waste each year (Fig. 4). Two types of waste can be distinguished: spoils and cleaning wastes. It is estimated that production of 1 tonne of hard coal is accompanied by the additional production of 0.4 tonnes of various wastes. Of this figure 46% remains underground and the rest (54%) is dumped on the surface. The wastes are carboniferous claystones, mudstones, sandstones and gravelstones/conglomerates, all of which contain variable amounts of heavy metal sulphides.

Ashes and slags from power stations constitute a relatively small percentage of wastes (2–3%). About 140 waste dumps are situated in the USCB, some of them being outside the mining areas. Although land rehabilitation has been

Table 1. *Soil contamination (mg kg⁻¹) with metals in the Katowice District*

Sampling site	Lead	Cadmium	Zinc
Bedzin	22–835	2–7	330–9000
Bobrowniki	74–1310	n.d.–54	109–6220
Bukowno	46–1520	1–42	90–9200
Bytom	129–2292	2–85	193–12595
Chorzów	117–600	5–23	539–3749
Chrzanów	24–1100	1–35	62–5660
Czechowice Dziedzice	8–95	n.d.–2	33–217
Czeladź	150–401	5–34	460–2645
Dabrowa Górnicza	10–890	1–17	37–3820
Gliwice	18–104	n.d.–4	37–289
Godów	n.d.–246	n.d.–1	1–102
Jaworzno	22–2150	1–36	73–5700
Katowice	20–1050	n.d.–20	61–2110
Klucze	4–3500	1–9	33–540
Krzyżanowice	n.d.–112	n.d.–2	n.d.–185
Libiaż	22–255	1–7	47–510
Lubomia	n.d.–58	n.d.–1	24–313
Laziska Górne	22–80	n.d.–3	82–265
Lazy	8–294	n.d.–5	34–760
Mierzecice	35–610	n.d.–10	62–1690
Mikolów	25–126	n.d.–4	26–387
Myslowice	8–940	n.d.–7	43–1740
Ogrodzieniec	16–122	n.d.–6	39–319
Olkusz	12–820	n.d.–17	24–1400
Orzesze	15–221	n.d.–16	20–900
Piekary Śląskie	131–3500	2–72	400–10400
Psary	40–960	n.d.–69	82–6840
Pszczyna	5–136	n.d.–4	11–159
Ruda Śląska	21–278	1–9	90–700
Siemianowice Śląskie	86–576	2–27	417–3807
Siewierz	17–260	n.d.–50	39–7720
Slawków	37–352	2–16	148–2960
Sosnowiec	45–537	2–31	120–2690
Suszec	9–130	n.d.–4	5–244
Świerklaniec	98–388	2–8	167–1640
Świetochlowice	122–1320	8–61	580–8050
Tarnowskie Góry	26–8200	1–143	103–13250
Trzebinia	24–1250	n.d.–80	73–9700
Tworóg	14–211	n.d.–7	8–620
Tychy	10–190	n.d.–7	22–580
Zabrze	30–158	1–5	5–550
Zawiercie	4–260	n.d.–5	24–359

n.d., non-detectable. After Kucharski *et al.* 1992.

completed for most of the dumps, the pollution of groundwaters due to the leaching of soluble components by percolating meteoric waters, still continues. Thus, despite their age, the dumps are classified as potential pollution sources for surface and groundwaters.

In 1989 industrial plants generated about 170 million tonnes of waste, of which 43%, including 2–3 million tonnes of hazardous waste, were dumped at waste disposal sites. Almost half of the total industrial wastes at waste disposal sites in Poland (i.e. 1500–2500 million tonnes) is accumulated in the small area of Katowice province (Fig. 5).

One of the most dangerous environmental hazards caused by waste dumps is the pollution of surface and groundwaters by soluble compounds leached from the wastes. This conclusion is supported by studies of pore solutions derived from the carboniferous spoils and of waters seeping from the dumps (Twardowska *et al.* 1988). Aqueous solutions which are highly contaminated with the products of sulphide decomposition occur throughout the full thick-

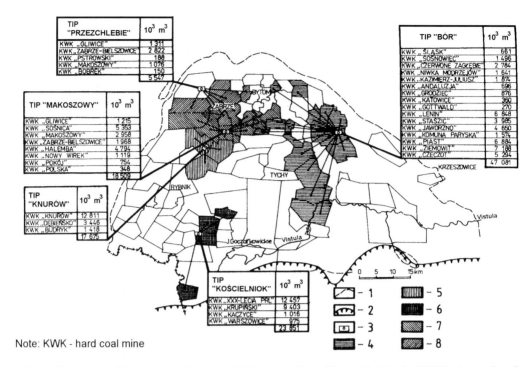

Fig. 4. Sketch map of the central coal mine spoil tips in the Upper Silesian Coal Basin (USCB): 1, margins of USCB; 2, margin of the Carpathian overthrust; 3, central coal mine spoil tips; 4–8, coal mines supplying spoils in the central tips: 4, Bór; 5, Knurów; 6, Kościelniok; 7, Makoszowy; 8, Przechlebie; 9, coal mine spoil tip Smolnica (Szczyglowice colliery). Tables contain volumes of spoils deposited in the period 1984–1990 (Twardowska et al. 1988).

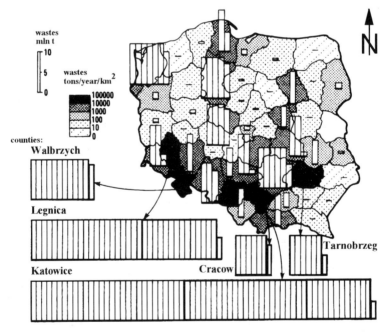

Fig. 5. Accumulated industrial wastes by counties in Poland (GUS 1991).

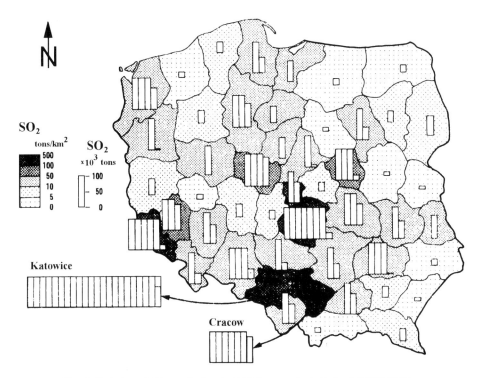

Fig. 6. Emission of SO$_2$ equivalent in 1990 by counties in Poland (GUS 1991).

nesses of typical spoil dumps. In solutions collected from beneath 7 to 15-year old spoils, sulphate ion concentrations exceeded by 10–80 times the quality standards for potable waters. Formation of especially dangerous, highly acid waters with pH below 4 appears to be a common effect. The leaching time of sulphides from fine-grained, oxygenated spoils is estimated to be about 11 years on average. In mine rock waste dumps, this process is much slower due to the coarse particle-size of the spoils and the limited volumes of meteoric waters available. However, some spoil dumps can remain a persistent source of pollution as a result of sulphide oxidation, and this may affect the environment over a period of decades.

Lower Silesian region

Of the mineral commodities which have been exploited for long periods in the Lower Silesia region the most important are: (1) hard coal in the Walbrzych and Nowa Ruda areas of the Lower Silesian Coal Basin (LSCB); (2) brown coal in the Turoszów area; and (3) copper ores in the Lubin Copper District (LCD) (see Fig. 1).

In the Lower Silesian Coal Basin (LSCB), the total volume of soluble solids released into surface water systems from mine waters reached 180 tonnes per day in 1989. The mine waters are polyionic (SO$_4$-HCO$_3$-Mg-Ca-Na type) but locally contain high concentrations of Cl$^-$. The high contents of sulphate in groundwaters result from infiltration of industrially contaminated, meteoric waters, derived from the leaching of numerous spoil dumps and also from the long-term residence of groundwaters in abandoned mine workings. Potential pollution sources of waters infiltrating into the LSCB mines include 17 dumps containing various types of mine spoils.

The total area affected by damage from brown coal open pit exploitation is 60 000 hectares and is the largest area affected by open pit mining in Poland. Beside the Turoszów brown coal mining and power station industry district there are two others in operation in Poland: Belchatów, and mid-Poland (Konin, Turek). The adverse environmental effects caused by brown coal open pit mining are irreversible and include extensive wasteland areas. Apart from geomechanical processes which lead to the complete destruction of soils and irreversible changes to the landscape, the drainage of open pits influences hydrological systems. Both surface water and shallow and deep groundwater, are affected by: (1) formation of depression cones which result in the lowering of shallow and deep water tables, and (2) changes in watersheds and divides. Dewatering of open pits results in a deficiency of water in farm lands and forest districts. Minewaters from brown coal pits are, however, generally uncontaminated and commonly meet the standards for potable waters.

Environmental problems caused by the brown coal mining industry are linked to those produced by the adjacent power stations which utilize most of the brown coal. Combustion of brown coal results in massive emissions of gases (CO_2, SO_2, nitrogen oxides) and dusts which are composed mostly of Al_2O_3 and Fe_2O_3 together with an admixture of heavy metals (Zn, Pb, Cd). Brown coals mined in Poland have sulphur contents of between 0.5 and 1.1%. As this sulphur is organically-bound, its removal before combustion is impossible. Huge amounts of SO_2 are thus released to the atmosphere (Fig. 6) giving rise to 'acid rain' and, consequently, to the degradation of soils by acidification.

Within a small area adjacent to the Polish–German–Czech border is situated the largest basin for brown coal extraction, with a combined annual production of c. 200 million tonnes of which about 20 million tonnes are extracted within Poland. This represents about 25% of the total brown coal extraction in Europe and it is used within the same area in 12 big power stations (including one in Poland at Turów). The SO_2 emission from this area, called 'the Black Triangle', is very high accounting for 20% of the total SO_2 emissions in Europe. Deposition of sulphur compounds is more than $5000\,mg/m^2$ per year and the pH is below 3.0. This has caused the complete destruction of forests in the Izerskie Mountains (EMEP 1992; Nowicki 1993).

In the Lubin Copper District (LCD) mine and industrial waters are purified before they are discharged into the Odra River thus preventing deterioration of the water quality class. Periodically, during high river flows, waters with increased TDS ($14.5\,g/dm^3$) are released at rates of less than $70\,m^3$ per minute. The load supplied to the Odra River in 1984 included 66 000 tonnes of chlorides, 21 000 tonnes of sulphates, 3.6 tonnes of Cu, 3.0 tonnes of Pb and 3.2 tonnes of Zn (see Fig. 3; Wilk et al. 1990).

Metal ore mines in the LCD are accompanied by processing plants and smelters which contribute significantly to the overall pollution. The soils in the vicinity of copper smelters are rich in heavy metals. Cu and Pb in the upper soil layer may exceed 0.7% and 0.2%, respectively (Helios Rybicka et al. 1994).

The Polish base metal ores are relatively low-grade so the wastes, which are mostly tailings, constitute 90–98% of the total output. The volume of wastes produced places the base metal industry in second position, just below the coal and power industry, in terms of waste production. The wastes are deposited in dumps and tailings ponds. The latter especially are sources of environmental hazards caused by the elevation of the groundwater level in the adjacent areas – in some cases up to several metres. This results in excessive moisture in soils and formation of marshes and ponds. The waters percolating from tailings ponds are highly polluted and may affect both groundwater and soils. In the LCD, waters in tailings ponds are highly contaminated with Cu, Zn and Pb sulphates and chlorides as well as organic compounds. Where groundwaters are contaminated they cannot be used for drinking water and, moreover, can also be harmful to vegetation. Dispersion of contaminants from huge tailings ponds in the LCD (Gilów & Żelazny Most areas) has become a serious problem with pollution extending over an area of about a dozen square kilometres.

Cracow urban–industrial area

Cracow is located within a large urbanized area which contains a number of polluting industrial plants, of which the most dangerous are a metallurgical factory, an aluminium plant and coal power and heating stations. These sources of gas and dust pollution are principally responsible for smog incidents in Cracow city. More than 47% of the area of Cracow province is threatened with environmental degradation (Michna 1991).

Beside the steelworks and power heating stations in Cracow, the major sources of air pollution are domestic coal-fired stoves and local boiler houses. The annual mean concentrations of suspended particulates, SO_2 and fluorine in the air in the years 1981–1991 were very high (55–95, 60–105 and 1–4 $\mu g/m^3$ respectively), although these decreased from 1988. The highest emission of particulates (above 140 000 tonnes per year) was observed in 1985, of which the steelworks' contribution made up about 50% (Rymont 1992).

The main river in the Cracow region is the Vistula River which is very polluted (see Fig. 2). The river water, which makes up 80% of the region's water resources, does not comply with any of the quality classes (Fig. 8). Due to the enhanced salinity of the Vistula River caused by

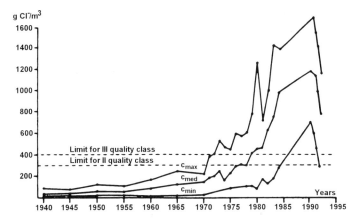

Fig. 7. Concentrations of Cl⁻ ions in the Vistula River water in 1940–1993 in Bielany/Cracow.

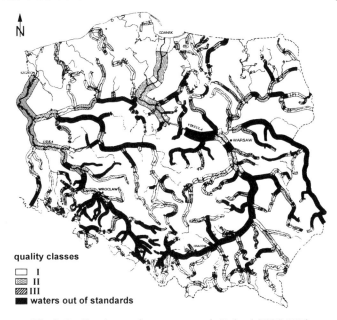

Fig. 8. Quality classes of stream waters in Poland (GUS 1991).

Table 2. *Total amount of main pollutants transported to the Baltic Sea by rivers from the Polish drainage basin (averages in tonnes per year estimated for 1988–89)*

phosphorus	nitrogen	zinc	lead	copper	cadmium	mercury	chromium	nickel
15 584	207 326	3331	448	515	40	90	443	284

Data from Main Statistical Office, GUS 1991.

the Silesian mines, water which reaches Cracow province is already too contaminated to be used as drinking water or for industrial and agricultural purposes. The tributaries of the Vistula River in the Cracow area are also heavily polluted. Significant contamination by Cd, Cr, Ni and Pb has been detected in allotment gardens in the Cracow conurbation (18, 86, 64 and 226 mg kg^{-1} respectively; Grodzińska *et al.* 1987).

In 1991 c. 3 100 000 tonnes of industrial waste were generated in the Cracow region of which only c. 2 200 000 tonnes were utilized. The rest were disposed of in dumps which occupy a total area of about 438 ha. The major wastes were metallurgical, including c. 1 400 000 tonnes from the steelworks and c. 800 000 tonnes of ash and slags from power stations.

Conclusions and recommendations

The mining and smelting industry causes tremendous devastation of the natural environment in Poland. The most endangered surface water flows are the upper courses of the Vistula River, including its Upper Silesian tributaries and the Odra River, in which the increase of contaminants, especially in recent years, must be regarded as alarming. Most of the polluting substances originate from highly mineralized waters pumped from the Upper Silesian coal mines. One example of a drastic increase of pollutants can be seen in the dispersion of chloride concentrations in the Vistula River near Cracow in the years 1940–1992 (Fig. 7).

Both the Vistula and Odra Rivers are important sources of water for domestic and industrial purposes. The Vistula River supplies about 0.5 billion m^3 of water along its 550 km long upper and middle courses which provide nearly 35% of the total water consumption in Poland. The Odra River yields an additional 300 million m^3 downstream to the mouth of the Nysa Łużycka River (i.e. about 2% of the total consumption). Progressive pollution of both rivers and their tributaries is a real danger and may lead to an environmental catastrophe.

Of the surface water flows included in the Polish water quality classification, about 50% do not meet the standards even for quality class III (Fig. 8). The total load of domestic and industrial sewage transported to the Baltic Sea by rivers from the Polish drainage basin includes, among others, the compounds listed in Table 2.

Other important sources of pollution are the wastes produced by the mineral industry. In 1990 the volume of such wastes included 660.5 million tonnes of spoils, of which 31.3% were disposed of in dumps, and 490.7 million - tonnes of tailings and cleaning wastes derived from coal, barite, native sulphur and base metal processing plants, of which 69.3% were dumped.

Hard-coal mining yields the largest volume of waste. It is estimated that in the period 1984–2000 c. 900 million m^3 of spoils will be dumped within the area of the Upper Silesian Coal Basin (Witczak & Szczepańska 1987). The annual load of chlorides in these spoils is up to 40 000 tonnes. Such a volume is sufficient to contaminate 133 million m^3 of water at concentrations exceeding the quality standards.

Many of the pollutants originate from the weathering of sulphides contained in spoils. Under full oxygenation, the decomposition of sulphides in these spoil tips may yield 450 000 tonnes of sulphates per year (Twardowska et al. 1988) – a high enough volume to pollute 2300 million m^3 of water (i.e. twice the annual consumption of the whole Katowice province) at a level exceeding quality standards. Many dumps produce highly acid waters (pH < 4).

A combination of hydrological, chemical and geochemical investigations can provide much additional valuable information on concentrations of pollutants, including potentially harmful trace elements such as As, Cd, Cu, Pb and Zn. Such data would facilitate the evaluation of the environmental impact of mining and processing activities on the river, groundwater and soil systems.

To evaluate and control the environmental impact of mining and smelting industries it is recommended that geochemical investigations should be undertaken to (1) estimate the extent of local and regional pollution, (2) quantify the sources of heavy metals, chlorides, sulphates, nitrates, total acid mine drainage, and other contaminants, and (3) help define and control the major sources of pollution, i.e. waste rock heaps, tailings, waste dumps and air pollution.

References

EMEP. 1992. *Calculated budgets for airborne acidifying components in Europe 1985–91.* (European Ministry for Environmental Protection)/MSC, Report, 1/92, August 1992.

GRODZIŃSKA, K., GODZIK, P. & SZAREK, G. 1987. Vegetables and soil contamination by heavy metals in allotment gardens in Cracow agglomeration (S. Poland). *Bulletin of the Polish Academy of Sciences, Biological Sciences,* **35**, 111.

GUS. 1991. *Statistical yearbook. Environmental Protection, 1991.* Warsaw, Main Statistical Office.

GUZIEL, A. 1988. *Ochrona i ksztaltowanie środowiska w rozwoju górnictwa w Polsce. Parts I & II, No. 2.* CPBP (Polish Central Programme of Fundamental Research) 04.10.

HELIOS RYBICKA, E. 1992. Heavy metal partitioning in polluted river and sea sediments: clay minerals effects. *Mineralogia et Petrografia Acta,* **XXXV-A.** 297–305.

—— 1993. Phase-specific bonding of heavy metals in sediments of the Vistula River, Poland. *Applied Geochemistry,* **2**, 45–48.

——, WILSON, M. J. & MCHARDY, W. J. 1994.

Chemical and mineralogical forms and mobilization of copper and lead in soils from Cu-smelting area in Poland. *Journal of Environmental Science and Health*, Part A Environmental Science and Engineering, 531–546.

KABATA-PENDIAS, A. & PENDIAS, H. 1992. *Trace Elements in Soil and Plants*. 2nd ed., CRC Press, Boca Raton, Ann Arbor, London.

KOZLOWSKI, S. 1983. *Przyrodnicze uwarunkowania gospodarki przestrzennej Polski*. Wszechnica Polish Academy of Sciences, **10**.

KUCHARSKI, R., MARCHWIŃSKA, E. & GZYL, J. 1992. Agricultural policy in polluted areas. *Science of the Total Environment*, 61–67.

MACKLIN, M. G. & KLIMEK, K. 1992. Dispersal, storage and transformation of metal-contaminated alluvium in the upper Vistula basin, southwest Poland. *Applied Geography*, **12**, 7–30.

MICHNA, W. 1991. *Ekosystemy żywicielskie i żywność; zagrożenia i problemy ochrony*. [Food-producing ecosystems and food: threats and protection issues]. IOS (Polish Institute of Environmental Protection), Warsaw.

NEY, R. 1988. *Kierunki zagospodarowania zasolonych wód kopalnianych z Górnoslaskiego Zaglebia Weglowego*. Polish Academy of Sciences.

NOWICKI, M. 1993. *Environment in Poland. Issues and Solutions*. Kluwer.

PRZYBYLSKI, T. 1991. *Zagrożenie środowiska przyrodniczego w województwie katowickim*. [Threats to the environment in Katowice province.] Fundacja Silesia, Katowice.

RYMONT, A. 1992. *Environmental Protection in Cracow Region*. Malopolska Poligrafia, Cracow.

TWARDOWSKA, I., SZCZEPAŃSKA, J. & WITCZAK, S. 1988. The effects of coal mine spoils on the water environment. *In:* GODZIK, S. (ed.) *The estimation of environmental contamination, prognosis, prevention*. Prace i Studia, Polish Academy of Science, Wroclaw, **35**.

VERNER, J., RAMSEY, M., HELIOS RYBICKA, E. & JEDRZEJCZYK, B. 1994. Heavy metal contamination of soils in Bukowno, Poland. Abstr. *In: 3rd International Symposium on Environmental Geochemistry. 12–15 Sep. 1994 Cracow, Poland*. 426–427.

WILK, Z., ADAMCZYK, A. & NALECKI, T. 1990. *Impact of mining activities on aquatic environment in Poland*. CPBP 04.10 (Polish Central Programme of Fundamental Research), **27**. Warsaw, Agricultural Academy.

WITCZAK, S. & SZCZEPAŃSKA, J. 1987. *Acidification of infiltration waters as a result of the seepage through Carboniferous hostrocks deposited on the surface*. Proceedings of the International Symposium on Acidification and Water Pathways, Bolkesjo, Norway.

Lacustrine sediment geochemistry as a tool in retrospective environmental impact assessment of mining and urban development in tropical environments: examples from Papua New Guinea

KEITH NICHOLSON

Environment Division, School of Applied Sciences, The Robert Gordon University, Aberdeen AB1 1HG, UK

Abstract: Lake sediments represent an archive on chemical inputs into the environment. Baseline data for natural or background inputs can be obtained by analysis of deep (> 1 m) cores in conjunction with rigorous statistical procedures such as probability plots. Surface sediments illustrate the geographical distribution of the contamination, while shallow cores (< 1 m) permit estimates of sedimentation rate from the onset of contamination discharge. In Papua New Guinea lakes are particularly important as fish stocks which represent a vital source of protein, and any impact which may diminish these stocks is a cause for concern. Two lakes considered are respectively impacted by mining and urban development and show significant enrichments in metals and enhanced sedimentation rates. Surface enrichments in Cu in the mining-impacted lake attain levels of 348% over baseline; while the lake impacted by urban run-off and sewage discharge displays enrichments of 450% for Pb, 500% for Mn and 1100% for P. Sedimentary geochemistry is a potential tool in determining background or 'natural' metal inputs into a lacustrine system. Such data can permit environmental impact evaluations and legislative guidelines to be developed retrospective to development. In developing countries where urban and industrial growth has preceded environmental legislation, this technique may be employed to provide baseline data on which future discharge limitations or environmental controls may be built.

Lake sediments are well known for their ability to record palaeo-environmental changes, and such changes can be reflected in the geochemistry of the sediments (Engstrom & Wright 1984). Sedimentary geochemistry has also been employed on a century-decade timescale to identify industrial pollution impacts (e.g. Salomons & Forstner 1984). The elevated erosion–sediment transport regime present in tropical climates may permit environmental impacts to be recorded in lake sediments on a finer, possibly annual, timescale. As such, lake sedimentary geochemistry could be a valuable tool in determining baseline metal inputs permitting retrospective environmental impact evaluations and legislative guidelines to be developed. This may be of particular importance in developing countries where urban and industrial growth have outstripped environmental controls which may need to be implemented at a later date.

Chambers (1987) catalogued the lakes of Papua New Guinea and reviewed limnological studies to 1984. Many of these investigations discuss physical or biological characteristics of the lakes. Only a few studies on water chemistry are noted, and these commonly deal with levels of nutrients such as phosphate and nitrate. Work on the sedimentary geochemistry of Papua New Guinea lakes published to date is limited to that of Osborne and co-workers (Osborne et al. 1988; Polunin et al. 1988; Osborne & Totome 1992). Lacustrine systems in Papua New Guinea are particularly important as fish stocks represent a vital source of protein and fish forms a large part of the diet, being eaten two–three times daily (Kyle & Ghani 1982). Any impact which may cause a decline or reduction in quality of the fish is therefore of concern.

This study considers examples of mining and urban development in Papua New Guinea, and draws on unpublished data and literature reports to illustrate the potential use of lacustrine sedimentary geochemistry in environmental impact assessment.

Impact of gold–copper mining

The Ok Tedi Au–Cu mine is situated in the west of Papua New Guinea near the Irian Jaya border (Fig. 1). Mining commenced in 1984 using open pit methods, extracting 150 000 t per day of ore and overburden by 1989 (Salomons & Eagle 1990). The watershed of the mine area drains into the Fly River via streams such as the Ok Tedi (Fig. 1) and is one of the wettest regions on

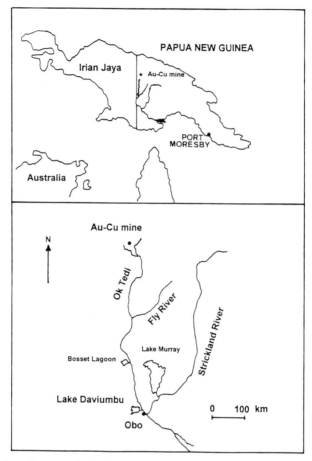

Fig. 1. Location map of Fly River and the Ok Tedi Au–Cu mine, Papua New Guinea.

earth, with annual rainfall ranging from 4.7–11.9 m in the upper catchment. Mining has increased the sediment load of the Fly River by 55%, adding up to 70 million tonnes per year to the total suspended load (Alongi et al. 1992), with copper enrichments at concentrations in excess of 6000 mg kg^{-1} in the fine fractions (Salomons & Eagle 1990). The original environmental study, undertaken before mining commenced, identified increased sedimentation in the Fly River and associated swamps and lateral lakes as a potential impact. However, on the basis of clay mineralogy Salomons & Eagle (1990) determined that a substantial input of sediments from the Fly River into Bosset Lagoon (a lateral lake) could be discounted.

Sampling and analytical methods

To investigate the impact of the mine on the lateral lakes, samples of surface sediments and cores were collected from Lake Daviumbu, a lateral lake situated above the confluence of the Srickland and Fly Rivers and fished by the villagers of Obo (Fig. 1). Surface sediment samples were collected from 32 sites across the lake. Sediment cores were collected from 22 sites by manually forcing an aluminium corer (internal diameter 7.3 cm) into the sediment, a pile driver was used for deep cores. The corer had been previously sawn in half longitudinally and bound with waterproof tape. Cores were sliced into 2 cm sections, which were then stored in polyethylene bags at 4°C for 2–3 days in the field and subsequently frozen. Samples were dried at 100°C and then ashed in a muffle furnace at 550°C for 3 hours to determine organic matter content. Samples for XRF analysis were subsequently ashed at 900°C for 3 hours. A total of 32 surface sediments, 3 river sediments, 9 shallow cores (< 1 m depth) and 13 deep cores (to 4 m depth) were collected. All surface sediments were

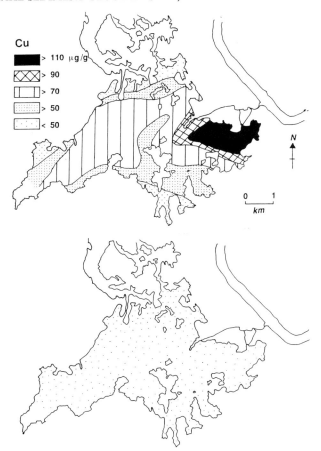

Fig. 2. Lake Daviumbu copper concentrations in (**a**) surface sediments and (**b**) core sediments from c. 1 m depth.

analysed by XRF, samples were also taken from three shallow cores and nine deep cores and similarly analysed. The surface sediments and shallow cores, collected in December 1984, permit the areal extent and depth of contamination to be assessed. The deep cores were taken to determine background or 'natural' levels of copper input into the sediments, and thereby define a baseline from which the level of contamination can be evaluated. Such a baseline level must be defined if an environmental impact assessment is to be made; lake sediments provide the necessary archive to enable threshold concentrations to be determined.

Results

Data for copper concentrations in surface sediments, shallow cores and deep cores are shown in Figs 2, 3 and Table 1 respectively. These data were obtained by XRF analyses of the total sediment sample; no size fraction distribution of the Cu was examined since it was the total Cu content of the sediments which was relevant to the establishment of baseline data and evaluation of the extent of any contamination. The baseline value for background copper was evaluated in three ways: visual inspection of the data with median Cu content determined on non-enriched samples; median Cu content derived from the whole dataset; and probability plots derived from the whole dataset.

Visual inspection of the shallow core profiles (Fig. 3) shows that copper is enriched in the upper 50 cm of the cores. Excluding data for samples in this enriched zone (n = 9) and using the remaining Cu data from both the shallow and deep cores (n = 42; Table 1, Fig. 3) gave a median value of 37.0 mg kg^{-1} Cu. However, taking the full dataset of all surface and core samples (n = 90) gave a median of 42.0 mg kg^{-1} Cu. Finally, a probability plot of all samples (n = 90) was employed. Probability plots have

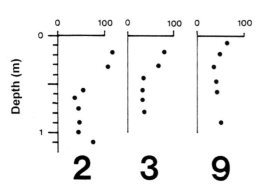

Fig. 3. Copper concentrations in shallow (< 1 m) cores from Lake Daviumbu.

Table 1. *Copper concentrations in deep cores from Lake Daviumbu, Papua New Guinea*

	Depth (m)	Cu (mg/kg)
Core 1	1.2	38
	2.0	44
	2.8	35
	3.6	40
Core 2	0.4	33
	0.8	31
	1.2	39
Core 3	0.6	28
	1.2	36
	2.0	33
	2.8	42
	3.6	41
Core 5	0.6	28
	1.0	29
	1.4	29
Core 7	0.8	31
	1.6	39
	2.4	34
	3.2	41
	4.0	35
Core 10	0.8	45
	1.8	43
	2.6	34
	3.4	24
Core 11	2.2	33
	3.0	23
	3.8	42
Core 12	2.6	29
	3.0	41
	3.4	32
	4.0	41
Core 13	2.0	48

been commonly employed in geochemical exploration (Sinclair 1986) but may equally be used for environmental applications such as this. Using this technique the dataset may be divided into a series of sample populations each with a range of concentrations. In this application it is the upper concentration of the low-Cu population that is required; Sinclair (1986) provides details of the procedure. This method yielded a value of 48.0 mg kg^{-1} Cu as the threshold value for background Cu levels. This latter figure was defined in a more statistically rigorous manner than either of the median-based background levels, and was taken as the baseline concentration for copper. It is also the highest background concentration value determined and therefore provides the most optimistic assessment of any contamination.

Discussion

Using the defined baseline of 48 mg kg^{-1} Cu, 68.6% of the surface sediments show enriched Cu concentrations up to 348% over baseline. In the shallow cores, Cu contents are in excess of baseline values in the upper 25 cm (Core 9) to 60 cm (Core 2), attaining enrichments of up to 244% (Core 2). Since these samples were collected less than one year after mining commenced, the data imply sedimentation rates of up to 60 cm per year, assuming the elevated Cu is all due to mine contamination. Such high sedimentation rates have obvious implications for the preservation of the lake if bed erosion does not occur. Contrary to the conclusion of Salomons & Eagle (1990), these data provide compelling evidence for a significant contribution of sediments from the Fly River to the lateral lakes such as Daviumbu, with concomitant implications for the metal enrichments in the lake sediments, and the potential for loss of the lacustrine environment due to greatly elevated sedimentation rates.

Impact of urban development

Port Moresby (Fig. 1) is the capital of Papua New Guinea and has seen increased urbanization in post-war years. The city lies in the catchment of Waigani Lake, and the geochemistry of the sediments of this lake provides a record of this urban expansion. Since 1965 sewage and urban run-off has been discharged into the lake, which now receives over 80% of the sewage effluent from Port Moresby (Osborne & Polunin 1986). Discharge of sewage and urban run-off into the lake has been cited by Osborne & Polunin (1986) and Osborne & Leach (1983) as a factor involved in a change in the aquatic flora of the lake and the onset of open water conditions from a swamp environment. Polunin et al. (1988) collected a 1 m deep core from the lake close to the sewage outfall. Sediment slices 2 cm thick were analysed by XRF for major and trace element constituents. Although no baseline data were defined, visual inspection of the core chemistry shows near-surface enrichments, commencing at c. 30 cm depth, for both P and Mn. Concentrations of these elements range up to c. 7500 mg kg^{-1} Mn and c. 3 wt% P. These enrichments could be caused by run-off and sewage discharge and/or by redox-based remobilization of these species. However, Zn and Pb are also at higher concentrations in the upper 30 cm zone, to 150 mg kg^{-1} and 90 mg kg^{-1} respectively and it is likely that these enrichments are anthropogenic. These concentrations represent significant increases over those found at the base of the core, with enrichments of 450% for Pb, 500% for Mn and 1100% for P (Polunin et al. 1988). The change in the lake flora was mapped through dated aerial photographs by these authors who were then able to date the core correlating changes in the floral remnants with the recorded physical and biological evolution of the lake. From this dating they estimated sedimentation rates of 2.6 cm per year for the period 1966–1975 and 3.2 cm per year for 1975 to present, an increase which reflects the enhanced sewage discharge. Although less detailed than the mining study, the work by Polunin and Osborne further illustrates the application of lake sediment geochemistry to documenting the impact of recent environmental change.

Conclusions

Lacustrine sedimentary geochemistry can be an effective tool for the retrospective determination of the extent of environmental impact due, for example, to mining and urban growth. In Papua New Guinea lakes represent valuable foodsource ecosystems and are vulnerable to such development through excessive heavy metal and nutrient loadings and excessive water and sediment discharge which can change the nature of the lake environment.

Surface sediments permit the areal or geographical extent of the contamination to be mapped, while shallow cores can be used to provide estimates of sedimentation rate and to illustrate the depth of enrichment. Fundamental to such investigations, however, is a reliable estimate of the natural or background levels of input into a lake. Baseline levels are archived in the lake sediments and can be accessed through analysis of deep cores (> 1 m depth) and the application of statistically rigorous methods such as probability plots to the chemical data. Establishment of baseline data permits the extent of contamination to be evaluated and may form a framework for future environmental monitoring or legislation.

References

ALONGI, D. M., CHRISTOFFERSEN, P., TIRENDI, F. & ROBERTSON, A. I. 1992. The influence of freshwater and material export on sedimentary facies and benthic processes within the Fly Delta and adjacent Gulf of Papua (Papua New Guinea). *Continental Shelf Research*, **12**, 287–326.

CHAMBERS, M. R. 1987. The freshwater lakes of Papua New Guinea: an inventory and limnological review. *Journal of Tropical Ecology*, **3**, 1–23.

ENGSTROM, D. R. & WRIGHT, H. E. 1984. Chemical stratigraphy of lake sediments as a record of environmental change. *In:* HAWORTH, E. Y. & LUND, J. W. G. (eds) *Lake Sediments and Environmental History*. Leicester University Press.

KYLE, J. H. & GHANI, N. 1982. Methylmercury in human hair: A study of a Papua New Guinean population exposed to methylmercury through fish consumption. *Archives of Environmental Health*, **37**, 266–271.

OSBORNE, P. L. & LEACH, G. J. 1983. Changes in the distribution of aquatic plants in a tropical swamp. *Environmental Conservation*, **10**, 323–329.

—— & POLUNIN, N. V. C. 1986. From swamp to lake: recent changes in a lowland tropical swamp. *Journal of Ecology*, **74**, 197–210.

—— & TOTOME, R. 1992. Influences of oligomixis on the water and sediment chemistry of Lake Kutubu, Papua New Guinea. *Archives of Hydrobiology*, **124**, 427–449.

——, POLUNIN, N. V. C. & NICHOLSON, K. 1988. Geochemical traces of riverine influence on a tropical lateral lake. *Verh. Internat. Verein. Limnol.*, **23**, 207–211.

POLUNIN, N. V. C., OSBORNE, P. L. & TOTOME, R. 1988. Environmental archive: Topical urban development reflected in the sediment geochemistry of a flood plain lake. *Archives of Hydrobiology*, **114**, 199–211.

SALOMONS, W. & EAGLE, A. M. 1990. Hydrology, sedimentology and the fate and distribution of copper in mine-related discharges in the Fly River System, Papua New Guinea. *Science of the Total Environment*, **97/98**, 315–334.

—— & FORSTNER, U. 1984. *Metals in the Hydrocycle*. Springer, Berlin.

SINCLAIR, A. J. 1986. Statistical interpretation of soil geochemical data. *In:* FLETCHER, W. K., HOFFMAN, S. J., MEHRTENS, M. B., SINCLAIR, A. J. & THOMSON, I. *Exploration Geochemistry: Design and Interpretation of Soil Surveys*. Society of Economic Geologists, El Paso, Reviews in Economic Geology, Vol. 3, 97–115.

Geochemistry of iodine in relation to iodine deficiency diseases

RON FUGE

Institute of Earth Studies, University of Wales, Aberystwyth SY23 3DB, UK

Abstract: Seawater is the most important reservoir for terrestrial iodine (mean concentration 58 μg l^{-1} I); this is a major influence on iodine distribution in the secondary environment. Volatilization of iodine from the oceans, possibly as elemental iodine or as an organically-bound species, is the main source of the element in the environment. The distribution of iodine in the secondary environment is, therefore, largely controlled by proximity to the oceans, with rainwater and surface run-off relatively enriched in iodine in near-coastal regions. Soil iodine content is also strongly influenced with coastal soils being much enriched and central continental soils being depleted. Topography has a marked influence with soils in rain shadow areas being relatively depleted in iodine. While iodine input is a major controlling influence on its geographical distribution in soils, the soil's ability to retain iodine is also an important factor. Organic matter together with iron and aluminium oxides and clays are the important sinks of soil iodine.

An additional factor in the distribution of iodine in the secondary environment, and its subsequent availability to the biosphere, relates to its speciation in soils. In acid oxidizing conditions iodine is likely to be present as the I$^-$ ion and as such liable to be volatilized as I$_2$. In near-neutral or alkaline conditions iodine is likely to be present as the IO$_3^-$ ion which is not volatilized. Soils in limestone areas, with pH values of 7 and above, have thus been found to be much enriched in iodine compared to acidic soils in neighbouring areas.

It is suggested that iodine from oceanic sources migrates in a series of 'steps' across landmasses by deposition followed by revolatilization. High pH soils and organic-rich soils then act as a migration barrier for iodine. In addition, it is suggested that volatilized iodine is bioavailable and any geochemical barrier to such volatilization deprives the biosphere of a major source. In this context it is interesting to note that several goitre endemias occurred in areas with limestone bedrock.

Iodine was the first element recognized as being essential to human health, with goitre being the first endemic disease found to be related to environmental geochemistry. Despite the early recognition of the importance of iodine to human health the problems of goitre and other iodine deficiency disorders are still a cause of concern. According to some sources (Hetzel 1991), of more than one billion people who are potentially at risk from some form of iodine deficiency disease, some 200 million have goitre and as many as 20 million have brain damage due to iodine deprivation during foetal development and infancy. Gillie (1973) has described endemic goitre as a disease of the poor and Hetzel (1991) points to the major areas of iodine deficiency occurring in remote parts of the world such as the Himalayas, Indonesia, New Guinea, Zaire and parts of China. However, it is of interest to note that even some supposedly affluent areas of Europe are affected to some extent by iodine deficiency problems (Thilly *et al.* 1992; Galvan 1993; Nohr *et al.* 1993).

While it is generally accepted that several dietary factors, including goitrogens, can exert an influence on iodine deficiency disorders, the primary cause is the lack of iodine in the diet (Thilly *et al.* 1992). For this reason there is a need to clarify the general geochemistry of iodine but more particularly to explain its distribution and behaviour in the secondary environment.

Much of the recent interest in the geochemistry of iodine in the secondary environment is related to concern about the fate of radioactive iodine released from nuclear installations (Chamberlain & Dunster 1958; Paquette *et al.* 1986; Rucklidge *et al.* 1994). As a radioactive hazard it also poses a threat to health (Prisyazhiuk *et al.* 1991; Kazakov *et al.* 1992).

The present paper reviews the current state of knowledge regarding iodine geochemistry, concentrating on the secondary environment. Attempts are made to correlate the geochemical behaviour of iodine in the secondary environment with the distribution of iodine deficiency disorders (IDD).

Iodine in the primary environment

Igneous and metamorphic rocks

Iodine with its ionic radius of 220 pm and electronegativity of 2.5 eV is not generally concentrated in primary minerals, having a relatively uniform content in all mineral groups (Fuge & Johnson 1986). There is, however, some suggestion that it is relatively concentrated in the chlorine-containing minerals eudialyte and sodalite where it presumably replaces the chloride ion (Kogarko & Gulyayeva 1965). The low concentration of iodine in the primary minerals is reflected in the low and generally uniform levels in all igneous rock groups (Table 1), there being also no difference between intrusive and extrusive types. The mean value for igneous rocks is 0.24 ppm (Fuge & Johnson 1986).

Similarly, the limited data available suggest that iodine has a uniformly low concentration in the various metamorphic rocks, with concentrations being similar to those in igneous rocks (Fuge & Johnson 1986).

Table 1. *Iodine in igneous rocks*

Rock type	Mean I content (ppm)
Granites	0.25
All other intrusives	0.22
Basalts	0.22
All other extrusives	0.24
Volcanic glasses	0.52

Data mainly based on compilations in Johnson (1980)

Sedimentary rocks

The distribution of iodine in sedimentary rocks is more variable, with higher concentrations in argillaceous than in arenaceous rocks. Some limestones are also relatively enriched in iodine. However, iodine concentrations are only greatly enhanced in organic-rich sediments, with values of over 40 ppm having been recorded. Marine derived sediments are richer in iodine than are those of non-marine origin, with recent marine sediments being particularly enriched. Fuge & Johnson (1986) have suggested a mean value of 2 ppm I in sedimentary rocks which are not bituminous or associated with oilfield brines. They further suggested the following mean values for the main sedimentary rock groups: carbonates, 2.7 ppm; shales, 2.3 ppm; sandstones, 0.8 ppm; with recent sediments having a range of 5–200 ppm.

Iodine in the secondary environment

The hydrosphere and atmosphere

The world's oceans are the most important reservoir of iodine. While there are some differences in the iodine content of the various oceans, together with variations reflecting differing salinity and depth, the mean iodine content of sea water has been estimated as $58\,\mu\text{g}\,\text{l}^{-1}$ (Fuge & Johnson 1986). The thermodynamically stable form of iodine in sea water is the IO_3^- anion (Tsunogai & Sase 1969) which, therefore, constitutes the major occurrence of the element in this medium.

The oceanic reservoir has the largest influence on iodine distribution in the environment, as iodine from this source passes into the atmosphere and thence onto the landmass. The mechanism for iodine release from the sea to the atmosphere has been the subject of some debate. While it is likely that some is released in spray and is subsequently carried as spray droplets and salt particles onto the land surface, this is unlikely to be the major mechanism for its transfer. In the sea the I/Cl ratio is 3×10^{-6} while in rain it has been found to be as high as $> 2 \times 10^{-3}$ (Fuge *et al.* 1987) and in snow $> 5 \times 10^{-4}$ (Heumann *et al.* 1987). In addition, it has been shown that gaseous iodine generally accounts for a large percentage of total atmospheric iodine (Duce *et al.* 1973; Gabler & Heumann 1993). Thus it seems likely that appreciable quantities of the iodine released from the oceans is volatilized.

The form of the volatile iodine and the mechanism for its release from the oceans is also the subject of much debate with Miyake & Tsunogai (1963) suggesting that I_2 is released by the action of UV light on iodide ions in the surface sea waters, the iodide having been derived biologically (Tsunogai & Sase 1969). Garland & Curtis (1981) pointed out that ozone could interact with iodide ions to produce elemental iodine and Heumann *et al.* (1987) suggest that NO_2 could also bring about this reaction.

Other workers (Lovelock *et al.* 1973; Rasmussen *et al.* 1982) have proposed that iodine is volatilized from the oceans as methyl iodide which is produced by microorganisms, while Heumann *et al.* (1990) found CH_3I in sea water and in air in Antarctica. Heumann *et al.* (1987) have suggested that several mechanisms are important in the transfer of iodine from the marine environment to the continent of Antarctica. They propose that short distance transfer of iodine is effected by formation of HI and I_2 and by transfer of sea spray, while for longer distance transport CH_3I is responsible.

More recently, Gabler & Heumann (1993) have found organoiodine to be the most abundant species of atmospheric iodine in

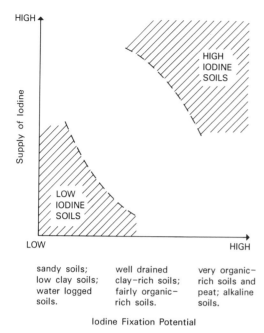

Fig. 1. A simple model for the iodine content of soils based on supply and fixation potential (after Fuge & Johnson 1986).

European samples while HI and I_2 are most abundant in the Antarctic atmosphere.

Iodine is transferred from the atmosphere to the land surface in both wet and dry precipitation, with some workers suggesting the latter as being the more important process (Chamberlain & Chadwick 1966; Sugawara 1967). Whitehead (1984) and Fuge et al. (1987) believe that wet precipitation adds considerably more iodine to the terrestrial environment than does dry precipitation. Whitehead (1984) has estimated that 9.6 g I per year are added to each hectare of soil by dry deposition but this will vary markedly depending on proximity to the oceans. Similarly, the terrestrial iodine input from wet deposition varies considerably depending on the locality. The concentration of iodine in rainwater is generally up to $5 \mu g\,l^{-1}$ (Fuge & Johnson 1986). However, samples of rain from an upland area 12 km from the coast in mid Wales have been found to contain between 5 and $6 \mu g\,l^{-1}$ compared to $2.0 \mu g\,l^{-1}$ in samples 84 km from the sea (Fuge et al. 1987). Iodine occurs in rainwater as I^- and IO_3^- (Jones 1981; Muramatsu & Ohmomo 1988); however, it has been suggested by Luten et al. (1978) that the amount of IO_3^- in rainfall decreases with distance from the coast.

In addition, Muramatsu & Ohmomo (1988) have shown that IO_3^- is converted to I^- on storage of rainwater. It is likely also that some organoiodine occurs in rain (Dean 1963). It is perhaps of note that Heumann et al. (1987) suggested that CH_3I in the atmosphere could be converted to I_2 by photolysis; this would explain the preponderance of I^- and IO_3^- in rainfall.

The iodine content of surface waters is very variable but generally does not exceed $15 \mu g\,l^{-1}$ (Fuge & Johnson 1986). There is considerable regional variation related to proximity of the sea, with surface run-off considerably enriched in iodine in near coastal environments (Fuge 1989), and geology; Konovalov (1959) noted that waters draining marine-derived sediments are relatively enriched in iodine. In addition, various contaminant sources of iodine, such as sewage effluent (Vought et al. 1970; Whitehead 1979), agricultural practices (Fuge 1989) and drainage from disused metalliferous mines (Fuge et al. 1978; Johnson 1980) can add significant amounts of the element to surface waters. While Sugawara & Terada (1957) have suggested that 90% of the iodine in surface waters occurs as the I^- ion, it is likely that in alkaline waters there will be appreciably more IO_3^- as this ion is stable in high pH environments.

Sub-surface waters are generally richer in iodine than surface waters (Fuge 1989).

Soils

The iodine content of soils is very variable but is generally controlled by the quantity of the element supplied coupled with the ability of the soil to fix iodine, this latter property, which is dependent on a large number of factors, being termed the 'Iodine Fixation Potential' by Fuge & Johnson (1986) (Fig. 1).

From reliable literature data for soil iodine it is apparent that there is a very wide range from < 0.1 to > 100 ppm. Given such a broad spread it is difficult to suggest a mean value but some authors have attempted to provide such a figure. Vinogradov (1959) suggests a mean value of 5 ppm and Kabata-Pendias & Pendias (1984) 2.8 ppm, with Fuge & Johnson (1986) suggesting a value of between 4 and 8 ppm. In a major survey of 399 soils of the USA Shacklette & Boerngen (1984) quote a mean value of 1.2 ppm I and a geometric mean of 0.75.

The iodine content of soils is generally considerably higher than their parent materials, with little iodine being added to soils from rocks during weathering (Goldschmidt 1954). The major supplier of soil iodine is the atmosphere, by way of wet and dry precipitation. Therefore,

Table 2. *Iodine in soils*

Soil, source etc. (No. of samples)	Iodine content (ppm)		Source of data
	Range	Mean	
Peats (>70% organic matter), UK (10)	28–98	56	Johnson (1980); Fuge (unpublished data)
Coastal soils NW Norway (13)	5.4–16.6	9.0	Låg & Steinnes (1976)
Inland soils E. Norway (16)	2.8–7.6	4.4	Låg & Steinnes (1976)
Soils, within 25 km of coast, Wales (424)	1.5–149	14.7	Johnson (1980); Al Ajely (1985) Fuge (unpublished data)
Soils, Welsh borders and central England >60 km from coast (28)	1.8–10.5	4.2	Fuge (unpublished data)
Continental soils, Missouri, USA (92)	0.12–5.6	1.26	Fuge (1987)

as might be expected, soils in coastal localities are much enriched in iodine when compared with those from environments more remote from the sea (Table 2; also Fig. 2). The relatively low value for USA soils, as discussed above, possibly reflects the large area of continental versus marine influenced soils. In this context, data for soils from the former Soviet Union, again a major continental area, are relatively low in iodine when compared with strongly marine-influenced areas such as Britain (see Fuge & Johnson 1986).

In addition, in many upland regions where there is enhanced rainfall and consequent greater input of iodine, there is frequently strong enrichment of the element (Fig. 2), though this is also dependent on the ability of the soils in these upland regions to retain iodine (see below and Fig. 1). In contrast, the rain shadow areas beyond the upland regions are generally depleted in iodine. This behaviour of iodine is demonstrated for the mid Wales and Welsh borderland region, where the predominant wind direction is from sea to land (Fig. 2).

Many factors have been implicated in the retention of iodine by soils. Several workers have commented on the enrichment of iodine in organic-rich soils (Vinogradov 1959; Sazonov 1970; Johnson 1980; Sheppard & Thibault 1992; see also Fuge & Johnson 1986), with peaty soils

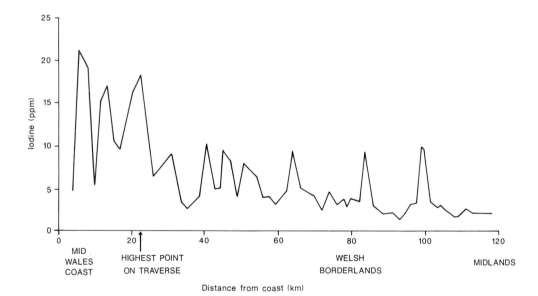

Fig. 2. Iodine in topsoils from a traverse from the west coast of Wales across the Cambrian Mountains, to the Welsh Borderlands and the English Midlands (highest point on traverse c. 350 m OD).

being particularly enriched (see Table 2). Johnson (1980) has shown that top soil (0–20 cm) iodine contents show strong correlation with the quantity of organic matter, while subsoils (> 20 cm) showed no correlation.

While organic matter plays a central role in the retention of soil iodine, Whitehead (1974, 1979, 1984) has suggested that iron oxide/hydroxide and hydrated aluminium oxide are also important for sorption of soil iodine. This author (Whitehead 1974) has demonstrated that sorption of iodine by aluminium and iron oxides is markedly influenced by pH with greater sorption resulting from increasing acidic conditions and no sorption under neutral conditions. It is also of interest to note that iron-rich horizons in soils have been found to concentrate iodine while Heumann et al. (1990) have found that the weathered surfaces of Antarctic iron meteorites are strongly enriched in iodine.

Some workers have pointed to the importance of clay minerals in the retention of soil iodine (Hamid & Warkentin 1967; De et al. 1971; Vinogradov & Lapp 1971). Prister et al. (1977) found that the adsorption of iodine on clays is pH dependent with increasing pH resulting in lower adsorption. Whitehead (1973, 1978, 1984) has suggested that sorption of iodine on soil clays is relatively unimportant when compared with the sorptive capacity of organic matter and the hydrated aluminium and iron oxides.

It is also of note that neutral to alkaline soils over limestones are much richer in iodine when compared with more acid soils occurring in neigbouring areas (Fuge & Long 1989). This would suggest that such neutral to alkaline soils tend to retain iodine (see next section).

From the above discussion, it is apparent that much soil iodine is fixed by soil components, the rate of sorption of soluble iodine added to soils being variable and controlled by several factors (Sheppard & Thibault 1992; Sheppard et al. 1994). It has been shown that, in general, only a small percentage of soil iodine is water soluble (Johnson 1980; Fuge & Johnson 1986). The water soluble component is likely to be I^-, IO_3^-, I_2 and soluble organically-bound iodine. Several authors have suggested that the I^- ion is the most important form of soluble iodine in acidic soils (Whitehead 1984; Muramatsu et al. 1989; Sheppard et al. 1994). However, Yuita (1992) found that while I^- was the major form of soluble iodine in flooded Japanese soils, IO_3^- was the major form in non-flooded soils, suggesting that Eh is an important control on the form of soil iodine. This author, however, further points out that in the non-flooded soils a far greater proportion of the total soil iodine is in a bound form.

Volatilization and iodine geochemistry

Volatilization of iodine in the secondary environment

Volatilization of iodine from the marine hydrosphere and its subsequent transfer to the land surface is the most important stage of the geochemical cycle of iodine. As previously stated this transfer of iodine from the oceans results in near coastal environments being much enriched in iodine while central continental areas contain considerably less. However, the zone of strong iodine enrichment would seem to be fairly narrow, as evidenced in the UK, which has a relatively small landmass and a strong maritime influence resulting from the southwesterly airstream. Soils in the coastal region of mid Wales (within 20 km of the sea) tend to have iodine concentrations within the range 10–25 ppm (Fuge 1987; see also Fig. 2). The values for iodine in soils from the region of the Welsh borderlands and the English Midlands (> 70 km from the sea) are generally below 5 ppm (Fuge unpublished data; also Fig. 2). In addition, in Derbyshire, northern England, c. 80–100 km from the coast in the direction of the prevailing wind, over half of the topsoil samples analysed (28) contained < 4 ppm iodine (Fuge & Long 1989). While the iodine concentrations in the soils of these regions of England and Wales are greater than might be expected in central continental areas such as the USA. (mean = 1.2 ppm, Shacklette & Boerngen 1984; mean = 1.26 ppm, Fuge 1987), they are not greatly so and certainly indicate that a marked depletion of soil iodine content can occur over relatively short distances from the coast.

It is apparent then, that while gaseous iodine is found in continental atmosphere (Gabler & Heumann 1993), a relatively low percentage of the directly marine-derived iodine reaches central continental areas. This has led Cohen (1985) to suggest that the iodine found in continental soils is derived from weathering of the lithosphere. While in some continental areas, geologically derived sources of iodine may be important (Låg & Steinnes 1976) and anthropogenic sources such as combustion of fossil fuels (Schroll & Krachsberger 1970; Vought et al. 1970) and sulphide ore smelting (Fuge et al. 1988) may have some localized influence, the major source is still likely to be marine. However, the initially marine-derived iodine

Table 3. *Iodine in topsoils, Derbyshire, UK*

Bedrock (No. of samples)	Iodine content (ppm)		pH
	Range	Mean	
Limestone (12)	2.58–26.0	8.2	6.0–6.9
Dolomite ⎫ Sandstone ⎬ (15) Shale ⎭	1.88–8.53	3.44	4.2–5.5

After Fuge (1990).

may have experienced revolatilization from soils during its transport from the marine environment.

The possible volatilization of iodine from soils was first suggested by Fellenberg *et al.* (1924) who suggested that I_2 would be lost under acidic oxidizing conditions, while McClendon (1939) was of the opinion that iodine added to soils as KI was lost following the conversion of the iodide ion to I_2. However, Whitehead (1981) demonstrated that loss of iodine by volatilization from KI added to surface soils was negligible apart from an acid sandy podsol. Fuge (1990) has proposed that volatilization of iodine from soils plays an important role in iodine's secondary environment geochemistry.

Reference to the Eh–pH diagram for iodine (Fig. 3) suggests that in an oxidizing acid environment iodide is readily converted to I_2. This elemental iodine, being gaseous, would be easily lost. In addition, Perel'man (1977) further suggests that the Fe^{3+} and Mn^{4+} cations would also play a part in the oxidation of the iodide ion in both acid and alkaline media. From the Eh–pH diagram (Fig. 3) it is also apparent that under alkaline conditions the iodate anion is the stable form of iodine.

Fuge (1990) is of the opinion that Eh–pH conditions control the form of iodine in the soil and this in turn exerts an important influence on its redistribution in the secondary environment. Thus, in acidic oxidizing conditions any unbound iodine in the soil is present as the iodide ion and is therefore likely to be converted to volatile I_2. Conversely, under alkaline conditions where the iodate ion is the stable form, iodine is immobilized. This phenomenon of iodine concentration under alkaline conditions has long been known and is evidenced in the iodate-rich caliche deposits of the Atacama desert and in alkaline lakes (Perel'man 1977). It has been demonstrated that neutral to alkaline soils occurring over limestones are enriched in iodine compared to neighbouring acidic soils occurring over other lithologies (Fuge 1990; see also Table 3).

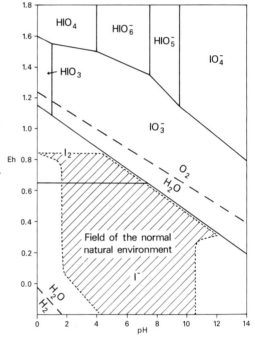

Fig. 3. Eh–pH diagram for iodine (after Vinogradov & Lapp 1971).

Similarly, in acidic reducing media, the iodide ion is stable and is, therefore, not lost by conversion to volatile I_2. This phenomenon, it has been suggested (Fuge 1990), is responsible for the enriched iodine concentrations found in soils over sulphide mineral deposits (Andrews *et al.* 1984; Fuge *et al.* 1986, 1988).

Volatilization and the iodine cycle in soils

Iodine added to soils from precipitation is in both the I^- and IO_3^- forms. Conditions in the soil will determine the fate of these ions. They will be sorbed in part by the organic matter, iron and aluminium oxides and clays, the degree of sorption being controlled by several factors such

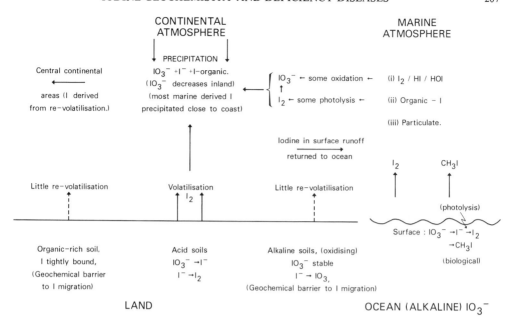

Fig. 4. A simplified model showing part of the iodine cycle involving transport from the marine to the terrestrial environment.

as pH (Sheppard *et al.* 1994). Some of the added iodine will remain in a mobile form, the quantity remaining in this form being dependent on the number of binding sites available in the soil. In acid soils with oxidizing conditions the IO_3^- will be converted to I^- and all of the unbound I^- will be available to be converted to volatile I_2, while in alkaline soils the added I^- will be converted to IO_3^- and no volatilization will occur.

Thus the degree of volatilization will, as suggested by Whitehead (1981), be controlled in the first instance by the degree of binding exerted in the soil, so governing the fixation of iodine. However, the secondary controlling influence will be the Eh–pH regime of the soil.

In addition to volatilization to I_2 in soil it has been suggested that some iodine in soils is present as CH_3I (Reiniger, 1977) and that as such this compound could also be volatilized and added to the atmospheric iodine flux (Sheppard & Thibault 1991).

It is likely then that directly marine-derived iodine is mainly precipitated in the near coastal environment and most of the iodine reaching central continental areas is that revolatilized from soils. This revolatilization is envisaged as being a series of stepwise movements across the terrestrial environment (see Fig. 4).

If volatilization from soils does constitute the major source of iodine in the continental interior then the stepwise movement across the terrestrial landscape could be strongly influenced by geochemical barriers. Thus if an organic-rich soil is encountered then iodine becomes tightly bound in the soil and is not available for further progress. Alkaline soils would act in similar fashion so causing iodine to be fixed in the soil and therefore unavailable for revolatilization. Thus geochemical barriers to iodine migration could have a marked effect on the secondary environmental geochemistry of iodine (Fig. 4).

Iodine geochemistry in relation to IDD

Some aspects of the aetiology of IDD

It has long been recognized that severe goitre and cretinism endemias generally occur in high mountain ranges, central continental regions and rainshadow areas (Kelly & Sneddon 1960). The geographical distribution of these diseases correlates well with the behaviour of iodine in the secondary environment, as outlined above. However, it has been pointed out by several workers that there are anomalies within this general distribution. Some of the anomalies have been related to goitrogenic substances in the diets of those affected. Gaitan (1973) and coworkers (Gaitan *et al.* 1969, 1972, 1986, 1993) have invoked geological sources of some dietary goitrogens, such as sulphur containing hydrocarbons. Other goitrogens occur in foodstuffs including some vegetables of the genus *Brassica* (family *Cruciferae*), (Greer & Astwood 1948: Greer 1950) and staple diet items like cassava (Gaitan 1980).

Other elements have also been suggested as being involved in the aetiology of IDD. These include fluorine (Wilson 1941), arsenic (Scott 1938), zinc (Koutras 1980), magnesium (Day & Powell-Jackson 1972) and manganese and cobalt (Blokhina 1970). More recently selenium has been strongly implicated in some IDD problems (Vanderpas et al. 1990; Thilly et al. 1992).

Of the other elements which have been implicated in IDD calcium would seem to be of most interest. The possible importance of this element in the aetiology of endemic goitre has been explored by McCarrison (1926) and Turner (1954), with the latter suggesting calcium to be a goitrogen. However, experimental work by Harrison et al. (1967) has failed to prove the goitrogenic nature of calcium in humans. The possible importance of calcium in goitre aetiology stems from suggestions that consumption of hard water could be a factor in the occurrence of the disease (Murray et al. 1948; Day & Powell-Jackson 1972). It has also been shown that areas underlain by limestone are prone to endemic goitre (Boussingault 1831; Turner 1958; Carter et al. 1959; Perel'man 1977).

Geology and IDD

Many workers have commented on possible geological influences on IDD. Some, as mentioned above, have pointed to the geological sources of some goitrogens (Gaitan 1980; Stewart 1990), while others have suggested that variable soil types may influence uptake of goitrogens (Thilly et al. 1972). However, it is apparent that geological influences may also control the movement of iodine in the secondary environment and thereby its availability to the biosphere.

The example of England and Wales, which exhibits a large range of geology over a comparatively small area, is of particular interest. Thus some areas of northern England were reported by Inglis (1832) to be as badly affected by endemic goitre as were the Alps. The areas affected in the Yorkshire Dales and the Derbyshire Peak District were essentially underlain by limestone. In the case of the Debyshire Peak District, where the disease was known as 'Derbyshire Neck', it has been found that iodine concentrations in soils over carboniferous limestone are considerably richer than are soils over other rock types in adjacent areas (Fuge & Long 1989; see also Table 3). Indeed, soils from the area of the goitre endemia are not significantly depleted in iodine when compared to British soils in non-goitrous areas. This is also true of the area of north Oxfordshire, around the village of Hook Norton where goitre was particularly prevalent (Kelly & Snedden 1960) and where iodine deficiency in schoolchildren was still being recorded in the late 1950s (Hughes et al. 1959). Soils over Jurassic limestones here also contain appreciable quantities of iodine (5–10 ppm).

In south Wales also the major goitre endemia occurred in areas underlain by limestone bedrock (Kelly & Snedden 1960).

In southwest England Kelly & Snedden (1960) indicated a fairly extensive area of endemic goitre. This area dominated by granitic bedrock, is predominantly moorland with quite extensive areas of peat development. Limited data for this area suggest iodine contents of 10–20 ppm, which does not suggest an area of iodine depletion.

The occurrence of IDD in areas of relatively iodine enriched soils suggests that the problem may be one of non-bioavailability rather than depletion. Thus rather than calcium being a goitrogen, its presence in soils could reflect relatively high pH soils. It is possible that the high calcium content, or high pH of the soils decreases the uptake of iodine by plants (Katalymov & Churbanov 1960; Katalymov & Shirshov 1960).

Another possible explanation for the non-bioavailability of iodine could stem from its method of uptake by plants. It has been shown that there is little correlation between the iodine content of plants and the soils in which they grow (Al-Ajely 1985). It seems likely that absorption of iodine by leaves is an important mechanism for iodine uptake by plants (Chamberlain & Chadwick 1953). Indeed, this is possibly the most important source (Shacklette & Cuthbert 1967; Whitehead 1984). Additionally, airborne iodine may be an important source for humans by inhalation.

Therefore, if iodine is strongly bound in soils it is not available for volatilization and an important source of the element in the biosphere will be lost. Geochemical barriers to iodine migration could therefore be a major influence in the aetiology of IDD. In an area where iodine is in relatively short supply the non-bioavailability of some of the soil iodine could then 'tip the balance' in favour of iodine deficiency.

Conclusions

Iodine in the secondary environment is essentially derived from the oceans by volatilization. However, this oceanic iodine is likely to be revolatilized from soils several times in its progress across the terrestrial environment in a 'step-wise' fashion. Any barrier to such migration is likely to result in relative depletion of bioavailable iodine and is likely to result in problems of IDD, particularly where iodine supply is marginal or goitrogens interfere with its metabolism.

References

AL-AJELY, K. O. 1985. *Biogeochemical prospecting as an effective tool in the search for mineral deposits.* PhD Thesis, University of Wales, Aberystwyth.

ANDREWS, M. J., BIBBY, J. M., FUGE, R. & JOHNSON, C. C. 1984. The distribution of iodine and chlorine in soils over lead-zinc mineralisation, east of Glogfawr, mid Wales. *Journal of Geochemical Exploration*, **20**, 19–32.

BLOKHINA, R. I. 1970. Geochemical ecology of endemic goitre. *In:* MILLS, C. F. (ed.) *Trace Element Metabolism in Animals*. Livingstone, Edinburgh, 426–432.

BOUSSINGAULT, J. B. 1831. Recherches sur la cause qui produit le Goitre dans les Cordileres de la Novelle-Grenade. *Annal Chimie (Phys)*, **48**, 41–69.

CARTER, C. W., COXON, R. V., PARSONS, D. S. & THOMPSON, R. H. S. 1959. *Biochemistry in Relation to Medicine*. 3rd edition, Longman, London, 165–166.

CHAMBERLAIN, A. C. & CHADWICK, R. C. 1953. Deposition of airborne radio-iodine vapour. *Nucleonics*, **11**, 22–30.

—— & —— 1966. Transport of iodine from atmosphere to ground. *Tellus*, **18**, 226–237.

—— & DUNSTER, H. J. 1958. Deposition of radioactivity in N.W. England from the accident at Windscale. *Nature*, **182**, 629–630.

COHEN, B. L. 1985. The origin of I in soil and the ^{129}I problem. *Health Physics*, **49**, 279–285.

DAY, T. K. & POWELL-JACKSON 1972. Fluoride, water hardness and endemic goitre. *The Lancet*, 1135–1138.

DE, S. K., RAO, S. S., TRIPATHI, C. M. & RAI, C. 1971. Retention of iodide by soil clays. *Indian Journal of Agricultural Chemistry*, **4**, 43–49.

DEAN, G. A. 1963. The iodine content of some New Zealand drinking waters with a note on the contribution of sea spray to the iodine in rain. *New Zealand Journal of Science*, **6**, 208–214.

DUCE, R. A., ZOLLER, W. H. & MOYERS, J. L. 1973. Particulate and gaseous halogens in the Antarctic atmosphere. *Journal of Geophysical Research*, **78**, 7802–7811.

FELLENBERG, T. VON, GEILINGER, H. & SCHWEIZER, K. 1924. Untersuchungen uber das Vorkommen von Jod in der Natur, 8. Uber das Freiwerden Element Jods Erde. *Biochemische Zeitschrift*, **152**, 172–184.

FUGE, R. 1987. Iodine in the environment: Its distribution and relationship to human health. *In:* HEMPHILL, D. D. (ed.) *Trace substances in Environmental Health-XXI*. University of Missouri, Columbia, 74–87.

—— 1989. Iodine in waters: possible links with endemic goitre. *Applied Geochemistry*, **4**, 203–208.

—— 1990. The role of volatility in the distribution of iodine in the secondary environment. *Applied Geochemistry*, **5**, 357–360.

—— & JOHNSON, C. C. 1986. The geochemistry of iodine – a review. *Environmental Geochemistry and Health*, **8**, 31–54.

—— & LONG, A. M. 1989. Iodine in the soils of North Derbyshire. *Environmental Geochemistry and Health*, **11**, 25–29.

——, ANDREWS, M. J. & JOHNSON, C. C. 1986. Chlorine and iodine, potential pathfinder elements in exploration geochemistry. *Applied Geochemistry*, **1**, 111–116.

——, JOHNSON, C. C. & PHILLIPS, W. J. 1978. An automated method for the determination of iodine in geochemical samples. *Chemical Geology*, **23**, 255–265.

——, LAIDLAW, I. M. S., ANDREWS, M. J. & JOHNSON, C. C. 1987. Aspects of the atmospheric contribution of iodine to the environment. *In:* HEMPHILL, D. D. (ed.) *Trace Substances in Environmental Health-XXI*. University of Missouri, Columbia, 113–120.

——, ANDREWS, M. J., CLEVENGER, T. E., DAVIES, B. E., GALE, N. L., PAVELEY, C. F. & WIXSON, B. G. 1988. The distribution of chlorine and iodine in soil in the vicinity of lead mining and smelting operations, Bixby area, S.E. Missouri, U.S.A. *Applied Geochemistry*, **3**, 517–521.

GABLER, H.-E. & HEUMANN, K. G. 1993. Determination of iodine species using a system of specifically prepared filters and IDMS. *Fresnius Journal of Analytical Chemistry*, **345**, 53–59.

GAITAN, E. 1973. Water-borne goitrogens and their role in the etiology of endemic goiter. *In: World Review of Nutrition and Dietetics*, vol 17. Karger, Basle, 53–90.

—— 1980. Goitrogens in the etiology of endemic goiter. *In:* STANBURY, J. B. & HETZEL, B. S. (eds) *Endemic Goiter and Cretinism Iodine Nutrition in Health and Disease*. Wiley, New York, 219–236.

——, COOKSEY, R. C. & LINDSEY, R. H. 1986. Factors other than iodine deficiency in endemic goiter: goitrogens and protein calorie malnutrition (PMC). *In:* DUNN, J. T., PRETELL, E. A., DAZA, C. H. & VITERI, F. E. (eds) *Towards the Eradication of Endemic Goiter, Cretinism, and Iodine Deficiency*. Pan American Health Organisation, Washington, DC, 28–45.

——, ISLAND, D. P. & LIDDLE, G. W. 1969. Identification of a naturally occurring goitrogen in water. *Transactions of the American Association of Physicists*, **42**, 141–152.

——, MACLENNAN, R., ISLAND, D. P. & LIDDLE, G. W. 1972. Identification of water-borne goitrogens in the Cauca Valley of Columbia. *In:* HEMPILL, D. D. (ed.) *Trace Substances in Environmental Health - V.* University of Missouri, Columbia, 55–66.

——, COOKSEY, R. C., LEGAN, J., CRUSE, J. M., LINDSAY, R. H. & HILL, J. 1993. Antithyroid and goitrogenic effects of coal-water extracts from iodine-sufficient areas. *Thyroid,* **3,** 49–53.

GALVAN, G. 1993. Iodine nutrition and goiter in Austria. *Acta Medica Austriaca,* **20,** 3–5.

GARLAND, J. A. & CURTIS, H. 1981. Emission of iodine from the sea surface in the presence of ozone. *Journal of Geophysical Research,* **86,** 3183–3186.

GILLIE, R. B. 1973. Endemic goitre. *In: Readings From Scientific American, Food.* Freeman, San Francisco, 63–71.

GOLDSCHMIDT, V. M. 1954. *Geochemistry.* Oxford University Press, London.

GREER, M. A. 1950. Nutrition and goiter. *Physiological Reviews,* **30,** 513–548.

—— & ASTWOOD, E. B. 1948. The antithyroid effect of certain foods in man determined with radioactive iodine. *Endocrinology,* **43,** 105–119.

HAMID, A. & WARKENTIN, B. P. 1967. Retention of I-131 used as a tracer in water-movement studies. *Soil Science,* **104,** 279–282.

HARRISON, M. T., HARDEN, R. McG. & ALEXANDER, W. D. 1967. Effect of calcium on iodine metabolism in man. *Metabolism,* **16,** 84–86.

HETZEL, B. S. 1991. The conquest of iodine deficiency: A special challenge to Australia from Asia. *In: Proceedings of the Nutrition Society of Australia,* **16,** 69–78.

HEUMANN, K. G., GALL, M. & WEISS, H. 1987. Geochemical investigations to explain iodine-overabundances in Antarctic meteorites. *Geochimica et Cosmochimica Acta,* **51,** 2541–2547.

——, NEUBAUER, J. & REIFENHAUSER, W. 1990. Iodine overabundances measured in the surface layers of an Antarctic stony and iron meteorite. *Geochimica et Cosmochimica Acta,* **54,** 2503–2506.

HUGHES, D. E., RODGERS, K. & WILSON, D. C. 1959. Thyroid enlargement in schoolchildren of north Oxfordshire. *British Medical Journal,* 280–281.

INGLIS, J. 1838. *Treatise on English Bronchocele with a few remarks on the use of iodine and its compounds.* London.

JOHNSON, C. C. 1980. *The geochemistry of iodine and a preliminary investigation into its potential use as a pathfinder element in geochemical exploration.* PhD Thesis, University of Wales, Aberystwyth.

JONES, S. D. 1981. *Studies on the speciation of iodine in rain and freshwaters.* PhD Thesis, University of Wales, Bangor.

KABATA-PENDIAS, A. & PENDIAS, H. 1984. *Trace Elements in Soils and Plants.* CRC, Boca Raton.

KATALYMOV, M. V. & CHURBANOV, V. M. 1960. (Effect of liming of the soil and use of chlorine-containing fertilisers on entry of iodine into plants.) *Doklady Akademii Nauk SSSR,* **131,** 1191–1193. [In Russian.]

—— & SHIRSHOV, A. A. 1960. (Iodine contents of plants and dependence on species characteristics and soil properties.) *Trudy Izvestkovaniya Derno-Podzolistykh Pochvovedenie,* 267–271. [In Russian.]

KAZAKOV, V. S., DEMIDCHIK, E. P. & ASTAKHOVA, L. N. 1992. Thyroid cancer after Chernobyl. *Nature,* **359,** 21.

KELLY, F. C. & SNEDDEN, F. W. 1960. Prevalence and geographical distribution of endemic goitre. *In: Endemic Goitre.* World Health Organization, Geneva, 27–233.

KOGARKO, L. N. & GULYAYEVA, L. A. 1965. Geochemistry of the halogens in the alkalic rocks of the Lovozero Massif (Kola Peninsula). *Geochemistry International,* 1965, 729–740.

KONOVALOV, G. S. 1959. (Removal of trace elements by rivers of the USSR.) *Doklady Akademia Nauk SSSR,* **129,** 912–915. [In Russian.]

KOUTRAS, D. A. 1980. Trace elements, genetic and other factors. *In:* STANBURY, J. B. & HETZEL, B. S. (eds) *Endemic Goiter and Endemic Cretinism Iodine Nutrition in Health and Disease.* Wiley, New York, 255–268.

LÅG, J. & STEINNES, E. 1976. Regional distribution of halogens in Norwegian forest soils. *Geoderma,* **16,** 317–325.

LOVELOCK, J. E., MAGGS, R. J. & WADE, R. J. 1973. Halogenated hydrocarbons in and over the Atlantic. *Nature,* **241,** 194–196.

LUTEN, J. B., WOITTIEZ, J. R. W., DAS, H. A. & DE LIGNY, C. L. 1978. The determination of iodate in rain-water. *Journal of Radioanalytical Chemistry,* **43,** 175–185.

MCCARRISON, R. 1926. Effects of excessive ingestion of lime, on thyroid gland and influence of iodine in counteracting them. *Indian Journal of Medical Research,* **13,** 817–821.

MCCLENDON, J. F. 1939. *Iodine and the Incidence of Goitre.* Oxford University Press.

MIYAKE, Y. & TSUNOGAI, S. 1963. Evaporation of iodine from the ocean. *Journal of Geophysical Research,* **68,** 3989–3993.

MURAMATSU, Y. & OHMOMO, Y. 1988. Tracer experiments for the determination of the chemical forms of radio-iodine in water samples. *Journal of Radioanalytical and Nuclear Chemistry,* **124,** 123–134.

——, UCHIDA, S., SUMIYA, M., OHMOMO, Y. & OBATA, H. 1989. Tracer experiments on transfer of radio-iodine in the soil-rice plant system. *Water, Air and Soil Pollution,* **45,** 157–171.

MURRAY, M. M., RYLE, J. A., SIMPSON, B. M. & WILSON, D. C. 1948. *Thyroid enlargement and other changes related to the mineral content of drinking water (with a note on goitre prophylaxis).* Medical Research Council, Memorandum, **18,** HMSO, London.

NOHR, S. B., LAURBERG, P., BORLUM, K. G., PEDERSEN, K. M., JOHANNESEN, P. L., DAMM, P., FUGLSANG, E. & JOHANSEN, A. 1993. Iodine deficiency in pregnancy in Denmark – regional variations and frequency of individual iodine supplementation. *Acta Obstetrica et Gynecologica Scandinavica,* **72,** 350–353.

PAQUETTE, J., WREN, D. J. & FORD, B. L. 1986. Iodine chemistry. *In:* TOTH, L. M. (ed.) *The Three Mile Island Accident: Diagnosis and Prognosis.* American Chemical Society, Symposium Series, **293**, 193–210.

PEREL'MAN, A. J. 1977. *Geochemistry of Elements in the Supergene Zone*. Keterpress Enterprises, Jerusalem.

PRISTER, B. S., GRIGOR'EVA, T. A., PEREVEZENTSEV, V. M., TIKHOMIROV, F. A., SAL'NIKOV, V. G., TERNOVSKAYA, I. M. & KARABIN, T. 1977. (Behaviour of iodine in soils). *Pochvovedenie*, **6**, 32–40. [In Russian.]

PRISYAZHIUK, A., PJATAK, O. A., BUZANOV, V. A., REEVES, O. K. & BERAL, V. 1991. Cancer in the Ukraine, post Chernobyl. *Lancet*, **338**, 1334–1335.

RASMUSSEN, R. A., KHALIL, M. A. K., GUNAWARDENA, R. & HOYT, S. D. 1982. Atmospheric methyl iodide. *Journal of Geophysical Research*, **87**, 3086–3090.

REINIGER, P. 1977. Transfer of iodine in terrestrial ecosystems. In: *Iodine-129, Proceedings of NEA Specialist Meeting*. OECD, Paris, 117.

RUCKLIDGE, J., KILIUS, L. & FUGE, R. 1994. ^{129}I in moss down-wind from the Sellafield nuclear fuel reprocessing plant. *Nuclear Instruments and Methods in Physics Research Section B*, **92**, 417–420.

SAZONOV, N. N. 1970. (Contents and migration of iodine in central Yakutia soils.) *Referatwnyi Zhurnal Geologiia*, V 1970, Abstract No. IV36. [In Russian.]

SCHROLL, E. & KRACHSBERGER, H. 1970. Geochemistry of impurities in atmospheric precipitation in the Vienna metropolitan area. *Radex Rundsch*, 334–340.

SCOTT, M. 1938. The possible role of arsenic in the etiology of goiter, cretinism and endemic deaf-mutism. *Transactions of the Third International Goitre Conference and American Association for Study of Goitre*, 34–49.

SHACKLETTE, H. J. & BOERNGEN, J. G. 1984. *Element concentrations in soils and other surficial materials of the conterminous United States*. U.S. Geological Survey Professional Paper, **1270**.

—— & CUTHBERT, M. E. 1967. Iodine content of plant groups as influenced by variation in rock and soil types. In: *US Geological Survey Special Paper*, **90**, 30–46.

SHEPPARD, M. I. & THIBAULT, D. H. 1991. A four-year mobility study of selected trace elements and heavy metals. *Journal of Environmental Quality*, **20**, 101–114.

—— & —— 1992. Chemical behaviour of iodine in organic and mineral soils. *Applied Geochemistry*, **7**, 265–272.

——, ——, MCMURRY, J. & SMITH, P. A. 1994. Factors affecting the soil sorption of iodine. *Water, Air and Soil Pollution*, in press.

STEWART, A. G. 1990. For debate: Drifting continents and endemic goitre in northern Pakistan. *British Medical Journal*, **300**, 1507–1512.

SUGAWARA, K. 1967. Migration of elements through phases of the hydrosphere and atmosphere. In: VINOGRADOV, A. P. (ed.) *Chemistry of the Earth's Crust, 2. Program for Scientific Translations*, Jerusalem, 501–510.

—— & TERADA, K. 1957. Iodine distribution in the western Pacific Ocean. *The Journal of Earth Sciences, Nagoya University*, **5**, 81–102.

TAYLOR, S. 1954. Calcium as a goitrogen. *Journal of Clinical Endocrinology*, **14**, 1412–1422.

THILLY, C. H., DELANGE, F. & ERMANS, A. M. 1972. Further investigations of iodine deficiency in the etiology of endemic goiter. *The American Journal of Clinical Nutrition*, **25**, 30–40.

——, VANDERPAS, J. B., BEBE, N., NTAMBUE, K., CONTEMPRE, B., ET AL. 1992. Iodine deficiency, other trace-elements, and goitrogenic factors in the etiopathogeny of iodine deficiency disorders (IDD). *Biological Trace Element Research*, **32**, 229–243.

TSUNOGAI, S. & SASE, T. 1969. Formation of iodide-iodine in the ocean. *Deep Sea Research*, **16**, 489–496.

—— 1958. The thyroid nodule. *Lancet*, 751–754.

VANDERPAS, J. B., CONTEMPRE, B., DUALE, N. L., GOOSSENS, W., BEBE, N. G. O., ET AL. 1990. Iodine and selenium deficiency associated with cretinism in northern Zaire. *American Journal of Clinical Nutrition*, **52**, 1087–1093.

VINOGRADOV, A. P. 1959. *The Geochemistry of Rare and Dispersed Elements in Soils*. 2nd edition. Consultants Bureau, New York.

—— & LAPP, M. A. 1971. (Use of iodine haloes to search for concealed mineralisation.) *Vestnik-Leningradskii Universitat Seriia Geologii i Geografii*, No 24, 70–76. [In Russian.]

VOUGHT, R. L., BROWN, F. A. & LONDON, W. T. 1970. Iodine in the environment. *Archives of Environmental Health*, **20**, 516–522.

WHITEHEAD, D. C. 1973. Studies on iodine in British soils. *Journal of Soil Science*, **24**, 260–270.

—— 1974. The influence of organic matter, chalk and sesquioxides on the solubility of iodide, elemental iodine and iodate incubated with soil. *Journal of Soil Science*, **25**, 461–470.

—— 1978. Iodine in soil profiles in relation to iron and aluminium oxides and organic matter. *Journal of Soil Science*, **29**, 88–94.

—— 1979. Iodine in the U.K. environment with particular reference to agriculture. *Journal of Applied Ecology*, **16**, 269–279.

—— 1981. The volatilisation, from soils and mixtures of soil components, of iodine added as potassium iodide. *Journal of Soil Science*, **32**, 97–102.

—— 1984. The distribution and transformation of iodine in the environment. *Environment International*, **10**, 321–339.

WILSON, D. C. 1941. Fluorine in the aetiology of endemic goitre. *Lancet*, 211–212.

YUITA, K. 1992. Dynamics of iodine, bromine, and chlorine in soil II. Chemical forms of iodine in soil solutions. *Soil Science and Plant Nutrition*, **38**, 281–287.

Iodine in the environment and endemic goitre in Sri Lanka

C. B. DISSANAYAKE[1] & R. L. R. CHANDRAJITH[2]

[1] *Institute of Fundamental Studies, Hantana Road, Kandy, Sri Lanka*
[2] *Department of Geology, University of Peradeniya, Peradeniya, Sri Lanka*

Abstract: The prevalence of endemic goitre is extremely high in certain parts of Sri Lanka where rates as high as 44% have been observed. With nearly 10 million people at risk the aetiology of endemic goitre in Sri Lanka needs to be clearly ascertained. The endemic goitre belt of Sri Lanka coincides with the wet climatic zone, indicating an apparent relationship of iodine geochemistry with climatic factors. However, the fact that there is a correlation of only −0.64 between the iodine content and the prevalence of goitre suggests the existence of other factors such as humic substances in the soil, clay minerals and soil pH that exert an influence on the bioavailability of iodine.

Iodization of salt used in food is the most convenient method of goitre prevention.

It has been estimated that out of a population of 17 million in Sri Lanka, nearly 10 million people are at risk of goitre (Fernando *et al.* 1987). The association between iodine and goitre was one of the earliest demonstrated links between a trace element in the environment and human health and nutrition. However, there is still very little information on the geochemistry of iodine and its effect on the prevalence of endemic goitre in Sri Lanka. The few studies carried out (Mahadeva & Shanmuganathan 1967; Weerasekera *et al.* 1985; Chandrajith 1987; Fernando *et al.* 1987, 1989; Balasuriya *et al.* 1992; Dissanayake & Chandrajith 1993) have shown that the endemic goitre belt of Sri Lanka lies in the wet climatic zone and that the climate has a marked influence on the geochemical distribution of iodine in the environment.

Geochemistry of iodine

Iodine with a large univalent anion does not usually enter crystal lattices of minerals on account of its large size. Ionic substitution of iodine is therefore unlikely and its geochemistry is mainly governed by the surface chemistry of geological materials. Fuge & Johnson (1986) in a review of the geochemistry of iodine have noted that iodine is particularly concentrated in the biosphere with strong affinities for organic matter. The easy transfer of iodine to the atmosphere and the mechanical and chemical removal from soil by groundwater tend to deplete soil of iodine. The geochemical cycle of iodine in the soil is critically important in the geographical distribution of goitre.

Levels of iodine in soil and water in Sri Lanka

Due to errors in analytical procedures, accurate data on iodine in soils and water are lacking. The earlier surveys used tedious colorimetric methods. Balasuriya *et al.* (1992) studied 609 samples of drinking water collected from scattered sources of the eight districts of Kandy, Matale, Kalutara, Anuradhapura, Polonnaruwa, Colombo, Puttalam and Gampaha using the Orion electrode method of analysis. Figure 1 illustrates the locations of these districts and the endemic goitre region of Sri Lanka as known at present. Table 1 shows the data obtained by them for iodine in water. It is seen that the Anuradhapura and Polonnaruwa Districts had higher water iodine levels while the Gampaha and Kalutara Districts had the lowest levels. It was also noted that tube wells had in general more iodine than dug wells.

Table 1. *Mean and median concentrations of water iodide by district (after Balasuriya* et al. *1992)*

District	Mean $\mu g\, l^{-1}$	Median $\mu g\, l^{-1}$
Kandy	30.96	19.1
Matale	16.91	11.1
Kalutara	15.50	12.2
Anuradhapura	119.03	101.6
Polonnaruwa	47.21	33.0
Colombo	16.86	11.4
Puttalam	34.68	21.6
Gampaha	11.90	5.0

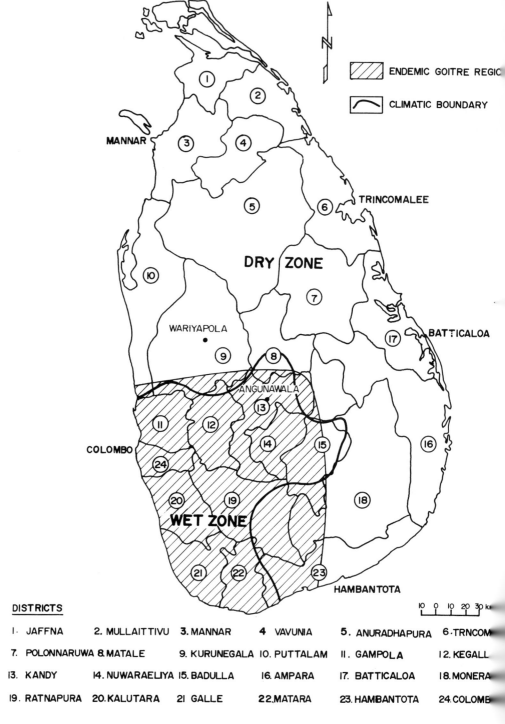

Fig. 1. The endemic goitre belt of Sri Lanka.

Table 2. *Geochemical data for water in the endemic goitre area around Angunawela–Daulagala in the Kandy District (No. samples = 60)*

Parameter	Minimum	Maximum	Mean
pH	5.85	8.2	7.7
Alkalinity (mg l^{-1})	30	420	138
F (μg l^{-1})	44	700	297
Cl (mg l^{-1})	6	108	35
I (μg l^{-1})	15	150	55
NO$_3$ (mg l^{-1})	1.5	15	8.5
Na (mg l^{-1})	27	1016	512
K (mg l^{-1})	0.55	9.4	7.3
Ca (mg l^{-1})	1.35	1616	50
Mg (mg l^{-1})	0.23	16.64	5.9
Mn (μg l^{-1})	1	208	53
Fe (μg l^{-1})	520	2430	1166
Hardness (mg l^{-1})	7	341	82
Co (μg l^{-1})	1	23	11

*Sample size, 11.

Table 3. *Geochemical data on the soils in the endemic goitre area around Angunawela–Daulagala in the Kandy District*

Parameter	Minimum	Maximum	Mean	Standard deviation
pH	3.8	6.8	5.2	0.7
E.C. (mhos cm^{-1})	159	310	236	28
F (mg kg^{-1})	0.4	44.9	10.2	6.5
Cl (mg kg^{-1})	0.6	4.3	1.6	0.8
I (mg kg^{-1})	0.04	6.6	1.9	1.2

Dissanayake & Chandrajith (1993) studied the geochemistry of endemic goitre in a village in the Central province near Kandy where a 40% rate of endemic goitre had been reported. Chandrajith (1987) analysed a total of 120 soil and water samples collected from 60 locations using the Orion electrode for iodine. The authors also analysed the samples for pH, alkalinity, F$^-$, Cl, NO$_3^-$, Na$^+$, Ca^{2+}, Mg^{2+}, Mn^{2+}, Fe and total hardness.

Table 2 shows the data obtained in the study of Dissanayake & Chandrajith (1993). A comparison of the iodine levels obtained from the two studies mentioned above shows that in general the water iodine levels of Sri Lanka are higher than the value of 10 μg l^{-1} regarded as the value below which goitre becomes endemic for Sri Lanka as suggested by Mahadeva & Shanmuganathan (1967). They are also considerably higher than those found in other endemic goitre regions such as northern England, southwest England and Wales as reported by Fuge (1989).

Figure 2 shows the spatial distribution of iodine in soils in the Angunawela–Daulagala area, a region of high endemic goitre in the Kandy District (Fig. 1). The soils for the endemic region have an average of 1.9 mg kg^{-1} iodine (Table 3) with 27% below 1.0 mg kg^{-1}. Only 15% of the soil samples analysed had more than 3 mg kg^{-1} iodine.

In contrast, soils in Wariyapola, a non-goitre region in the Dry Zone of northwest Sri Lanka (Fig. 1) had on the average more than 9 mg kg^{-1} of iodine. The frequency histogram for both areas is shown in Fig. 3. The distribution of iodine in the soils is unimodal with peaks at 32% for both areas.

The content of iodine in the soil and other factors such as geology and climate show a clear relationship with the incidence of goitre. The Wariyapola region located in the Dry Zone of Sri Lanka consists mostly of reddish-brown earth, and is characterized by flat topography with a few inselbergs. The area is underlain mainly by biotite gneisses, for which no iodine data are available. Fuge & Johnson (1986), however, quote a figure of 0.04 mg kg^{-1} for biotite gneisses and 0.38 for amphibolites. The iodine content of soils is expected to be greater than that in rocks and Fuge & Johnson (1986)

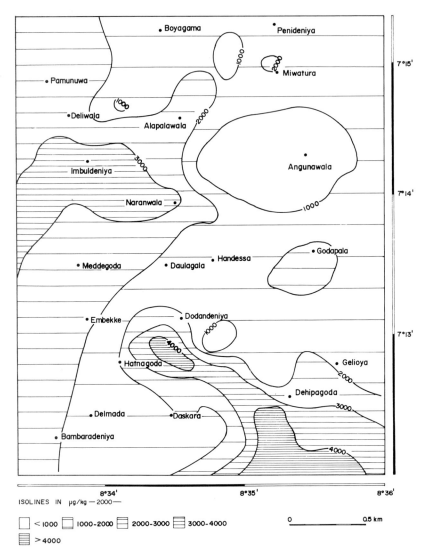

Fig. 2. The spatial distribution of iodine in soils in the endemic goitre area around Angunawela–Daulagala in the Kandy District.

record a range of 4–8 mg kg^{-1} as a mean value for iodine in soils.

Soils are known to contain very low percentages of water soluble iodine. Magomedova et al. (1970) observed that for some soils in the USSR, only 1–12% of the total iodine was water soluble. Whitehead (1973) noted that only 24% of the total iodine could be extracted by boiling water while Fuge & Johnson (1986) recorded that 80% of the soils analysed, had less than 10% cold water extractable iodine. The iodine content of groundwater is therefore expected to be very low and this is seen in the case of the endemic goitre region of Angunawela.

The high rainfall of the Wet Zone takes away the iodine in the groundwater to deeper levels, while in the Dry Zone evaporation tends to bring up the iodine to surface soil layers, aided by capillary action. Some island-wide goitre surveys carried out earlier by Wilson (1954) and Mahadeva & Shanmuganathan (1967) had indicated that goitre is prevalent in areas where the rainfall is highest. Table 4 shows the mean iodine content of soils in relation to the incidence of endemic goitre.

Salt and seafood including dry fish are the main sources of iodine in the diet. According to Mahadeva et al. (1968) the iodine content of

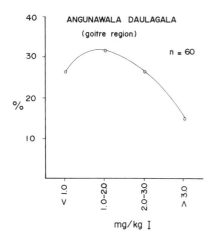
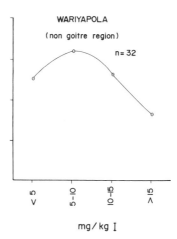

Fig. 3. Frequency histograms for the distribution of iodine in soils in an endemic goitre region and a non-goitre region in Sri Lanka.

Table 4. *Incidence of endemic goitre in relation to the iodine content in soils*

Area	Population suffering from goitre (%)	Mean iodine content (mg kg^{-1})	Reference
Taranaki (New Zealand)	4	14	Hercus
Auckland (New Zealand)	4	12	et al.
Dunedin (New Zealand)	19	3.2	(1931)
Clutha Valley (New Zealand)	40	0.4	
South Canterbury (New Zealand)	62	0.3	
Angunawala–Daulagala (Sri Lanka)	45*	1.9	present
Wariyapola (Sri Lanka)	12*	9.4	study

*Dr K. B. Heath, pers. comm.

Table 5. *Consumption of salt in different countries*

Country	Salt consumption (g per person per day)	Reference
South India	18	Pasricha (1966)
Taiwan	10	Kung-pei Chen (1964)
USA	10	Johnson (1951)
USA	20	Black (1952)
UK	15–20	Davidson et al. (1975)
Sri Lanka	07	Mahadeva & Karunanayake (1970)

local food varies from one area to another, the Dry Zone crops having higher quantities of iodine.

The iodine content in the local salt used for cooking is low when compared to that in other countries (Table 5). The salt used for cooking in Sri Lanka is known to contain about 498 mg 100 g^{-1} wet weight iodine (Mahadeva et al. 1968). The washing of the salt prior to cooking and excessive heat applied during the cooking result in major losses of iodine.

Factors affecting the endemicity of goitre in Sri Lanka

The analysis of the data for water as obtained by Balasuriya et al. (1992) show that the rank order correlation between iodine and the prevalence of goitre was only −0.64. The fact that the Kalutara District which has a very high prevalence of goitre (≃44%) is located near the sea further suggests that some factors other than the iodine content may contribute to the aetiology of goitre

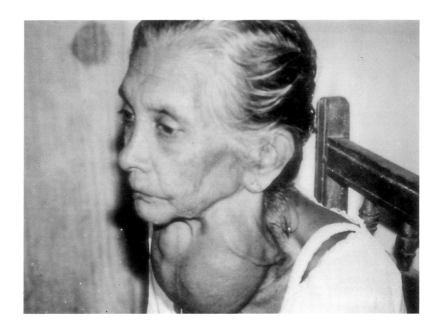

Fig. 4. Typical case of goitre in Sri Lanka.

in Sri Lanka, bearing in mind that the sea is a rich source of iodine.

From among the factors other than nutritional iodine deficiency, goitrogens, which affect the utilization of iodine by the thyroid gland, are of extreme importance. The high rate of goitre among certain villagers (Fig. 4), whose lifestyles and dietary patterns are no different from those who live in non-goitre regions, point to a major role played possibly by goitrogens yet unknown in Sri Lanka. It is apparent that the utilization of iodine by the thyroid gland is obstructed by a goitrogen that may well be environmental in origin.

Even though iodine may be found in sufficient concentrations in soil and water, its bioavailability may be seriously hampered if the iodine is fixed by substances prevalent in the environment.

Organic matter

The ability of the soil to retain iodine is related to a large number of complex interrelated soil characteristics and described by Fuge & Johnson (1986) as Iodine Fixation Potential (IFP). They defined IFP of a soil as the total amount of iodine that can be fixed by the various fractions in a soil and which is a characteristic of the soil in a given environment.

Organic matter in soils is known to trap and absorb iodine resulting in higher concentrations of iodine in soils rich in organic matter. This fact was demonstrated by Johnson (1988) who observed that surface soil samples (0–20 cm) show good correlation between organic content and total iodine ($r_{sp} = 0.70$).

Humic substances in the organic matter play a major role in the speciation and geochemical mobility of chemical elements (Dissanayake 1991). Iodine in view of its large ion could well be fixed by the humic substances resulting in a lowering of the bioavailability. The nature and extent of organic matter in the soils of the endemic areas are therefore worthy of further detailed investigation.

Soil pH

It has been suggested that highly acidic soils tend to be low in iodine due to the I^- ion being easily oxidized to I_2 which is then easily volatilized (Fuge 1990). The soil pH also influences the soil anion exchange capacity, which directly affects the iodine content. The study of Dissanayake & Chandrajith (1993) also showed that in the endemic goitre region around Kandy the soil pH was low ranging from 3.8–6.7 with a mean of 5.2. Figure 5 illustrates the correlation of I^- with other chemical parameters in hand pump wells,

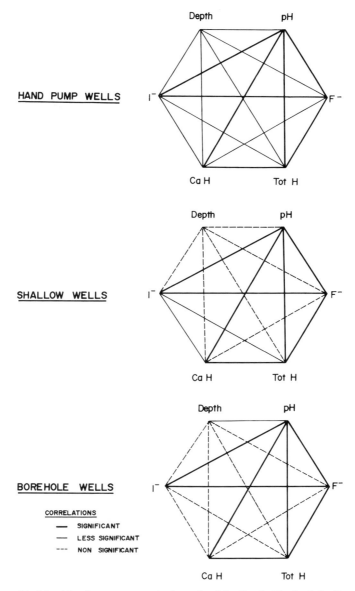

Fig. 5. Correlations of iodide with other parameters in the wells of the Kandy District (after Perera et al. 1992).

shallow wells and boreholes in the Kandy District as observed by Perera et al. (1992), from a study of about 200 wells. It is seen that the pH and the F^- correlate significantly with that of the I^- ion in the water.

Climatic influence

The geographical distribution of endemic goitre in Sri Lanka is clearly influenced by the climate. The high prevalence of goitre in the wet climatic zone appears to indicate a strong leaching effect of iodine from soils due to the heavy rainfall. The higher levels of iodine as found in the Dry Zone exemplified by the Anuradhapura District indicate a process of concentration.

Effect of clays

As in the case of the organic matter present in the soil, clays also have a marked tendency to adsorb ions. Iodine is strongly sorbed in or on clays and colloids and the presence of clays in the soils markedly affects the bioavailability of iodine. Whitehead (1979) has suggested that iodine in soils is retained mainly due to Al and Fe oxides and organic matter based on a good correlation between iodine and the sesquioxides present in soil.

Interestingly, the different clay mineral provinces of Sri Lanka also coincide with the climatic zones. The wet zone consists mainly of gibbsite and kaolinite while the Dry Zone has the montmorillonite type of clays. The geochemistry of iodine in the soils is therefore governed to a marked extent by the combined effect of the climate and the clays bearing in mind that a low pH favours the uptake of iodine by iron and aluminium oxides.

Prevention of goitre

The most convenient method of prevention of goitre is the iodization of the salt used in food. This practice has now been introduced in Sri Lanka, particularly in the endemic areas. However, due to the fact that the vast majority of the population that appears to undergo iodine deficiency disorders (IDD) in Sri Lanka are among the low income rural groups, a major awareness programme on the use of iodized salt is necessary. Such a venture must necessarily involve a concerted and coordinated effort by public health workers, manufacturers of salt, distributors and finally consumers. It is also of vital importance to monitor the progress of such a community health programme continuously.

Conclusions

Several factors govern the geochemical pathways of iodine and the incidence of goitre. Among the factors most clearly seen in Sri Lanka are climate, geology, nature of soil inclusive of organic matter, clay content, pH and food. The fact that the iodine contents of soil and water do not show marked correlation with the endemicity of goitre in Sri Lanka indicates a multifactorial aetiology which may include goitrogens, which affect the utilization of iodine by the thyroid gland.

Grateful thanks are due to Sandra Paragahawewa, Anne George and Harshini Aluwihare for their assistance.

References

BALASURIYA, S., PERERA, P. A. J., HERATH, K. B., KATUGAMPOLA, S. L. & FERNANDO, M. A. 1992. Role of iodine content of drinking water in the aetiology of goitre in Sri Lanka. *Ceylon Journal of Medical Science*, **35**, 45–51.

BLACK, D. A. K. 1952. *Sodium Metabolism in Health and Disease*. Blackwell, Oxford.

CHANDRAJITH, R. L. R. 1987. *The geochemical behaviour of iodine and some chemical parameters in high endemic goitre areas in Sri Lanka*. BSc Thesis, University of Peradeniya, Sri Lanka.

DAVIDSON, S., PASSMORE, R., BROCK, J. F. & TRUSWELL, A. S. 1975. *Human Nutrition and Dietetics*, 6th edition. Chap. 10. Churchill Livingstone, Edinburgh.

DISSANAYAKE, C. B. 1991. Human substances and chemical speciation – implications on environmental geochemistry and health. *International Journal of Environmental Studies*, **37**, 247–258.

—— & CHANDRAJITH, R. L. R. 1993. Geochemistry of endemic goitre, Sri Lanka. *Applied Geochemistry*, Supplementary Issue **2**, 211–213.

FERNANDO, M. A., BALASURIYA, S., HERATH, K. B. & KATUGAMPOLA, S. L. 1987. Endemic goitre in Sri Lanka. *In*: DISSANAYAKE, C. B. & GUNATILAKA, A. A. (eds) *Some Aspects of the Chemistry of the Environment of Sri Lanka*. Sri Lanka Association for the Advancement of Science, Colombo, Sri Lanka, 46–64.

——, ——, —— & —— 1989. Endemic goitre in Sri Lanka. *Asia–Pacific Journal of Public Health*, **3**, 11–18.

FUGE, R. 1989. Iodine in waters: possible links with endemic goitre. *Applied Geochemistry*, **4**, 203–208.

—— 1990. The role of volatility in the distribution of iodine in the secondary environment. *Applied Geochemistry*, **5**, 357–360.

—— & JOHNSON, C. C. 1986. The geochemistry of iodine – a review. *Environmental Geochemistry and Health*, **8**, 31–54.

HERCUS, C. E., AITKEN, H. A. THOMPSON, H. M., & COX, G. H. 1931. Further observations on the occurrence of iodine. *Journal of Hygiene*, **24**, 321–402.

JOHNSON, C. C. 1988. *The geochemistry of iodine and a preliminary investigation into its potential use as a pathfinder element in geochemical exploration*. PhD Thesis, University College of Wales, Aberystwyth, UK.

JOHNSON, D. 1951. *Modern Dietetics*. Putnam and Sons, New York.

KUNG-PEI CHEN. 1964. *Second Far East Symposium on Nutrition*. Taiwan. Publication of the US Government Printing Office.

MAGOMEDOVA, L. A., ZYRIN, N. G. & SALOMONOV, A. B. 1970. Iodine in the soils and rocks of mountainous Dagestan. *Agrokhimiya*, **1**, 117–125. [In Russian.]

MAHADEVA, K. & SHANMUGANATHAN, S. 1967. The problem of goitre in Sri Lanka. *British Journal of Nutrition*, **21**, 341–352.

—— & KARUNANAYAKE, E. 1970. Salt intake in Ceylon. *British Journal of Nutrition*, **24**, 811–814.

——, SENEVIRATNE, D. A., JAYATILAKE, B., SHANMUGANATHAN, S. S., PREMACHANDRA, P. & NAGARAJA, M. 1968. Further studies on the problem of goitre in Ceylon. *British Journal of Nutrition*, **22**, 525–537.

PASRICHA, S. 1966. Endemic goitre in South India. *Journal of Nutritional Dietetics*, **3**, 160–167.

PERERA, P. A. J., WELIGAMA, K. N. H., NADARAJAH, B. & PADMASIRI, J. P. 1992. *A study of fluoride and iodide levels in hand pump wells in Kandy District.* Unpublished report prepared by Kandy District Water Supply and Sanitation Project and University of Peradeniya.

WEERASEKERA, D. A., BANDARA, H. M. N. & PIYASENA, R. D. 1985. Iodine content of drinking water and goitre endemicity in Sri Lanka. *Ceylon Medical Journal*, **30**, 117–123.

WHITEHEAD, D. C. 1973. Studies on iodine in British soils. *Journal of Soil Science*, **24**, 260–270.

——— 1979. Iodine in the U.K. environment with particular reference to agriculture. *Journal of Applied Ecology*, **16**, 269–279.

WILSON, D. C. 1954. Goitre in Ceylon and Nigeria. *British Journal of Nutrition*, **8**, 90–99.

Clinical and epidemiological correlates of iodine deficiency disorders

A. G. STEWART & P. O. D. PHAROAH

Department of Public Health, University of Liverpool, PO Box 147, Liverpool L69 3BX, UK

Abstract: The iodine deficiency disorders, endemic goitre, stillbirths, abortions, congenital abnormalities, endemic cretinism, and impaired mental function, are found worldwide. Known for millenia, they currently put 29% of the world's population at risk. The most important is cretinism, found in two forms. Neurological cretinism possibly develops due to an insult in early pregnancy, hypothyroid cretinism to an insult late in pregnancy or in early life.

The classical medical causes of environmental iodine deficiency, glaciation and leaching, do not adequately explain the prevalence and distribution of iodine deficiency disorders, nor have the new insights into the environmental chemistry of iodine been assimilated properly with medical knowledge. Joint studies would be of help.

The presence of various factors other than iodine deficiency in the causation of the disorders is recognized. These are often thought to be geologically related. Recent work suggests plate tectonics as the unifying factor, since iodine deficiency disorders are found along many plate collision zones. Concentration of major and trace elements as a result of plate subduction provides a possible source of goitrogens. A systematic search is needed.

The iodine goitrogen ratio could provide the link between environmental iodine and differing intensities of the iodine deficiency disorders found in affected communities.

Endemic goitre and its attendant disorders are a worldwide problem. The review by Kelly & Snedden (1958) is still the most detailed, though lacking in data for many parts of the world. Recently the WHO has published an up-to-date listing of the size of the problem (Micronutrient Deficiency Information System 1993), both nationally and by survey areas. Almost no country is exempt from the potential for iodine deficiency disorders but due to preventive programmes in many countries and an increased standard of living in others, some are now goitre free. However, there are now an estimated 1570 million people at risk (29% of the world's population). This is due to both the ever increasing population of the world as well as a reduction in the acceptable level of goitre from 10% to 5%, this latter redefinition being a result of the growing awareness of the deleterious effects of iodine deficiency on the brain.

Areas with a high prevalence of goitre are often determined by certain common features. (1) Most are high mountain regions, including the Alps and its adjoining chains, the Himalayas and its extensions through Burma, Thailand, Malaysia, Indonesia to Papua New Guinea; the Andean Cordillera and its continuation north into Central America. Low mountains are also affected e.g. the Appalachians and the Urals. (2) Rain shadow areas, e.g. Rocky Mountains, North America. (3) Alluvial plains, e.g. the Great Lakes basin of North America, Finland, the low lying Netherlands, and the Indo-Gangetic plain. (4) Areas with water supplies percolating through limestone, e.g. the Peak District of England and parts of Colombia. Whether these are distinct areas or whether there is a common underlying cause is not clear.

Endemic cretinism

Endemic goitre is only the tip of the clinical iodine deficiency iceberg. The iodine deficiency disorders include stillbirths, abortions, congenital abnormalities, endemic cretinism (commonly characterized by mental deficiency, deaf–mutism, spastic diplegia and lesser degrees of neurological defect), and impaired mental function in children and adults with goitre associated with reduced levels of circulating thyroid hormones (Hetzel 1986). Other more controversial effects of iodine deficiency are breast disorders such as fibrocystic disease (Ghent et al. 1993). Goitre itself is rarely a major health problem in a community. It is the fetal wastage and the impaired mental functions associated with goitre that are so devastating to the community and the individual. But goitre remains the visible marker of such disorders.

By far the most important of the disorders is endemic cretinism (Pharoah 1985). There are two recognized types, the neurological and the hypothyroid. These are probably not distinct entities but the two ends of a clinical spectrum.

The prevalence of the major varieties varies worldwide and many affected individuals on close examination display a mixture of the two forms. The neurological variety is more often seen in the mountainous zones of iodine deficiency disorders, while the hypothyroid variety is common in Zaire in particular. The reason for this difference in distribution is thought to be due to the time of injury to the growing brain. An insult in early fetal life leads to neurological cretinism, whilst a later insult, towards the end of pregnancy or during early postnatal life, results in hypothyroid cretinism. The severity of the cretinism is correlated with the degree of the insult, which may be related to serum thyroid hormone levels: the greater the deficit the more severe the resulting cretinism in the child. In early fetal life the maternal thyroid hormone levels appear to be important, since at that stage the child's thyroid is not sufficiently developed to manufacture its own hormones (Pharoah & Connolly 1991a). The child's thyroid gland becomes functional later, and the fetal, or subsequently the neonatal, thyroid hormone levels are influential. Thyroid hormones are associated with cognitive and motor measures related to the developing brain (Pharoah & Connolly 1991b). The timing and the depth of the insult are therefore crucial. During adult life thyroid hormones are required for normal mental competency and hormonal deficiency may be epitomized by myxoedema madness. Seen uncommonly in the developed countries, this is a psychological state produced by a reduction in thyroid hormone levels caused by thyroid disease. It is eminently treatable by thyroid hormone replacement therapy, whereas the developmental problems in neurological cretinism are probably not reversible (Pharoah & Connolly 1987).

Because the brain is sensitive to thyroid hormone lack, cretinism itself is not a discrete disorder, but merges into normality. It may be correct to say that in severely affected areas where the overt cretinism rate may rise to 10–15% of the community, almost everyone will be affected and their brain development influenced to some extent. Any increase in mental performance in cretins with treatment during early childhood is probably a functional improvement rather than any change in the brain itself. It is the permanency of the neurological disorder which makes control of iodine deficiency disorders so important.

Iodine deficiency

Goitre has been known for many centuries, the Hindu and Chinese literature of 4000 years ago bearing witness to neck swellings (Langer 1960). A coin from Syracuse c. 425 BC shows a goitrous figure, Arethusa the city's patron nymph (Hart 1973). The first representation of cretinism is on a frieze from second or third century Ghandara, the great sub-Himalayan Buddhist kingdom (Merke 1971). Seaweeds such as *Laminaria* and *Sargassum* and preparations of pig and deer thyroids were used historically as treatments. In the first century Pliny knew of the efficacy of burnt seaweed for goitre (Cogswell 1837). When Courtois in 1811 isolated iodine from the seaweed *Fucus vesiculosus*, Prout in 1816 and Coindet in 1820 independently prescribed iodine as treatment for goitre, since this was the seaweed whose ashes had long been used empirically. Subsequently Coindet recognized the development of iodism and hyperthyroidism from large and toxic dosages and warned against such excess. Doses of the order of 5000 times the optimum were used (Matovinovic & Ramalingaswami 1960), and contributed to a decline in iodine therapy. But with the recognition of the concentration of iodine by the thyroid and its importance in the active hormone (Harington & Barger 1927) the use of iodine revived. The earlier unpalatable preparations have now largely been superseded and a variety of prophylactic compounds is available. The major challenge today is not the dose or the form of iodine but the distribution to the often remote and impoverished areas where the need is greatest.

Boussingault promoted the use of iodized salt for prevention 150 years ago (Kelly & Snedden 1958), but it was not until Marine & Kimball (1920) gave sodium iodide supplements to schoolgirls in Akron, Ohio that prophylaxis was successfully demonstrated. The use of iodinated oil injections was later pioneered in Papua New Guinea in the late 1950s (Hennessy 1964) and successfully eliminated endemic cretinism from a community (Pharoah & Connolly 1987).

Such studies, allied to the physiology of iodine in the thyroid and brain, are evidence in support of the iodine deficiency hypothesis. An inverse relationship between environmental iodine and goitre prevalence has been repeatedly demonstrated (Roche & Lissitzky 1960; Gutekunst & Scriba 1987). Urinary iodine excretion, which parallels dietary intake and is easier to measure, is often low in goitrous areas (Follis 1964; Bourdoux et al. 1985), and animals fed on a low iodine diet develop goitre (Axelrad et al. 1955). Nevertheless, there are anomalies in the correlation between iodine deficiency and the prevalence of goitre (e.g. Burma, Taiwan and Java (Kochupillai et al. (1980); Cuba (Pretell & Dunn 1987); Spain (Gutekunst & Scriba 1987); Missouri and northern England (Fuge 1989)), and adequate prophylaxis is not always successful in eradicating the disorder. Two recent studies have been unable to explain most of the variation in goitre rates. In Sri Lanka, Balasuriya et al. (1992) found that only 40% of the variability of goitre prevalence could be explained in terms of the iodide content of the drinking water, although drinking water only accounts for between 10 and 20% of the dietary iodine. In Bangladesh Filteau et al. (1994) could account for as little as 12% by iodine status and anthropometric variables. This suggests that there are other factors in the causation of iodine deficiency disorders, and many goitrogens have been implicated over the years but few have been adequately confirmed. There are even suggestions that iodine deficiency is not the cause of endemic goitre (Greenwald 1957).

Goitrogens

One goitrogen that has much support is thiocyanate, released from inadequately prepared cassava. Studies in Zaire show that a raised urinary iodine/thiocyanate ratio may be considered healthy, since no clinical problems are known and biochemical indices are normal. Low ratios are associated with goitre and very low ratios with cretinism (Delange et al. 1982). However, there are unanswered questions about the role of geology in some parts of Zaire. The clear cut difference in goitre prevalence in communities living on two different rock types on Idjwi Island in Lake Kivu, Zaire has been ascribed to cassava, but the original geological association (Delange et al. 1968) has never been explained adequately. The northern area of high goitre is based on a granite and gneiss basement while the non-goitrous southern end of the island has a basalt basement. Selenium deficiency has also recently been identified as a cause of goitre, but its role has yet to be clarified, although its involvement in goitrogenesis has been raised in these cassava areas of Zaire (Thilly et al. 1993).

McCarrison (1906) in the Karakoram noted the increase in goitre prevalence in successive villages down river, which he attributed to the progressive pollution of the water, and with other later observations concluded that a microorganism was involved. *E. coli* was later incriminated as the causative organism in other endemias (Vought et al. 1967; Malamos et al. 1971). Chapman et al. (1972) studying McCarrison's original work in the same Karakoram valley showed that the increase in goitre prevalence there was not due to bacterial contamination of the water. They suggested that the high silt load was absorbing the iodine from the water, but failed to explain why this should increase down river when the main source of iodine in the environment is the atmosphere, not bedrock.

The higher prevalence of goitre in hard water areas has long been appreciated. Prosser (1769) recognized that 'Derbyshire Neck' was associated with limestone regions. Experimental evidence for this association was seen in the observation (Taylor 1954) that calcium induces goitre in the iodine deficient rat though Harrison et al. (1967) and Malamos et al. (1971) dispute this. Day & Powell-Jackson (1972) found that the variation in goitre prevalence correlated in Nepal not only with the degree of hardness but also with the fluoride content of the water. The role of fluorine is not yet clear despite this last observation. Other areas of fluorosis do not have consistently high goitre prevalence and it must be asked whether or not in Nepal the fluorine, as well as the hardness factor, was acting as a marker for some other mineral which affects goitrogenesis and occurs in related geological formations. What this mineral could be is as yet unknown. The geology of the iodine deficiency disorders is still a poorly understood subject.

A possible correlation exists with plate subduction zones. In northern Pakistan, in the area where McCarrison made his original observations, there seems to be a clear cut geological differentiation of goitre prevalence into two groups north and south of the east–west trending Main Karakoram Thrust (Stewart 1990). Markedly different goitre prevalences are found and are shown in Fig. 1. As expected there is an age and sex variation but there is also an independent, statistically significant, difference between the two areas ($p < 0.01$). The Main Karakoram Thrust is the western continuation of the Indus–Tsangpo Line and part, therefore, of the collision boundary between the Indian and the Asian plates. Whether the difference is simply due to the differing geochemical signatures between the two plates or whether it is related to the collision is still unclear. However, several suspected goitrogens such as BF_4, SO_3F, molybdate, and Li are recognized to occur in the minerals north of the Main Karakoram Thrust in Pakistan and Afghanistan (Stewart 1990; Tahirkheli 1982; Quasin Jan, University of Peshawar, pers. comm.) and are believed to be concentrated in Asia, the non-subducting plate, by crustal melting and metallogenic processes

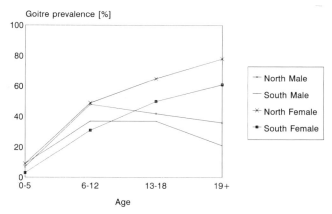

Fig. 1. Goitre prevalence across the Main Karakoram Thrust, Pakistan. Sample size: 1896 persons (3.4% total population) to the north and 4321 persons (2.7% of total population) to the south of the thrust.

associated with the plate boundary. Currently a search for possible goitrogens in northern Pakistan is underway, admitting the possibility of testing this hypothesis in other collision zones since it seems possible that there is a link between other collision zones and goitre too. The postulate that subduction concentrates possible goitrogens in the non-subducting plate is intriguing in the light of the distribution of goitre already noted earlier, in particular the mountainous regions. Goitre is particularly prevalent on the non-subducting plate of subduction zones associated with the Andes and the Himalaya–Indonesia–New Guinea belt. All these areas are noted for the severity of the whole spectrum of the iodine deficiency disorders. Other non-subducted plates such as the Philippines and Turkey also have iodine deficiency disorders associated with the plate boundary. Ordinary weathering will supply goitrogens to the food chain. In such situations iodine would act in a similar way to its action with cassava, the ratio of iodine to the goitrogen, rather than the absolute level of iodine, being important.

Environmental iodine

However, if our understanding of the geology of the disorders leaves a lot to be desired, our awareness of the links between environmental iodine levels and the disorders is not as clear as is often assumed.

Two medical red herrings concerning environmental iodine levels need to be exposed. The classical environmental explanations in the medical literature for goitre occurrence are that glaciation and leaching cause soil iodine in endemic areas to be low (Hetzel 1989). Neither of these is true as quoted although still repeated (Micronutrient Deficiency Information System 1993), and recent work by environmental scientists on the distribution of iodine is hardly known in the medical literature.

The glaciation hypothesis rests on the supposition that the great glaciations stripped soils from the underlying rocks and there has been insufficient time elapsed to raise soil iodine to prevent goitre developing. The hypothesis is usually ascribed to Goldschmidt (1954) though it occurs in the medical literature some 14 years earlier (Shee 1940). One other medical worker in Switzerland supports the hypothesis with evidence (Merke 1965, 1967). However, there are several objections to the hypothesis. Goitre is now known to have a much wider distribution than glaciation and continues beyond the glaciated boundaries of northern Europe and North America. The medical surveys of both Shee (1940) and Merke (1967) are open to doubt in the light of more recent findings. Shee was unaware of the influence of seaweed fertilizer as a source of iodine while Merke's analysis dismissed non-local foods as an important source of dietary iodine. Both of these facts are now well attested and raise doubts about the medical basis for the glaciation hypothesis. But glaciation is still assumed in the medical literature to be the local cause of environmental iodine deficiency (Micronutrient Deficiency Information System 1993). The medical aspects of the hypothesis have never been examined critically, although non-medical scientists have questioned it (Fleming 1980; Fuge & Johnson 1986). Unpublished observations from northern Pakistan found no correlation between glaciation and goitre prevalence.

Leaching is even less well understood in the

medical literature, there being no clear description of the concept but a wide acceptance of its validity. The assumption is made that the occurrence of endemic goitre *must* mean low soil iodine, and the further conjecture that leaching must have occurred is quietly added often without any environmental evidence. Water flowing through soil is expected to transport iodine from the topmost layers, either depositing it in the deeper levels where it lies unavailable to plants, or removing it completely in the run-off. The source of water is quoted variously as rain, river, flood or irrigation (Hetzel 1989). The assumption is that soil iodine is high initially and then reduced by the flow-through of water. Leaching by glacial melt may even be quoted as the mechanism for the efficiency of glaciation in lowering soil iodine levels (Koutrous et al. 1980). The idea that soil iodine may be tightly bound and not free to move through the groundwater into rivers is not known medically, though Fuge & Johnson (1986) have demonstrated that iodine distribution in soils is governed both by the supply of iodine and the ability of the soil to retain it.

Atmospheric input in wet and dry deposition, soil type (Zborishchuk & Zyrin 1974; Fuge & Johnson 1986), the state of growth of covering vegetation (Asperer & Lansangan 1986; Muramatsu et al. 1991) humus, animal biomass (Bors & Martens 1992) and oxides of Al and Fe (Whitehead 1978) have all been shown to affect the retention and distribution of iodine in the soil. It is no wonder that there is at times confusion over the role of environmental iodine in the causation of the iodine deficiency disorders. These various factors need to be taken into account in any future discussion of soil iodine and endemic goitre. In this context it would be good if the Iodine Fixation Potential of soils, as proposed by Fuge & Johnson (1986), was given a quantitative basis, to define its potential in characterizing areas where goitre occurs. A collaborative study between medical and earth scientists would be ideal. Simple soil iodine concentrations do not give enough information on the potential availability of iodine.

One point that has received little attention from either group is the definition of low environmental iodine, whether soil or water. For example, the mean river iodine level is $5 \mu g \, l^{-1}$ (Fuge & Johnson 1986), and values under this may be low environmentally, but whether or not they are low medically is a moot point. If goitrogens are as important as they appear to be then this could explain why water iodine levels sometimes do not correlate with goitre prevalence (Fuge 1989), and sometimes do; see for example the various figures given in Stanbury & Hetzel (1980), where in such diverse places as Algeria, Cameroon, Michigan, Tunisia and Iraq water iodine is inversely related to variations in goitre prevalence, while in Burma, Italy and Taiwan it is not (Table 1). There is also the impression of different goitre prevalences at the same environmental iodine levels, though this has not been formally studied. Perhaps the suggested joint study needs to cover water as well as soils to clarify the environmental factors in the aetiology of goitre.

Table 1. *Water iodine levels and goitre prevalence Adapted from, Stanbury & Hetzel 1980*

Place	Water iodine $\mu g \, l^{-1}$	Goitre prevalence %
Algeria	3.5	80
	4.5	70
Cameroon	2–4	58
	>4	non-endemic
Italy	0.3	more
	0.35	11
	0.55	49
	0.82	48
	1.75	less
	2.02	10
	2.26	49
	2.38	5–40
	9.09	38
	500	7
Michigan	0.0	65
	0.3	56
	7.3	33
	8.7	26
Tunisia	3.9	50
	3.9	36
	6.9	1

In examining terrestrial iodine the atmospheric input must not be ignored. The atmosphere is recognized as the source of soil and water iodine, transporting iodine from its sources in the sea to land. Figures quoted by Whitehead & Truesdale (1982) indicate that in Berkshire (UK) and central Russia the annual losses through run-off were probably similar to annual input from the atmosphere 20 and 45 g I ha^{-1} respectively). This would suggest that there are steady states in the iodine cycle which are set variously according to local factors. Distance from the sea is correlated with goitre prevalence, such as in a coastal country like the Netherlands (Kelly & Snedden 1958). No analysis over a greater distance has yet been published, but it is conceivable that this correlation is limited to the immediate coastal area in the light of the many complexities already outlined in the environmental control of iodine levels. Maritime influence on environmental iodine probably decreases exponentially with distance from the sea. Sadasivan (1980) reports an order of magnitude difference between rain iodine near the coast of India and inland during the monsoon, while Perel'man (1972) notes a similar difference between iodine in Black Sea air (10–20 $\mu g/m^3$) and that in the air of the Cherkassy region, former USSR, some 3000 km or more inland (1.28 $\mu g/m^3$). Låg & Steinnes (1976) studied soil iodine in Norway and commented that in the northern area nearer the sea the decrease in soil iodine with increasing distance from the sea far exceeded the decrease in annual precipitation, while in eastern Norway the drop in values was lower than the corresponding decrease in precipitation. This sounds like an exponential decrease, not unexpected, since many environmental parameters behave in this manner, and even mineral dust shows a probable exponential decrease in concentration as it is transported over the sea (Knap & Kaiser 1990). Atmospheric iodine occurs as gaseous inorganic, gaseous organic and particulate species. Each has a different residence time and may be removed by either wet or dry deposition (Whitehead 1984). The depositional velocity depends on the species, for example, elemental iodine settles at about 1 cm $\sec^{-1}$, methyl iodide has a velocity of less than 0.01 cm $\sec^{-1}$, while that of various atmospheric particles ranges from 1–5 cm $\sec^{-1}$, though no figure for particulate iodine is known (Whitehead 1984). It is likely that the summation of these various rates will give an exponential decrease, but this needs confirmation.

Postulate

We can postulate a correlation between these various ideas, the timing and depth of injury to the growing brain affecting the type of cretinism produced, the role of goitrogens in the aetiology of the iodine deficiency disorders, and the role of environmental iodine. Water iodine may be related to the atmospheric input but not the availability of iodine. The availability of environmental iodine to the diet depends on a variety of soil factors, such as atmospheric input, soil type, vegetation cover, humus, animal biomass, and Al and Fe oxides. A low dietary iodine on its own will produce the full range of iodine deficiency disorders in a community. However, in the presence of a goitrogen iodine will act only as a limiting factor, the ratio of iodine to goitrogen being the determinant of the outcome. The goitrogen may often be geologically related, and may be subject, as may iodine, to an annual cycle of concentration. This will give a varying iodine–goitrogen ratio causing differences of timing and depth of any insult to the growing brain. Thus we will find differing intensities of cretinism and the other disorders in an affected community.

Conclusions

There is a need for a new examination of the environmental cycle of iodine, particularly in plants, soil and water, with respect to the occurrence of iodine deficiency disorders. Secondly, a systematic search for mineral goitrogens in endemic goitre areas should be undertaken to identify other factors which affect the development and progress of iodine deficiency disorders.

AGS was supported in part by gifts from churches and individual Christians.

References

ASPERER, G. A. & LANSANGAN, L. M. 1986. The Uptake of I-131 in Tropical Crops. *Trace Substances in Environmental Health*, **20**, 457–465.

AXELRAD, A. A., LEBLOND, C. & ISLER, H. 1955. Role of iodine deficiency in the production of goiter by the Remington diet. *Endocrinology*, **56**, 387–403.

BALASURIVA, S., PERERA, P. A. J., HERATH, K. B., KATUGAMPOLA, S. L. & FERNANDO, M. A. 1992. Role of Iodine Content of Drinking Water in the Aetiology of Goitre in Sri Lanka. *The Ceylon Journal of Medical Science*, **35**, 45–51.

BORS, J. & MARTENS, R. 1992. The contribution of microbial biomass to the adsorption of radioiodide in soils. *Journal of Environmental Radioactivity*, **15(1)**, 35–49.

BOURDOUX, P., DELANGE, F., FILETTI, S., THILLY, C. & ERMANS, A. M. 1985. Reliability of the iodine/creatinine ratio: a myth? *In:* HAL, R. & KOBBERLING, J. (eds) *Thyroid Disorders Associated with Iodine Deficiency and Excess*. Raven, New York, 145–152.

CHAPMAN, J. A., GRANT, I. S., TAYLOR, G., MAHMUD, K., SARDAR-UL-MULK & SHAHID, M. A. 1972. Endemic goitre in the Gilgit Agency, West Pakistan. *Philosophical Transactions of the Royal Society of London, Biology*, **263**, 459–490.

COGSWELL, C. 1837. *An Experimental Essay on the Relative Physiological and Medicinal Properties of Iodine and its Compounds; being the Haverian Dissertation for 1837*. Adam & Charles Black, Edinburgh.

DAY, T. K. & POWELL-JACKSON, P. R. 1972. Fluoride,

water hardness and endemic goitre. *Lancet*, **1**, 1135–1138.
DELANGE, F., THILLY, G. & ERMANS, A. M. 1968. Iodine deficiency, a permissive condition in the development of endemic goitre. *Journal of Clinical Endocrinology & Metabolism*, **28**, 114–116.
——, ITEKE, F. B. & ERMANS, A. M. 1982. *Nutritional Factors Involved in the Goitrogenic Action of Cassava.* International Development Research Centre, Ottawa.
FILTEAU, S. M., SULLIVAN, K. R., ANWAR, U. S., ANWAR, Z. R. & TOMKINS, A. M. 1994. Iodine Deficiency Alone Cannot Account for Goitre Prevalence Among Pregnant Women in Modhupur, Bangladesh. *European Journal of Clinical Nutrition*, **48**, 293–302.
FLEMING, G. A. 1980. Essential micronutirents II: iodine and selenium. *In:* DAVIES, B. E. (ed.) *Applied Soil Trace Elements.* Wiley, London, 199–234.
FOLLIS, R. H. 1964. Patterns of urinary iodine excretion in goitrous and nongoitrous areas. *American Journal of Clinical Nutrition*, **14**, 253–268.
FUGE, R. 1989. Iodine in waters: possible links with endemic goitre. *Applied Geochemistry*, **4**, 203–208.
—— & JOHNSON, C. C. 1986. The geochemistry of iodine – a review. *Environmental Geochemistry and Health*, **8(2)**, 31–54.
GHENT, R., ESKIN, B. A., LOW, D. A. & HILL, L. P. 1993. Iodine Replacement in Fibrocystic Disease of the Breast. *Canadian Journal of Surgery*, **36**, 453–460.
GOLDSCHMIDT, V. M. 1954. *Geochemistry.* Clarendon, Oxford.
GREENWALD, I. 1957 The history of goiter in the Inca empire: Peru, Chile and the Argentine Republic. *Texas Reports on Biology and Medicine*, **15**, 874–889.
GUTEKUNST, R. & SCRIBA, P. C. 1987. Iodine deficiency disorders in Europe. *In:* HETZEL, B. S., DUNN, J. T. & STANBURY, J. B. *The Prevention and Control of Iodine Deficiency Disorders.* Elsevier, Amsterdam, 249–264.
HARINGTON, G. R. & BARGER, G. 1927. Thyroxine II: Constitution and synthesis of thyroxine. *Biochemical Journal*, **21**, 169–183.
HARRISON, M. T., HARDEN, R. M. & ALEXANDER, W. D. 1967. Effect of calcium on iodine metabolism in man. *Metabolism*, **16**, 84–86.
HART, G. D. 1973. The diagnosis of disease from ancient coins. *Archaeology*, **26**, 123–127.
HENNESSY, W. B. 1964. Goitre prophylaxis in New Guinea with intramuscular injections of iodized oil. *Medical Journal of Australia*, **1**, 505–512.
HETZEL, B. S. 1986. The concept of iodine-deficiency disorders (IDD) and their eradication. *In:* DUNN, J. T., PRETELL, E. A., DAZA, C. H. & VITERI, F. E. (eds) *Towards the Eradication of Endemic Goiter, Cretinism, and Iodine Deficiency.* Pan American Health Organization, Washington, 109–114.
—— 1989. *The Story of Iodine Deficiency.* Oxford University Press.

KELLY, F. C. & SNEDDEN, W. W. 1958. Prevalence and geographical distribution of endemic goitre. *Bulletin WHO*, **18**, 5–173.
KOCHUPILLAI, N., RAMALINGASWAMI, V. & STANBURY, J. B. 1980. Southwest Asia. *In:* STANBURY, J. B. & HETZEL, B. S. (eds) *Endemic Goiter and Endemic Cretinism.* Wiley, New York, 101–121.
KOUTROUS, D. A., MATOVINOVIC, J. & VOUGHT, R. 1980. The ecology of iodine. *In:* STANBURY, J. B. & HETZEL, B. S. (eds) *Endemic Goiter and Endemic Cretinism.* Wiley, New York, 185–195.
KNAP, A. H. & KAISER, M. S. 1990. *The Long Range Atmospheric Transport of Natural and Contaminant Substances.* Kluwer, Dordrecht.
LÅG, J. & STEINNES, E. 1976. Regional distribution of halogens in Norwegian forest soils. *Geoderma*, **16**, 317–325.
LANGER, P. 1960. History of goitre. *In: Endemic Goitre.* WHO, Monograph Series, Geneva, **44**, 9–25.
McCARRISON, R. 1906. Observations on endemic goitre in the Chitral and Gilgit valleys. *Lancet*, **1**, 1110–1111.
MALAMOS, B., KOUTRAS, D. A., RIGOPOULOS, G. A., PAPAPETROU, P. D., GOUGAS, E., ET AL. 1971. Endemic goiter in Greece: some new epidemiological studies. *Journal of Clinical Endocrinology*, **32**, 130–139.
MARINE, D. & KIMBALL, O. P. 1920. The prevention of simple goiter in man. *Archives of Internal Medicine*, **25**, 661–672.
MATOVINOVIC, J. & RAMALINGASWAMI, V. 1960. Therapy and prophylaxis of endemic goitre. *In: Endemic goitre.* WHO, Monograph Series, Geneva, **44**, 385–410.
MERKE, F. 1965. Eiszeit als primordiale Ursache des endemischen Kropfes. *Schweizerische Medizinixche Wochenschrift*, **95**, 1183–1192.
—— 1967. Eiszeit als primordiale Ursache des endemischen Kropfes: Eiszeit und Kropf im Wallis. *Schweizerische Medizinixche Wochenschrift*, **97**, 131–140.
—— 1971. *Geschichte und Ikonographie des endemischen Kropfes und Kretinismus.* Hans Huber, Bern.
MICRONUTRIENT DEFICIENCY INFORMATION SYSTEM. 1993. *Global Prevalence of Iodine Deficiency Disorders.* WHO, Geneva.
MURAMATSU, Y., UCHIDA, S. & YOSHIDA, S. 1991. Radiotracer experiments on the desorption of iodine from paddy fields with and without rice plants. *Radioisotopes*, **40(11)** 440–443.
PEREL'MAN, A. I. 1972. *Geochemistry of Elements in the Supergene Zone.* Translated from the Russian, Israel Program for Scientific Translations, Jerusalem.
PHAROAH, P. O. D. 1985. The epidemiology of endemic cretinism. *In:* FOLLET, B. K., ISHII, S. & CHANDOLA, A. (eds) *The Endocrine System and the Environment.* Springer, Berlin, 315–322.
—— & CONNOLLY, K. J. 1987. A controlled trial of iodinated oil for the prevention of endemic cretinism: a long-term follow up. *International Journal of Epidemiology*, **16**, 68–73.

—— & —— 1991a. Effects of maternal iodine supplementation during pregnancy. *Archives of Diseases of Children*, **66**, 145–147.

—— & —— 1991b. Relationship between maternal thyroxine levels during pregnancy and memory function in childhood. *Early Human Development*, **35**, 43–51.

PRETELL, E. A. & DUNN, J. T. 1987. Iodine deficiency disorders in the Americas. *In:* HETZEL, B. S., DUNN, J. T. & STANBURY, J. B. (eds) *The prevention and control of iodine deficiency disorders.* Elsevier, Amsterdam, 237–247.

PROSSER, T. 1769. Account and method of cure of the bronchocele of Derby Neck. London.

ROCHE, J. & LISSITZKY, S. 1960. Etiology of endemic goitre. *In: Endemic Goitre.* WHO, Monograph Series, Geneva, **44**, 351–368.

SADASIVAN, S. 1980. Trace Constituents in Cloud Water, Rainwater and Aerosol Samples Collected Near the West Coast of India During the Monsoon. *Atmospheric Environment*, **14**, 33–38.

SHEE, J. C. 1940. Soil and fresh-water iodine content in Ireland in relation to endemic goitre incidence. *Scientific Proceedings of the Royal Dublin Society*, **22**, 307–314.

STANBURY, J. S. & HETZEL, B. S. 1980. *Endemic Goiter and Endemic Cretinism.* John Wiley, New York.

STEWART, A. G. 1990. Drifting continents and endemic goitre in northern Pakistan. *British Medical Journal*, **300**, 1507–1512

TAHIRKHELI, R. A. K. 1982. Geology of the Himalayas, Karakoram and Hindukush in Pakistan. *Geological Bulletin, University of Peshawar.* 15 (Special issue), 1–15.

TAYLOR, S. 1954. Calcium as a goitrogen. *Journal of Clinical Endocrinology*, **14**, 1412–1422.

THILLY, C. H., SWENNEN, B., BOURDOUX, P., NTAMBUE, K., MORENO-REYES, R., GILLIES, J. & VANDERPLAS, J. B. 1993. The epidemiology of iodine deficiency disorders in relation to goitrogenic factors and thyroid stimulating hormone regulation. *American Journal of Clinical Nutrition*, **57** (2nd Supplement), 267s–270s.

VOUGHT, R. L., LONDON, W. T. & STEBBING, G. E. T. 1967. Endemic goitre in Northern Virginia. *Journal of Clinical Endocrinology & Metabolism*, **27**, 1381–1389.

WHITEHEAD, D. C. 1978. Iodine in soil profiles in relation to Fe and Al oxides and organic matter. *Journal of Soil Science*, **29**, 88–94.

—— 1984. The distribution and transformations of iodine in the environment. *Environment International*, **10**, 321–339.

—— & TRUESDALE, V. M. 1982. *Iodine and its movements in the environment with particular reference to soil and plants.* Grassland Research Institute, Maidenhead, UK.

ZBORISHCHUK, Y. N. & ZYRIN, N. G. 1974. Average B, Mn, Co, Cu, Zn, Mo and I content of the soils of the European USSR. *Soviet Soil Science*, **6**, 209–215.

The protective role of trace elements in preventing aflatoxin induced damage: a review

M. G. NAIR, S. M. MAXWELL & B. J. BRABIN

Tropical Child Health Group, Liverpool School of Tropical Medicine, Pembroke Place, Liverpool L3 5QA, UK

Abstract: Recent evidence has shown that there are important interactions in experimental models and human beings between trace element status and aflatoxin B_1 (AFB_1) exposure. The hypothesis is presented that these interactions operate through altered host antioxidant status leading to excess free radical production resulting in cell damage and clinical disease. Geographical differences in the geochemical environment could have a major influence on the pattern of disease. Appropriate nutritional intervention studies are required to determine potential benefits of trace element supplementation in aflatoxin exposed populations.

The geochemical environment has a profound influence on the level of health of man and animals; moreover, the geochemical environment, particularly as it affects intake and utilization of trace elements, plays a major causal role in many specific diseases (Hopps 1975). In this symposium Plant *et al.* (1996) have reported that, in developing countries, alterations in the geochemical environment not only caused trace element deficiencies but also depleted conservative elements such as iodine, the deficiency of which is associated with goitre and cretinism. Trace element interactions with mycotoxins in foods are an example of such an effect. Mycotoxins (fungal toxins) are often detected in foods, particularly in developing countries. Chemically, the mycotoxins are organic compounds with low molecular weight. The most important among these, aflatoxin B_1 (AFB_1) which is elaborated by *Aspergillus flavus* and *A. parasiticus* is a well known hepatotoxin, mutagen, immunosuppressant and carcinogen in animals and man (Busby & Wogan 1984). The adverse biological effects of AFB_1 are mani-

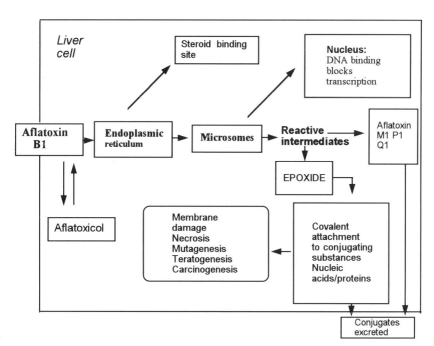

Fig. 1. Adverse biological effects of AFB_1.

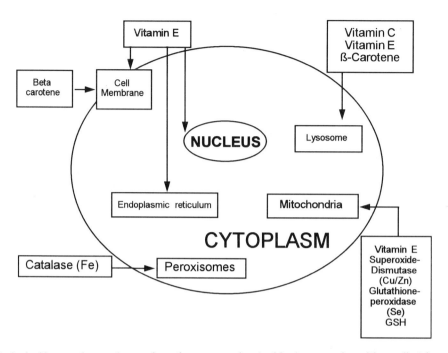

Fig. 2. Antioxidants and trace element dependent enzymes involved in the prevention of free radical formation.

fested after its metabolic activation by cytochrome P-450 into an active form AFB_1-8,9-epoxide and subsequent interaction with cellular macromolecules (Fig. 1).

Recent research has clearly indicated that drugs and toxins can interact with dietary nutrients in a number of ways. The mechanism of such interaction is poorly understood but the outcome of such encounters determines whether there will be no detectable effect or if an acute or chronic adverse effect will be experienced by the host. Several naturally occurring factors in the diet modify the metabolic activation of aflatoxins. Among these, trace elements play an important role in modulating the biological effects of AFB_1 and this review attempts to focus on possible mechanisms involved in this interaction.

Importance of dietary trace elements

Most of the trace elements are essential constituents of all living tissues with highly specific functions. Hopps (1975) highlighted the importance of trace elements in the geochemical environment in relation to health and disease and emphasized that specific requirements must be satisfied. The geochemical environment is an important factor in enabling the individual to meet these needs.

Of the trace elements, copper (Cu), selenium (Se), manganese (Mn), iodine (I) and zinc (Zn) are components of enzymes. Some are also bound to specific proteins, either for storage or transport purposes or for both, e.g. Cu in caeruloplasmin and Cu and Zn in metallothionein (Waterlow 1992). Trace element dependent enzymes are involved in the prevention of free radical formation. Free radicals are species containing one or more unpaired electrons. These radicals are chemically very active and have very short half-lives and are continuously produced *in vivo*. In consequence, organisms have evolved not only antioxidant systems to protect against them, but also repair systems that prevent the accumulation of oxidatively-damaged molecules. The chemistry of free radicals is reviewed in detail by Halliwell (1993). The trace element dependent enzymes (catalase, glutathione peroxidase, superoxide dismutase, and caeruloplasmin) are involved in the scavenging of reactive oxygen species (Fig. 2) and in this role have attracted considerable attention in recent years (Athar *et al.* 1993). One of the major antioxidant defence mechanisms in mammals is the sequestration of transition metal ions into forms incapable of stimulating free radical reactions (Halliwell 1993).

In vivo studies

Among the trace elements, selenium (Se) has been studied to the greatest extent. Under certain environmental conditions aflatoxin may be present in diets in which Se may be deficient or in excess.

Newberne & Conner (1974) observed that pretreatment of animals with sodium selenite in the diet (0.03–5 ppm) for a two to three week period prior to dietary administration of AFB_1, protected animals against the acute toxicity of aflatoxin at a dosage of 1.0 ppm even though this dose was mildly toxic. The dose levels of 0.03 and 5 ppm were found to be too low and too high respectively for their protective effects against aflatoxin toxicity. They attributed the protective mechanisms to the antioxidant effect of Se in quenching the free radicals produced in the tissues by AFB_1.

In rabbits orally dosed with sodium selenite ($50\,mg\,kg^{-1}$) and inoculated with *A. flavus* antigen, Alexsandrowicz (1976) observed that sodium selenite induced an accelerated response and significant increase in concentration of specific antibodies against this antigen.

In rats given 42 µg of AFB_1 for 27 days with sodium selenite in drinking water at doses of 3 and $6\,mg\,kg^{-1}$ for 79 weeks, Lei et al. (1990) observed an inhibitory effect of selenium on the development of preneoplastic liver lesions and this was more pronounced at the low than at the high dose of selenium.

Petr et al. (1990) recorded a protective effect of selenium against the mutagenic effect of AFB_1 in Chinese hamsters pretreated with 2 ppm sodium selenite.

Pretreatment of calves with a single intramuscular injection of selenium-vitamin E (5 mg of selenium and 68 IU of α-tocopherol per 60 kg of body weight for 14 days) before oral dosing with $0.1\,mg\,kg^{-1}$ AFB_1, improved the feed intake compared to that of the toxin only fed calves (Brucato et al. 1986). The protective effect of selenium was attributed to the stimulation of glutathione peroxidase with resultant conjugation of the reactive metabolites of aflatoxin and reduction in oxidative damage.

The exact mechanisms of protection are not known. The protective effect of Se against AFB_1 may be due to: (a) the antioxidant action of Se as an integral component of glutathione peroxidase protecting against lipid peroxidation and resultant peroxidative damage which may influence the carcinogenic process (Baldwin & Parker 1987); (b) reduction in the *in vivo* covalent binding of AFB_1 to macromolecules (Chen et al. 1982). (c) modulation of the hepatic and intestinal microsomal enzyme system (Lalor et al. 1978; Nyandieka et al. 1990) by dietary Se.

However, it is interesting that in another study (Burk & Lane 1983), Se deficiency in rats was shown to protect against the hepatotoxic effects of AFB_1 possibly by a mechanism related to more rapid glutathione synthesis and higher activity of glutathione S-transferase in Se deficient livers.

The selenium status of a population tends to be determined primarily by the geochemical environment and its interaction with diet and nutritional products. Because of geochemical differences, the estimates of adult human exposure to selenium via the diet range from 11 000 to 5000 $\mu g\,day^{-1}$, in different parts of the world; however, dietary intake more usually falls within the range of 20–300 $\mu g\,day^{-1}$ (WHO 1987). Low selenium in the soil results in low selenium levels in food and water, which has been associated with increased risk of cancer of various organs (Combs 1989) and also has been attributed to the high incidence of Keshan disease (endemic cardiomyopathy) and Kashin-Beck disease (an endemic osteoarthropathy) in China (WHO 1987).

The high incidence of hepatoma in the southern part of China has been directly related to the lifestyle and environment of the residents, particularly with regard to the contamination of their grain source by AFB_1 (Han 1993). Quidong County, Jiansu Province, has one of the highest incidence rates of hepatoma in China at 45.9/100 000. The high incidence of hepatoma is related to low levels of selenium in the blood of the residents in this region of China and low Se levels in their grain (Li et al. 1986).

Yu et al. (1985) have reported that the administration of a Se-enriched yeast suspension to Wistar rats and ducks, that also received AFB_1, reduced the incidence of hepatoma. Based on these observations an intervention study using Se-yeast (200 $\mu g\,day^{-1}$) was carried out on a high risk population with hepatitis B surface antigen. A significant decrease in the incidence of hepatoma was observed in the treated group during a two year follow-up (Han 1993). Yu et al. (1988) also observed a protective effect on reducing cellular DNA damage induced by AFB_1 in the lymphocytes from people receiving Se supplements.

Studies have also been carried out on other trace element interactions with AFB_1 such as copper, zinc, iron, manganese, fluorine and iodine. Acutely toxic AFB_1 doses ($5-7\,mg\,kg^{-1}$,

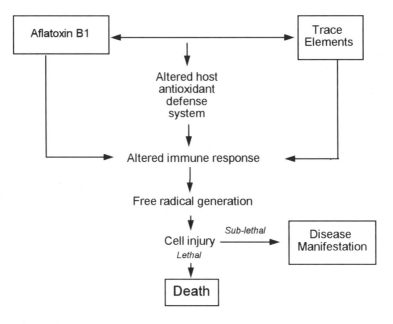

Fig. 3. The action of aflatoxins as noxae through interaction with trace elements.

intra-peritoneally) altered the distribution of Fe, Cu, and Mn in rat liver, kidney and spleen (Doyle et al. 1977). Zinc carbonate supplemented diets were found to prevent hepatocellular necrosis and cell proliferation in Syrian hamsters chronically dosed with aflatoxin (Llewellyn et al. 1986). Katzen & Llewellyn (1987) recorded the bio-interaction of Mn and AFB_1 in rats following the administration of AFB_1 and radioactive manganese chloride ($MnCl_2$). In pigs fed aflatoxins, a progressive reduction in Unsaturated Iron Binding Capacity (UIBC) and Total Iron Binding Capacity (TIBC) was observed and this was attributed to decreased transferrin levels in aflatoxicosis (Harvey et al. 1988). Increased molecular reactivity of AFB_1 leading to subsequent DNA adduct formation was evident in the liver microsomes of rats fed copper deficient diets for 9–13 weeks. The activities of microsomes returned to normal after supplementation of the same diet with a small level of copper emphasizing the role of copper in the activation, and hence in modulating the carcinogenicity of AFB_1 (Prabhu et al. 1989).

Maxwell et al. (1992) observed that dietary aflatoxin (25 μg of AFB_1 for 8 weeks) caused a significant decrease in serum copper and zinc levels in rats. The decrease in serum copper and zinc levels was attributed to the possible interaction of AFB_1 and these trace elements.

In vitro studies

Using *Salmonella typhimurium* strains TA100 and TA98, Francis et al. (1988) tested the inhibitory activity of various trace elements on mutagenesis induced by AFB_1 in the presence of a rat liver microsomal activation system and they observed that Cu, Mn and Se possessed exceptional inhibitory activity for AFB_1 whereas Zn and Fe produced moderate inhibition. The inhibition was competitive with regard to Se and Mn but non-competitive for Cu suggesting that these trace elements might interact with some components of the microsomal enzyme system, thus interfering with the bioactivation of AFB_1. The authors considered that the capacity of trace elements to inhibit AFB_1-induced bacterial mutagenicity reflects on their ability to afford protection against the development of AFB_1 induced neoplasia.

Outcome of interaction of aflatoxin and trace elements

Even though several mechanisms of interaction of AFB_1 with trace elements have been postulated it is most probable that free radicals are the net result of such interactions. AFB_1 is a known exogenous factor in the generation of reactive oxygen species (Busby & Wogan 1984) and the essential trace elements Cu, Mn, Se, Zn and Fe are important constituents of enzymes concerned with the scavenging of these species and provide protection against free radical-induced tissue damage (Athar et al. 1993). Interaction of AFB_1 with these elements results in free radical generation and cellular toxicity. In addition AFB_1 induced gastro-intestinal disturbances could interfere with the absorption of these trace elements. These trace elements are also required for host immunocompetence (Shronts 1993) as their deficiency has been shown to interfere with the antioxidant reactions related to phagocytic, cell-mediated and humoral immune functions. As aflatoxins are potent immuno suppressants as evident from extensive animal studies (Busby & Wogan 1984), this combination of factors in the malnourished individual may contribute to increased susceptibility to infections. This is exemplified in Kwashiorkor, a form of protein–energy malnutrition observed in children in the tropics in which the antioxidant nutrient status such as vitamin A, E, β-carotene, Zn, Se, Fe, Mn are low. The clinical symptoms of this disease have been considered to result from an impaired defence system unable to detoxify an increased production of free radicals (Golden & Ramdath 1987). The free radical theory in the causation of Kwashiorkor has been proposed by Golden et al. (1991). These authors argue that Kwashiorkor results from an imbalance between the production of free radicals and their safe disposal. The various noxae to which these children are exposed produce an oxidative toxic stress which leads to excess free radical generation.

Aflatoxins have been found to be an important etiological factor in Kwashiorkor based on epidemiological and clinical studies (Coulter et al. 1988; Hendrickse 1988). It is probable that aflatoxins could act as noxae through interaction with the antioxidant trace elements resulting in the generation of excess free radicals and also by directly or indirectly suppressing the immune response. This process is represented in Fig. 3.

The interaction of aflatoxin with trace elements is only one aspect of the multitude of interactions taking place in the biological system and further evidence that the geochemical environment can influence specific biological mechanisms of disease. There are significant interactions occurring between trace elements such as Zn, Cu and Fe (Davis 1980), toxic elements (Stanstead 1980) and other dietary factors. Understanding these complex interactions and characterizing them will involve a multidisciplinary approach involving biomedical scientists, nutritionists, epidemiologists, geologists and geochemists.

Conclusions

Aflatoxin B_1, a potent mycotoxin that occurs in the diet, interacts with trace elements as demonstrated by in vivo and in vitro studies. The interaction of AFB_1 with trace elements may lead to changes in host immunity and also favour free radical production. Aflatoxins have been implicated in humans and animals with malnutrition and presumably trace element deficiency. Intervention studies by supplementation with trace elements in aflatoxin endemic areas, as in the Quidong County in China, have significant potential in improving our understanding of the control of nutritional/environmental problems. The interaction of aflatoxin with trace elements is only a single aspect of the multitude of factors affecting trace element status in human beings and animals. Another example is Kashin-Beck disease, an endemic osteoarthropathy that occurs in eastern Siberia and certain parts of China which have been attributed to selenium deficiency. Although the etiology of this disease has not been fully established, selenium deficiency has been attributed as one of the main causative factors. Other factors such as mycotoxin contamination of cereals with certain Fusarium strains in China and high phosphate and manganese contents in the soil, food and drinking water in endemic areas in eastern Siberia have been suggested (WHO 1987).

Processes in the geochemical environment, in addition to causing trace element deficiencies, are also known to cause depletion of well recognized elements such as iodine which leads to goitre and cretinism (in some situations related to goitrogens in food). In certain situations geochemical alterations may also result in the escape of toxic metals such as cadmium, lead and arsenic from the soil–plant barrier which contaminate the food chain and cause toxicity to human beings and animals (Plant et al. 1993). As Hopps (1975) clearly points out 'Scientists are often divided as a result of an isolation that comes from concern with only the disciplinary area of their primary interest. All the advantages of each discipline are necessary to understand the complex interactions that reflect health or disease.' Primary prevention requires the application of highly sensitive methods for early detection of risk factors in the environment, food and water. Future progress depends on a multidisciplinary approach involving biomedical scientists, nutritionists, epidemiologists, geologists and geochemists and an integrated common corrective action.

References

ALEXANDROWICZ, J. 1976. The potential role of mycotoxins and of trace elements in prophylaxis of leukemia. *In:* HEMPHIL, D. D. (ed.) *Trace Substances in Environmental Health X.* 133–135.

ATHAR, M., ABDULLA, M., SULTANA, S., ET AL. 1993. Free Radicals and Trace Elements. *The Journal of Trace Elements in Experimental Medicine*, **6**, 65–73.

BALDWIN, S. & PARKER, R. S. 1987. Influence of dietary fat and selenium in initiation and promotion of aflatoxin B_1-induced pre-neoplastic foci in rat liver. *Carcinogenesis*, **8**, 101–107.

BRUCATO, M., SUNDLOF, S. F., BELL, J. U. & EDDS, G. T. 1986. Aflatoxin B_1 toxicosis in dairy calves pretreated with selenium-vitamin E. *American Journal of Veterinary Research*, **47**, 179–183.

BURK, R. F. & LANE, J. M. 1983. Modification of chemical toxicity by selenium deficiency. *Fundamental and Applied Toxicology*, **3**, 218–221.

BUSBY, W. F. & WOGAN, G. N. 1984. Aflatoxins. *In:* SEARLE, C. E. (ed.) *Chemical Carcinogens.* 2nd edn. American Chemical Society, 945–1136.

CHEN, J., GOETCHIUS, M. P., COMBS, G. F. JR. & CAMPBELL, T. C. 1982. Effects of dietary selenium and vitamin E on covalent binding of aflatoxin to chick liver cell macromolecules. *Journal of Nutrition*, **112**, 350–355.

COMBS, G. F. JR. 1989. Selenium. *In:* MOON, T. E. & MICOZZI, M. S. (eds) *Nutrition and Cancer Prevention – Investigating the Role of Micronutrients.* Marcel Dekker, Inc. New York, 389–419.

COULTER, J. B. S., SULIMAN, G. I., LAMPLUGH, S. M. ET AL. 1988. Protein-energy malnutrition in Northern Sudan: clinical studies. *European Journal of Clinical Nutrition*, **42**, 787–796.

DAVIS, G. K. 1980. Microelement interactions of zinc, copper, and iron in mammalian species. *Annals of New York Academy of Sciences*, **355**, 131–139.

DOYLE, J. J., STEARMAN, W. C., NORMAN, J. O. & PETERSON, H. D. V. 1977. Effects of aflatoxin B_1 on distribution of Fe, Cu, Zn, and Mn in rat tissues. *Bulletin of Environmental Contamination and Toxicology*, **17**, 33–39.

FRANCIS, A. R., SHETTY, T. K. & BHATTACHARYA, R. K. 1988. Modifying role of dietary factors on the mutagenicity of aflatoxin B_1: In vitro effect of trace elements. *Mutation Research*, **199**, 85–93.

GOLDEN, M. H. N. & RAMDATH, D. D. 1987. Free radicals in the pathogenesis of kwashiorkor. *Proceedings of the Nutrition Society*, **46**, 53–68.

——, —— & GOLDEN, B. E. 1991. Free radicals and malnutrition. *In:* DREOSTI, I. E. (ed.) *Trace elements, micronutrients and free radicals.* Humana, Totowa, NJ, 199–221.

HALLIWELL, B. 1993. The Chemistry of Free Radicals. *Toxicology and Industrial Health*, **9**, 1–21.

HAN, J. 1993. Highlights of the cancer chemoprevention studies in China. *Preventive Medicine*, **22**, 712–722.

HARVEY, R. B., CLARK, D. E., HUFF, W. E. et al. 1988. Suppression of serum iron-binding capacity and bone marrow cellularity in pigs fed aflatoxin. *Bulletin of the Environmental Contamination and Toxicology*, **40**, 576–583.

HENDRICKSE, R. G. 1988. Kwashiorkor and aflatoxins. *Journal of Paediatric Gastroenterology and Nutrition*, **7**, 633–635.

HOPPS, H. C. 1975. Geochemical environment related to health and disease. In: FREEDMAN, J. (ed.) *Trace Element Geochemistry in Health and Disease.* Geological Society of America, Special Paper, **155**, 1–8.

KATZEN, J. S. & LLEWELLYN, G. C. 1987. Further evidence supporting the concurrent influence of aflatoxin and manganese. *Veterinary and Human Toxicology*, **29**, 127–132.

LALOR, J. H., KIMBROUGH, T. D. & LLEWELLYN, G. C. 1978. Induction of duodenal serotonin production by dietary sodium selenite and aflatoxin B_1. *Food and Chemical Toxicology*, **16**, 611–613.

LEI, D. N., WANG, L. Q., RUEBNER, B. H. et al. 1990. Effect of selenium on aflatoxin hepatocarcinogenesis in the rat. *Biomedicine and Environmental Science*, **3**, 65–80.

LI, W. G., GONG, X. L, XIE J, R. ET AL. 1986. Regional distribution of liver cancer and its relation to selenium levels in Quidong County of China. *Chinese Journal of Oncology*, **8**, 262–264.

LLEWELLYN, G. C., HOKE, G. D., O'REAR, C. E. ET AL. 1986. Alteration of some toxic aflatoxin responses in Syrian hamsters fed zinc carbonate supplemented diets. *Journal of Industrial Microbiology*, **1**, 1–8.

MAXWELL, S. M., YOUNG, R. H., HARRISON, C. & HENDRICKSE, R. G. H. 1992. The effect of aflatoxin on serum zinc and copper levels. *Transactions of the Royal Society of Tropical Medicine and Hygiene*, **86**, 344.

NEWBERNE, P. M. & CONNER, M. W. 1974. Effect of selenium on acute response to aflatoxin B_1. In: HEMPHIL, D. D. (ed.) *Trace substances in Environmental Health VIII.* University Park Press, Columbia, 323–328.

NYANDIEKA, H. S., WAKHIS, J. & KILONZO, M. M. 1990. Association of reduction of AFB_1 induced liver tumours by antioxidants with increased activity of microsomal enzymes. *Indian Journal of Medical Research*, **92**, 332–336.

PETR, T., BARTA, I. & TUREK, B. 1990. In vivo effect of selenium on the mutagenic activity of aflatoxin B_1. *Journal of Hygiene Epidemiology Microbiology and Immunology*, **34**, 123–128.

PLANT, J. A., BALDOCK, J. W. & SMITH, B. 1996. The role of geochemistry in environmental and epidemiological studies in developing countries: a review. *This volume.*

PRABHU, A. L., ABOOBAKER, V. S. & BHATTACHARYA, R. K. 1989. In vivo dietary factors on the molecular action of aflatoxin B_1. *In vivo*, **3**, 389–392.

SHRONTS, E. P. 1993. Basic concepts of immunology

and its application to clinical nutrition. *Nutrition in Clinical Practice*, **8**, 177–183.

STANSTEAD, H. H. 1980. Interactions of toxic elements with essential elements: Introduction. *Annals of the New York Academy of Sciences*, **355**, 282–284.

WATERLOW, J. 1992. *Protein Energy Malnutrition*. Edward Arnold, London, 104–154.

WHO. 1987. Selenium. *In: Environmental Health Criteria 58*, World Health Organisation, Geneva, 13–306.

YU, S. Y., LI, W. G., ZHU, Y. L. *et al.* 1985. Regional cancer mortality incidence and its relation to selenium levels in China. *Biology of Trace Element Research*, **7**, 22–26.

———, CHU, Y. J. & LI, W. G. 1988. Selenium chemoprevention of liver cancer in animals and possible human applications. *Biology of Trace Element Research*, **15**, 231–241.

Preliminary studies of acid and gas contamination at Poas volcano, Costa Rica

R. A. NICHOLSON,[1] P. D. ROBERTS[1] & P. J. BAXTER[2]

[1]*British Geological Survey, Keyworth, Nottingham NG12 5GG, UK*
[2]*University of Cambridge, Department of Community Medicine, Fenner's, Gresham Road, Cambridge CB1 2ES, UK*

Abstract: Primary volcanic phenomena are potentially catastrophic, whilst those of a secondary nature may be merely considered as background activity but may be of sufficient intensity to cause long-term suffering to indigenous populations. The latter is the case at Poas volcano in Costa Rica, where a change in the quiescent state of the volcano has been shown by increasing activity during the last decade. Acid gases are apparently now being emitted more intensely than before, particularly during the dry season, to the detriment of the health of people and domestic livestock living in the surrounding countryside. Damage to crops and farm buildings is also evident.

A pilot geochemical study has been undertaken at Poas volcano to determine the principal constituents of the gas emissions, the form in which they are most likely to be transported and the extent of the area within which they can be easily measured.

The results obtained indicate that emissions of mixed volatile acids contribute to the problems encountered, and that, as might be expected, seasonal changes and distance from the source also exert control. Rainfall effectively scrubs the acid gases during the wet season, thereby considerably ameliorating the immediate effects of the emissions.

Poas is a composite stratovolcano rising 1400 m above its surroundings in the populated Cordillera Central of Costa Rica (Fig. 1). The last major eruptions occurred in 1952/53, when phreatic activity resulted in the loss of the crater lake and culminated in the growth of a 45 m high pyroclastic cone, which subsequently became the focus of continuous fumarolic activity. A new lake had formed by 1967 and reached a depth of about 50 m, circulating mildly acidic brines at about 40°C. Apart from minor fluctuations in level and temperature, and a brief period of

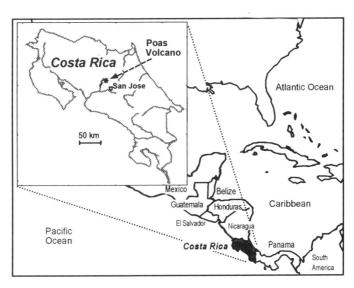

Fig. 1. Location of Poas Volcano, Costa Rica.

Table 1. *Sulphur dioxide concentrations from long-term diffusion tubes (24 hour average)*

Location	1991 ($\mu g\, SO_2/m^3$) Wet season conditions	1992 ($\mu g\, SO_2/m^3$) Dry season conditions
Control tubes (Bajos del Toro)	0.4	4
Cajon	250	249
San Luis	299	730*
Trojas	60	124
Farm near summit	314	499*
Public viewpoint	117	156

Sample locations shown on Fig. 2.
*Approximate data only due to normal tube capacity being exceeded.

geyser activity during 1978/79, the system appeared stable until early 1986 when a long-term decrease in the lake level and increase in acidity commenced. Intense geysering again occurred from June 1986 to July 1987, and these events heralded the start of a new magmatic cycle which has been studied by a variety of geophysical and geochemical techniques (Brown et al. 1989; Barquero & Fernández 1990; Rowe et al. 1992a, b). However, the environmental hazards posed by the escape of high temperature volcanic gases and aerosols are only now becoming apparent. The effects of acid gases can be seen on the western flank of Poas, which is completely denuded of vegetation on the inner-facing slopes for c. 1 km downwind, and only supports low scrub elsewhere.

Gas monitoring

A preliminary visit to Poas took place in late 1991 towards the end of the wet season in Costa Rica, during which the proposed sampling was curtailed because of the danger of slope instability in parts of the crater induced by heavy rain (Nicholson et al. 1992). As Poas volcano comprises andesitic lavas separated by layers of unconsolidated volcaniclastic materials, periods of exceptional rainfall cause some areas to be liable to spontaneous collapse. During 1991, parts of the crater area were therefore considered inaccessible for safety reasons, and it was decided to repeat the sampling exercise when drier conditions would be encountered. This would also allow comparison of analytical data obtained during wet and dry seasons. During this initial visit long-term diffusion tubes were emplaced for approximately two weeks to determine sulphur dioxide gas concentrations around the volcano. Figure 2 indicates the approximate locations of these sample sites in relation to the active crater.

A second visit took place a few months later in early 1992, when a marked difference was noted between the conditions encountered and those experienced at the end of 1991. Little or no rain fell throughout the entire visit, and there was only light precipitation from persistent afternoon cloud-cover. However, acid fall-out from the volcanic plume appeared to be more intense in the drier air, and this was evident by personnel experiencing dry throats and being able to smell 'sulphurous' fumes in the atmosphere. Brief interviews with local people confirmed this as typical of the dry season effects of the volcano (Nicholson et al. 1993).

Results

In the populated areas downwind, several kilometres from the summit, acid gases were detected in the atmosphere, although the highest concentrations of sulphur dioxide (SO_2) measured did not exceed 0.3–0.5 ppm over periods of a few minutes. These measurements were carried out using an MEI (Survey Plus) toxic gas monitor equipped with an electrolytic sensor for the detection of SO_2. Data subsequently obtained from the long-term diffusive gas samplers showed comparable values (Table 1).

During a visit to the crater rim on the northwestern side of the volcano (Fig. 2) in 1992, the highest temporal SO_2 concentration recorded using the toxic gas monitor was c. 35 ppm (Short Term Exposure Limit 5.0 ppm; HSE 1995). Stinging sensations caused by acid precipitation on exposed skin were pronounced. The acid gases present at this location were later confirmed by collecting the acid fall-out on filter papers impregnated with sodium formate, and measuring the sulphate and fluoride contents. On this later occasion low cloud within the crater obscured the view and variable air circulation ensured that the volcanic plume was constantly changing direction, and therefore levels of SO_2 measured with the toxic gas

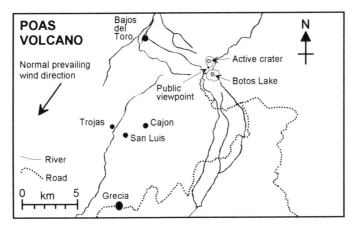

Fig. 2. Sampling locations, Poas Volcano.

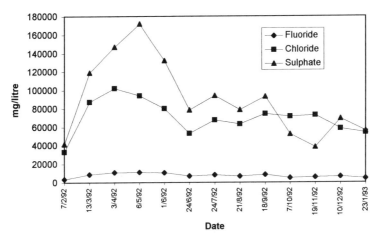

Fig. 3. Fluoride, chloride and sulphate concentrations measured in crater lake waters between February 1992 and January 1993.

monitor did not exceed 0.2 ppm at any one time. However, Draeger chemical diffusion tubes emplaced above open ground showed a total flux of 10 ppm SO_2 over a four-hour period, i.e. 2.5 ppm per hour. The use of similar tubes confirmed that there was no hydrogen chloride or hydrogen sulphide at this location. Studies carried out at the public viewing platform on the southern side of the main crater, which is much closer to the source but normal to the prevailing wind direction, did not detect measurable concentrations of any of the three acid gases, even after five hours of exposure.

In the crater bottom, values of 2.5 and 5.0 ppm per hour for hydrogen chloride and SO_2 respectively were obtained with the Draeger chemical diffusion tubes; there was no indication of hydrogen sulphide. However, atmospheric SO_2 concentrations measured using the toxic gas monitor ranged between 20 and 70 ppm. The wind direction was highly variable, and although positioned only 100 m–300 m from the source of the gases, it was often possible to work for prolonged periods without a respirator, because the gas plume was impacting throughout the caldera. This effect is very localized due to the nature of the terrain, and in no way affects the general southwesterly flow of the plume away from the crater (Fig. 2). It was not possible to approach any of the high temperature fumaroles venting through the remnants of the crater lake, due principally to the presence of acid waters and thixotropic sediments, but it was possible to collect lake water samples.

Crater lake water chemistry

Major and trace elements in the water samples were determined by a combination of ICP/AES, ion chromatography, flow injection analysis, selective ion electrode and titrimetric techniques.

The pH of the crater lake water samples was not determined but previous investigations have shown that acid concentrations were generally high, with pH values of $c.$ −0.5 being common (Rowe et al. 1992b).

Under such highly acid conditions it would be expected that any uncomplexed anionic species will be present as the free acids. Figure 3 shows the distribution of fluoride, chloride and sulphate in the lake waters throughout 1992/93, although practical difficulties associated with sampling from exactly the same location on each occasion may have slightly affected the overall results. These samples were collected by staff of OVSICORI–UNA (Observatorio Vulcanológico y Sismológico de Costa Rica–Universidad Nacional) as part of their routine sampling programme. In general much higher concentrations prevail throughout the drier months (December–April) indicating the gradual depletion by evaporation or drainage of the remnants of the lake, followed by dilution with rain water during the wet season. It was reported by local scientists that rainfall was much heavier than usual during the latter months of 1992 and early 1993. This obviously affected the concentrations of dissolved ions (Fig. 3), which would normally be expected to start increasing as the lake waters either evaporated or drained away (Brown et al. 1991).

Gas geochemistry

Diffusion samplers. To corroborate data on acid precipitation around the volcano, long-term diffusion tubes were emplaced at sites near the summit, and also at several locations on the downwind side, to collect SO_2. The village of Bajos del Toro some 6 km from Poas summit (Fig. 2) was chosen as the background location because there is no record of gases having impacted on this area on a regular basis, and it was also readily accessible by road. Other areas could equally have served as background monitoring sites, but were either more difficult to approach or records of gas dispersion were incomplete. The other diffusion samplers were placed in villages where they would receive constant exposure to the volcanic gases, subject to the prevailing winds. During the 1991 survey, tubes were emplaced for between 336 and 430 hours and during the 1992 survey for $c.$ 600 hours. The tubes were subsequently returned to the UK for analysis; all data are shown in Table 1 calculated as an average exposure time over 24 hours.

At some locations measurements indicated significantly increased levels of SO_2 during the dry season in early 1992, compared to those obtained under much wetter conditions in 1991. However, the order of magnitude increase shown by the control tubes placed in the Bajos del Toro is well within the unexposed control variation of the tubes. The release of acid gases and aerosols into the environment is much greater during the drier months, due principally to the loss of the scrubbing effects of rainfall near the summit source.

Impregnated papers. Simple chemical absorption techniques were used to monitor the short-term variation of gas dispersion at different locations around the crater (Fig. 2). During the 1992 visit, a site was chosen on the crater rim, due north of the public viewpoint and directly under the 'normal' direction of the gas plume. One litre volumes of air were pumped through individual filter papers impregnated with sodium formate, at predetermined intervals over a period of almost four hours, during which time the aerial SO_2 concentration was monitored using the toxic gas monitor. The papers were then sealed in polyethylene bags and returned to the laboratory for analysis. Each paper was cut into two pieces, one being leached with distilled water for sulphate determination, and the other with a special buffer solution to extract the fluoride. The results are shown in Table 2, and for comparison, also include data for 1 litre volumes of air sampled on the top of the lava dome within the crater, at atmospheric SO_2 concentrations of 10–20 ppm. These latter samples are significantly higher due to greater exposure to high gas concentrations. It is to be expected that there would be a general increase in both fluoride and sulphate content, with the accompanying temporal increase in the measured aerial SO_2 concentration.

Environmental and health effects

It has been shown that there is a wide variation in the amounts of acid contaminants dispersing into the air from Poas crater lake. The damage to vegetation is readily visible immediately downwind of the volcano, but there are undoubtedly potentially more serious effects at

Table 2. *Acid ion and SO_2 data from impregnated papers (1 l pumped samples)*

Sample no.	Time	F^- (mg l^{-1})	SO_4^- (mg l^{-1})	SO_2 (ppm)
1	12:15	0.15	0.64	3.0
2	13:00	0.11	0.53	0.4
3	13:15	0.10	0.42	0.4
4	13:30	0.09	0.87	0.6
5	13:45	0.10	0.54	0.2
6	14:00	0.10	0.66	0.2
7	14:15	0.15	0.54	0.4
8	14:30	0.22	0.56	0.3
9	14:45	0.20	0.55	0.6
10	15:00	0.25	1.79	1.5
11	15:15	0.31	1.63	1.0
12	15:30	0.29	1.76	0.8
13	15:45	0.32	1.99	1.8
Lava Dome		0.40	3.65	10.0
Lava Dome		0.50	7.87	15.0–20.0

greater distances from the summit, caused by the long-term low concentrations of these contaminants. The principal agricultural problems are likely to arise from general acid precipitation comprising halogen and sulphur acids, which readily disperse from the source during the dry season. Sulphur species measured as SO_2 have been monitored at least 7 km from the summit, and may cause health problems such as respiratory tract irritation and asthma attacks in susceptibles. The SO_2 results from higher strength sulphur acids being formed through evaporation of water during vapour transport (Nicholson et al. 1993). The formation of acid aerosols is suspected, and these preliminary measurements have indicated that although the amounts may be very small, they are a potential pathway through which acid gases escape to the environment. Further studies are needed to substantiate this.

The enhanced levels of SO_2 measured at some localities indicate the problems of acidic components in the atmosphere near the volcano, and are very conspicuous through visible damage to vegetation. This emphasizes the environmental problems created when a change in wind direction alters the fallout zone of the volcanic plume. During the wet season, many areas seem to be protected from the effects of the acidity, presumably because rainfall effectively scours the main acid components of the plume close to their source. The release of acid gases and aerosols into the environment is therefore apparently much greater during the drier months. Local farmers complained of aggravated respiratory problems during the dry season, and also damage to important (for Costa Rica) cash crops such as coffee. WHO guideline exposure values for SO_2 are $30\,\mu g/m^3$ for plants and $50\,\mu g/m^3$ for humans (annual average), and $100\,\mu g/m^3$ for plants and $125\,\mu g/m^3$ for humans (24 hour average). Therefore by reference to Table 1 it can be seen that the recommended exposure levels to SO_2 are being routinely exceeded, in some locations by substantial margins. Visible foliar damage to coffee plantations and other vegetation several kilometres from the summit of Poas was indeed apparent on each of the field trips, and therefore invisible damage to soil profiles may also be expected, with obvious consequences for plant health. It is hoped to undertake further studies at Poas involving health studies of the local population and air monitoring of SO_2 and airborne particulates with the objective of assessing the health impacts of gaseous emissions from Poas volcanic crater lake.

Only a few of the contributors to this work can be acknowledged individually. Special mention should be made of Eduardo Malavassi, Jorge Barquero and Erick Fernández from OVSICORI–UNA, Costa Rica, all of whom gave either assistance with field investigations and/or helpful advice. We also thank colleagues in the BGS Analytical Geochemistry Group who undertook many chemical determinations, and Martin Williams, Glen Campbell and Clare Downing of Warren Spring Laboratory who provided the SO_2 diffusion tubes and also carried out the subsequent analyses. The late Professor Geoff Brown, and his colleague Dr Hazel Rymer, from the Open University also assisted with an introduction to the field area, and by providing geological information based on their personal experiences over many years working at Poas.

References

BARQUERO, J. & FERNÁNDEZ, E. 1990. Erupciones de gases y sus consecuencias en el Volcan Poas, Costa Rica. *Boletin de Vulcanologia Universidad Nacional, Costa Rica*, **21**, 13–18.

BROWN, G. C., RYMER, H. & STEVENSON, D. 1991. Volcano monitoring by microgravimetry and energy budget analysis. *Journal of the Geological Society, London*, **148**, 589–593.

——, ——, DOWDEN, J., KAPADIA, P., STEVENSON, D., BARQUERO, J. & MORALES, L. D. 1989. Energy budget analysis for Poas crater lake: implications for predicting volcanic activity. *Nature*, **339**, **6223**, 370–373.

NICHOLSON, R. A., HOWELLS, M. F., ROBERTS, P. D. & BAXTER, P. J. 1992. Gas geochemistry studies at Poas Volcano, Costa Rica. British Geological Survey, Technical Report, **WC/92/10**.

——, ——, BAXTER, P. J., CLEGG, S. L. & BARQUERO, J. 1993. Gas geochemistry at Poas Volcano, Costa Rica. March 1992 and January 1993. British Geological Survey, Technical Report, **WC/93/21**.

HSE. 1995. *Occupational Exposure Limits*, 1995. Health and Safety Executive, Guidance Note, **EH 40**, HMSO, London.

ROWE, G. L., BRANTLEY, S. L., FERNANDEZ, M., FERNANDEZ, J. F., BORGIA, A. & BARQUERO, J. 1992a. Fluid-volcano interaction in an active stratovolcano: the crater lake system of Poas volcano, Costa Rica. *Journal of Volcanology and Geothermal Research*, **49**, 23–51.

——, OHSAWA, S., TAKANO, B., BRANTLEY, S. L., FERNANDEZ, J. F. & BARQUERO, J. 1992b. Using crater lake chemistry to predict volcanic activity at Poas volcano, Costa Rica. *Bulletin of Volcanology*, **54**, 494–503.

Kriging: a method of estimation for environmental and rare disease data

M. A. OLIVER

Institute of Public and Environmental Health, School of Chemistry, The University of Birmingham, Edgbaston, Birmingham B15 2TT, UK

Abstract: Most properties that are distributed in geographical space vary, and often in a complex way. The information available is usually fragmentary, but we often want to know the values at intermediate locations. This is the case for both environmental and disease data. However, the nature of the data and the information that we require from them ultimately can be quite different. For rare disease knowing its spatial distribution is often not enough, we also want to know the variation in the underlying risk. At present direct links between disease and factors in the environment are tenuous; however, there is increasing interest in obtaining more precise evidence. Geostatistics provides us with a means of analysing both kinds of data underpinned by the same body of theory, the Theory of Regionalized Variables. Two types of geostatistical estimation are described here. Disjunctive kriging for estimating the probabilities of exceeding particular threshold values in the environment: it is illustrated with an example of soil salinity in Israel. Standard geostatistical technique has been adapted to analyse the spatial distribution of a rare disease, and to estimate the risk of developing it: this is illustrated with an analysis of childhood cancer in the West Midlands Health Authority Region where the risk appears largest in rural areas. The aim in the future is to use the information from both approaches to identify areas for further investigation concerning the aetiology of certain diseases.

Most features of the environment vary spatially, often in a locally erratic and unpredictable way. In many instances investigators want to know how their data vary spatially, and this usually means analysing the variation and displaying the information as maps. Spatial pattern reflects the underlying structure in the variation, and it often provides clues as to its possible cause. Human diseases also vary spatially and the same observations apply. For some diseases the spatial pattern is well defined and there are clear links with the environment (Lovett & Gattrell 1988), but for others, especially rare diseases, the patterns, the relations with other factors and their causes are obscure. Walter (1993) has shown that identifying and assesssing spatial pattern by visual perception alone is not enough. It needs to be supplemented by some kind of statistical analysis of the data.

Geostatistics provides the tools to analyse spatial variation, and it is now well established in the Earth sciences, for instance in petroleum, mining, ground water, soil science, geology and so on. It embraces a large suite of analytical techniques, such as variography, coregionalization, kriging, simulation, fuzzy geostatistics, and so on (see Armstrong 1989; Soares 1993), all underpinned by a coherent and consistent theory. The purpose of this paper is to describe forms of kriging suitable for analysing environmental and disease data, and to suggest how they could be used in a complementary way in the future. The first example describes disjunctive kriging for estimating the probability of exceeding the threshold value of some concentration in the environment that could be harmful to crop, animal or human health. The other analyses the risk of childhood cancer by binomial kriging.

Governmental agencies, such as the European Union, are setting statutory limits for the concentration of certain substances in the soil, water, atmosphere and so on that might be harmful to plant and animal life. In addition land managers are also aware that if nutrient levels fall below certain critical concentrations plants and animals will fail to thrive. In both of these situations a method to aid the decision of whether or not to act to ameliorate the situation is needed. Disjunctive kriging provides estimates of the variable and of the probabilities that the threshold is or is not exceeded at each place of interest. In other words it assesses the risk of taking the estimate at face value. Yates & Yates (1988) and Webster & Oliver (1989) have shown that disjunctive kriging is well suited for decision making on environmental issues. The method is illustrated here with an example of soil salinity in Israel where crop yield is affected when the electrical conductivity of the soil solution exceeds 4 mS cm^{-1}. This method could also have potential in geographical epidemiology when epidemiologists have to decide which areas should be investigated in more detail.

Geographical epidemiology is concerned with identifying and analysing the spatial distribution of diseases. Epidemiologists want to know their overall regional pattern as well as those areas where the risk is unusually large. Historically geographical studies of diseases have provided important clues about their aetiology. For instance Snow's (1855) observations on the distribution of cholera in London, and Burkitt's (Burkitt & Wright 1970) mapping of the distribution of a lymphoma in Africa (Burkitt's lymphoma). These diseases had a fairly distinct pattern that could be explained; however, for others, especially rare ones, the patterns are more obscure, and their aetiology is only partly understood. Epidemiologists are also interested in aspects of the environment that could be a risk to human health. The apparent geographical variation of some diseases must be interpreted with caution: other factors could contribute to the variation in the recorded frequency apart from environmental ones. Consequently many statistical analyses have been applied to the spatial distributions of diseases to try to identify real differences in incidence from place to place (see Marshall 1991).

Oliver et al. (1992) have adapted standard geostatistical method to analyse the incidence of rare disease. The significance of this development is that both the disease and the environmental data can be analysed by techniques embraced by a single body of theory in the search for similar patterns. The ultimate aim is to gain insight into the aetiology of diseases.

Geostatistical theory

The spatial variation of most properties, regionalized variables, is so complex that it defies simple mathematical description. Therefore, Matheron (1965, 1971) suggested a stochastic approach and he brought together several isolated ideas in spatial analysis into a coherent body of theory, the Theory of Regionalized Variables. This treats any property distributed in geographical space, which may have one, two or three dimensions, as a random variable, $Z(\mathbf{x})$, where $\mathbf{x}$ denotes the spatial coordinates. However, their variation is not generally unstructured, it is almost always spatially dependent at some scale, i.e. points within a given distance apart depend on one another in a statistical sense. This structure may be overlain by more or less erratic local variation, or 'noise' at the working scale. Both of these components of spatial variation can be described by the variogram, which summarizes the variation succinctly.

The variogram

The variogram is central to geostatistics and essential for most other geostatistical analyses. It compares the similarity between pairs of points a given distance and direction apart (the lag), and expresses mathematically the average rate of change of a property with separating distance. If Matheron's intrinsic hypothesis holds, i.e. the local mean and the variance of the differences are constant over a given area, then the variogram exists. The usual computing formula is:

$$\hat{\gamma}(\mathbf{h}) = \frac{1}{2M(\mathbf{h})} \sum_{i=1}^{M(\mathbf{h})} \{z(\mathbf{x}_i) - z(\mathbf{x}_i + \mathbf{h})\}^2 \quad (1)$$

where $\hat{\gamma}$ is the experimental semivariance, $z(\mathbf{x}_i)$ and $z(\mathbf{x}_i + \mathbf{h})$ are the values at $\mathbf{x}_i$ and $\mathbf{x}_i + \mathbf{h}$, $\mathbf{h}$ is the separation, or lag, between any two places, and M is the number of pairs of comparisons at a given lag. The function that relates γ to $\mathbf{h}$ is the variogram.

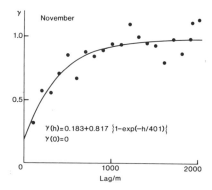

Fig. 1. The experimental variogram of the Hermite transformed electrical conductivities (EC): the points are the sample semivariances and the solid line the fitted exponential model. (Redrawn from Wood et al. 1990.)

Interpreting the variogram can sometimes provide insight into the structure of the variation. In most instances the variance increases with increasing separating distance as in Fig. 1. This corresponds with more or less strong correlation or spatial dependence at the shortest distances which weakens as the separation increases. Variograms often flatten when they reach a variance known as the 'sill' variance: they are bounded (Fig. 1). Such variograms are second order stationary and suggest that there are patches or zones with different kinds of soil for instance, whereas unbounded ones suggest continuous change over a region. The distance at which the sill is reached, the 'range', marks the limit of spatial dependence. The variogram often has a positive intercept on the ordinate, 'the nugget variance'. This is the part of the variation that we cannot predict. Much of it derives from spatially dependent variation within the smallest sampling interval, somewhat less from measurement error and purely random variation. A completely flat variogram, 'pure nugget', means that there is no spatial dependence in the data and that interpolation should not be attempted.

Kriging

The observations that describe many geographical properties are often fragmentary. To determine their spatial distribution we need to estimate their values at unsampled points. Kriging is the geostatistical technique for estimating values locally; it is based on a mathematical model of the spatial variation. It is an optimal estimator in the sense that the estimates are unbiased and have known estimation variances. The latter provide some information on the reliability of the estimates. Kriging is a generic term for a set of methods including ordinary kriging, disjunctive kriging, cokriging, Bayesian kriging and so on. It is essentially a local weighted averaging procedure, but one in which the data carry different weights according to their positions both in relation to the unknown point and to one another. The kriging weights are obtained from the variogram, or equivalently the covariance function. Estimates can be made for points or over blocks of land, water, etc.

Laslett et al. (1987) compared several methods of spatial estimation including global means and medians, moving averages, inverse squared distance, natural neighbour interpolation, quadratic trend surface analysis, Laplacian smoothing splines and ordinary kriging. They found that most of the methods were either poor predictors or suffered from theoretical drawbacks. Laplacian splines and ordinary kriging performed the best, but the splines over-smoothed the variation compared with kriging. In addition to being an optimal estimator kriging is based on sound theory, it honours the data and it is local. The reliability of kriging, however, depends on having a precise variogram and on satisfying the assumptions of the intrinsic hypothesis.

In ordinary kriging the estimate of our variable, Z, at a point, $\mathbf{x}_0$, is given by:

$$\hat{z}(\mathbf{x}_0) = \sum_{i=1}^{N} \lambda_i z(\mathbf{x}_i) \qquad (2)$$

where λ_i are the weights which sum to 1 to avoid bias, and subject to this are chosen to minimize the estimation variance. (Readers who want more information about the theory of kriging are recommended to read Chapter 14 of Webster & Oliver (1990) to start with, followed by Cressie (1993).) For environmental properties kriging has been used mainly for interpolation and mapping (Oliver & Webster 1990).

Disjunctive kriging for decision making

Ordinary kriging provides optimal estimates, but as with all estimates they embody some error because they derive from sample information. Decisions based on them with regard to the likelihood of toxicity or deficiency associated with a prescribed threshold value may be more or less sound in consequence. The true value at a place might exceed the threshold even though the estimate is less and *vice versa*. As a result an agency could fail to deal with pollution or a land manager might remedy a soil condition that did not exist. To aid decision there is a need to know the risk of taking the estimates at their face value, of incurring unnecessary expenditure or of perpetuating environmental damage and suffering if nothing is done. Disjunctive kriging was developed by Matheron (1976) for miners faced with this problem. It provides estimates of the property, the estimation variances and the probabilities that the true value exceeds or falls short of a critical threshold. Isarithmic maps of the probabilities enable those areas in most need of attention to be identified.

Theory

Rivoirard (1994) has given a clear and detailed description of the complex theory; this paper contains only a summary of the main points. To estimate the conditional probabilities requires stronger assumptions than those for linear kriging: that the values of the properties are the outcome of a second order stationary process, and that the bivariate probability distribution is known and that it too is stationary throughout the region of interest. To ensure this the variable must be transformed so that it has a standard normal distribution. Methods of transformation such as logarithms, square roots, logits, etc. change the general form of the distribution but not the detail. A more flexible method is needed, and the one that Matheron (1976) recommended for environmental data uses Hermite polynomials and has the form:

$$Z(\mathbf{x}) = \phi\{Y(\mathbf{x})\} = \sum_{k=0}^{\infty} Q_k H_k \{Y(\mathbf{x})\} \quad (3)$$

where $Y(\mathbf{x})$ is the transformed value, ϕ is a linear combination of Hermite polynomials and coefficients: H_k, $k = 1, 2, 3...$, is an infinite series of Hermite polynomials and Q_k are coefficients evaluated by Hermite integration. In general k is no more than order 10. Hermite polynomials are orthogonal polynomials related to the Normal distribution. In addition to providing a detailed transformation of the data the transformation is also invertible, hence the results are in the original units in which the property was measured.

The disjunctively kriged estimates are linear combinations of the estimates of the Hermite polynomials of the transformed sample values. In disjunctive kriging the weights have to be found k times per estimate instead of once as in ordinary kriging, and the kriging equations have to be solved k times per estimate also. The estimation variance is also determined for each estimate.

Once the Hermite polynomials have been estimated at an unknown point the conditional probability that the true value there exceeds the critical value, z_c, can be estimated by defining an indicator function, $\Omega[Z(\mathbf{x}) > z_c]$. Based on the transformation using Hermite polynomials the indicators are related and are given by:

$$\Omega[Z(\mathbf{x}) \geq z_c] = \Omega[Y(\mathbf{x}) \geq y_c]. \quad (4)$$

The probability of exceeding the threshold is written in terms of the Hermite polynomials with respect to the observed values in the neighbourhood. At the sampling points the probability is known, elsewhere it has to be estimated by:

$$\hat{\Omega}^{DK}[y(\mathbf{x}_0) > y_c] =$$

$$1 - G(y_c) - \sum_{k=1}^{L} \frac{1}{\sqrt{k}} H_{k-1}(y_c)g(y_c)\hat{H}_k \{y(\mathbf{x}_0)\} \quad (5)$$

where $G(Y_c)$ is the cumulative probability distribution, $g(y_c)$ is the probability density function, and L is the number of Hermite polynomials used, usually no more than ten.

Soil salinity in Bet Shean

The problem of soil salinity in the Bet Shean Valley, Israel (Wood et al. 1990) shows how disjunctive kriging can be used for decision making in situations where concentrations in the environment are likely to exceed a given tolerance which could limit crop growth or be harmful to human or animal health.

The salt concentration in soil is usually measured in terms of the electrical conductivity (EC) of the soil solution: it is greater the larger the salt concentration. An EC of $4\,\mathrm{mS\,cm^{-1}}$ in the soil is usually regarded as critical, marking the onset of salinization. In this example the aim was to use the probabilities from disjunctive kriging to advise farmers of the risk that they were taking by not ameliorating their soil by using gypsum or good quality irrigation water, improving drainage or having a fallow year, all of which are costly remedies.

Analysis and Results. Wood measured the electrical conductivity of 201 topsoil samples taken from an area of 2030 ha in November 1985 [Wood et al. (1990) give the details of the survey.] The measurements were strongly positively skewed and were transformed to a standard normal distribution using Hermite polynomials. The sample variogram was computed from the transformed values and an exponential model fitted to it. The model describes one form of second order stationarity, and so this assumption underlying disjunctive kriging is satisfied. Figure 1 shows the sample semi-variances plotted as points, the fitted model as the solid line, and its equation.

The electrical conductivities and probabilities were estimated at intervals of 100 m on a square grid using the variogram model, the Hermite transformed values, and a critical threshold of $4\,\mathrm{mS\,cm^{-1}}$. Figure 2a shows the isarithmic map of electrical conductivity. In much of the area the estimates exceed $4\,\mathrm{mS\,cm^{-1}}$ and in these areas the farmers could expect losses of yield unless they treat their land. The more interesting point relates to those parts of the region where the estimated EC is c. $4\,\mathrm{mS\,cm^{-1}}$ or less. In these areas we need to know the risk of taking these estimates at their face value. They covered about half of the area. Figure 2b shows the estimated conditional probabilities for the indicator $\Omega[EC(x) \geq 4]$ and they provide the answer. The probabilities exceed 0.4 over much of the area,

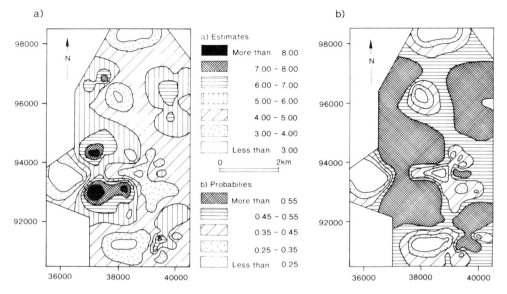

Fig. 2. (a) Map of the disjunctively kriged EC values for Bet Shean. (b) Map of the estimated conditional probabilities of EC exceeding a threshold of $4\,\mathrm{mS\,cm^{-1}}$. (Redrawn from Wood *et al.* 1990.)

only in a very small area are they less than 0.25. Thus the risks of salinity and of reduced yields are fairly large over a much greater area than the estimates suggested. The conditional probabilities identify those areas where there is an important risk of salinity even though the estimates suggest that the soil is not saline. Figure 2b provides a means of deciding where remedial action is most needed and where there is no need to act.

Analysing the spatial pattern of childhood cancer

This analysis addressed the problem of determining whether there were any real differences geographically in the risk of a child's developing cancer in the West Midlands Health Authority Region (WMHAR) for 1980 to 1984 inclusive, and of estimating it. If there were real and evident regional patterns in the risk then its solution could guide epidemiologists to search for the reasons, perhaps in the environment, and to concentrate on those areas where it is greatest.

Several investigators have devised methods for estimating and mapping the risk of disease. Clayton & Kaldor (1987), Besag *et al.* (1991) and Bernardinelli & Montomoli (1992) have used Bayes estimation to map the risk of various malignancies, including lip, thyroid and breast cancers, respectively, and Bithell (1990) used kernel density estimation to determine the risk of childhood cancer in Cumbria. These methods seemed not to make the best use of known spatial autocorrelation in the data. Carrat & Valleron (1992) used ordinary kriging to map the spread of a virus in an epidemic in France. The standard technique does take into account the autocorrelation, but it is not suited to estimating the risk of a rare disease, such as childhood cancer. If the rates of incidence of a rare disease are estimated by standard kriging or any other interpolator, such as inverse squared distance or splines, the result will show their variation from place to place. However, the rates are poor estimates of the risk and are unreliable. This is because there is a large error in estimating the risk from the incidences arising from the small risk and the limited population in the ward. Apparent pattern in the spatial distribution of the incidences is no guarantee that there is structure in the risk: the risk could be completely random.

The data

The West Midlands Health Authority Region (WMHAR) covers about $25\,000\,\mathrm{km^2}$. It includes rural, urban, industrial and suburban environments. During 1980 to 1984 inclusive there were 605 cases of cancer among the 1.13 million children under fifteen years of age living there. The cases are distributed among 345 of the total 840 electoral wards (the areas for which population is recorded). When the cases were mapped

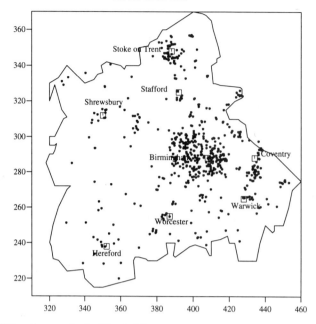

Fig. 3. Cases of childhood cancer in the West Midlands Health Authority Region from 1980 to 1984 inclusive. Each dot represents a diagnosed case, and its position is that of the child's home.

(Fig. 3) their density varied enormously: it is much greater in the towns than in the rural areas, if only because many more children live there. However, this apparent pattern in the cases may not reflect the true pattern in the underlying risk. My aim was, therefore, to explore the data for spatial autocorrelation with a view to identifying meaningful pattern in it.

Theory and analysis

Following Lajaunie (1991) the present author assumed that there is an underlying risk, $R(x)$, of a child's developing cancer and to which all children are exposed: this is the regionalized variable. As a first step to estimating the risk locally the local rate of incidence, $F(x_i)$, was calculated i.e. the ratio of the number of cases within each ward, $L(x_i)$, to the number of children living there, $n(x_i): F(x_i) = L(x_i)/n(x_i)$, where x_i, $i = 1, 2, \ldots$, denotes the centroids of the wards. It was also assumed that the different cases occur independently, so that R is the only possible source of correlation among them. Hence the observed rates, $F(x_i)$, are drawn from a binomial distribution that depends on the risk, $R(x_i)$, and the number of children, $n(x_i)$. The rates of incidence were plotted with the ward boundaries for the WMHAR (Fig. 4). This map gives a different impression of the apparent risk from that of the cases (Fig. 3): the higher rates are in the rural and suburban areas rather than the urban areas. This map must still be treated with caution, however, because the rates of incidence are poor estimates of the risk.

Estimating the risk variogram. The first step in estimating the risk variogram is to compute an experimental variogram (Fig. 5a) of the frequencies, γ_F, with the usual formula equation (1). The full derivation of the risk variogram is given in Oliver et al. (1993a). It can be estimated by:

$$\hat{\gamma}_R(\mathbf{h}) = \hat{\gamma}_F(\mathbf{h}) - \frac{1}{2}\{\overline{F}(1 - \overline{F})\} - \hat{\sigma}_R^2$$

$$\overline{\left\{\frac{n(x_i) + n(x_i + h)}{n(x_i).n(x_i = h)}\right\}} \quad (6)$$

in which the quantity beneath the bar is the average over all pairs of wards involved in calculating $\hat{\gamma}_F(\mathbf{h})$. It takes into account the variation in the numbers of children from ward to ward. The semivariances, $\hat{\gamma}_R(h)$ where $h = |\mathbf{h}|$ for isotropic variation, were calculated from the rates of incidence in this way, and the results are shown by the set of points in Fig. 5b. Since the true variogram is continuous a model was fitted by weighted least squares approximation

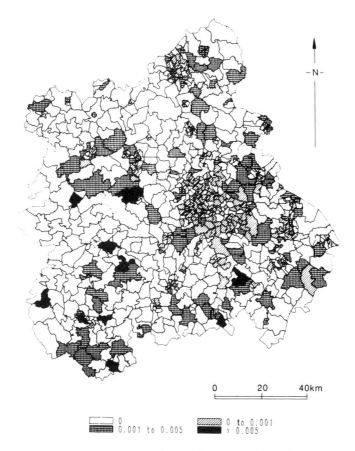

Fig 4. Map of the rates of incidence of childhood cancer in the West Midlands Health Authority Region for the period 1980 to 1984 inclusive plotted by electoral ward.

using Ross's (1987) maximum likelihood program. The solid line in Fig. 5b is Whittle's (1954) elementary correlation. In its isotropic form the function is:

$$\gamma(h) = c\left\{1 - \frac{h}{r}K_1\left(\frac{h}{r}\right)\right\} \qquad (7)$$

where c is the sill estimating the *a priori* variance of the process, r is a distance parameter, and K_1 is the modified Bessel function of the second kind.

Binomial kriging. The rates of incidence, $F(\mathbf{x}_i)$, and the risk variogram are used to estimate the risk, R, of a child's developing cancer at points $\mathbf{x}_0$ throughout the Region, and to map it. Our procedure is effectively an extension of ordinary cokriging (Oliver *et al.* 1993*a*). Assuming the mean to be unknown the risk at a place, $\mathbf{x}_0$, is

$$\hat{R}(\mathbf{x}_0) = \sum_{i=1}^{N} \lambda_i F(\mathbf{x}_i) \qquad (8)$$

where N is the number of data and λ_i are weights. It involves solving the kriging system

$$\sum_{i=1}^{N} \lambda_i C_F(\mathbf{x}_i, \mathbf{x}_j) + \psi = C_{FR}(\mathbf{x}_0, \mathbf{x}_j) \; \forall j$$

$$\sum_{i=1}^{N} \lambda_i = 1 \qquad (9)$$

where ψ is a Lagrange multiplier, the $C_F(\mathbf{x}_i,\mathbf{x}_j)$ are the covariances of the rates, and the $C_{FR}(\mathbf{x}_0,\mathbf{x}_j)$ are the covariances between the rates of incidence and the risk. The covariances of the risk C_R are obtained from the risk variogram and from these the other covariances are derived

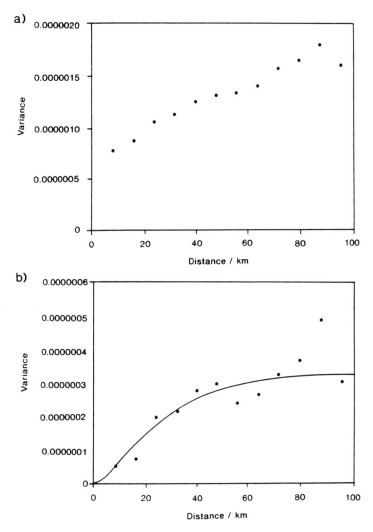

Fig 5. Variograms of childhood cancer in the West Midlands. (a) The experimental variogram of the rates of incidence, $\hat{\gamma}_F(\mathbf{h})$, computed using the usual formula (1). (b) The experimental variogram of risk, $\hat{\gamma}_R(\mathbf{h})$, with the fitted model shown as the solid line. (Redrawn from Oliver et al. 1992.)

(see Oliver et al. 1993a). The risk of a child's developing cancer and the estimation variances were estimated by solving the kriging equations (9) at 2 km intervals on a square grid throughout the region. In most situations a kriging neighbourhood containing 16–25 points is sufficient, but in this adaptation of kriging we found (Oliver et al. 1993b) that more distant data carry sufficient weight to be included in the equation and N was set to 100. The estimates of the underlying risk and the estimation variances were then 'contoured' to produce isarithmic maps, (Fig. 6a, b).

Results and discussion

Both the variogram of rates of incidence, $\hat{\gamma}_F(h)$, and that of the risk, $\hat{\gamma}_R(h)$, are bounded (Fig. 5). This suggests that the risk of a child's developing cancer has a coarse patchy distribution; i.e. wards with large risk in general occur near to others with large risk, and similarly those with small risk occur close to others where it is small. The distance parameter of the risk variogram suggests that the pattern or autocorrelation in the risk extends to c. 47 km. The semivariances of the rates are much larger than those of the

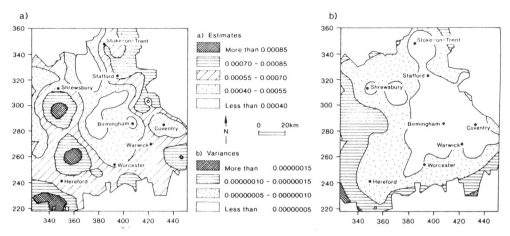

Fig. 6. (a) Map of the estimated risk of childhood cancer in the West Midlands made by binomial kriging. (b) Map of the estimation variances of the risk of childhood cancer in the West Midlands. (Redrawn from Oliver *et al.* 1992.)

risk, (Fig. 5a) and the projection of the empirical values to the ordinate has an appreciable nugget variance. The latter is the principal difference between the variograms; it represents the error in estimating the risk from few cases among the finite population. There is no nugget variance in the risk variogram because the wards, which form the support of our sample, are contiguous.

The kriged map of the risk, Fig. 6a, supports the interpretation of the variogram: the risk has a patchy distribution. The rural areas and the fringes of the conurbation appear to have a large risk, whereas it is small in the urban areas in general. This contrasts markedly with the impression gained from the distribution of cases, (Fig. 3) and it is also somewhat different from the map of the rates (Fig. 4). Figure 6b displays the estimation variances, showing that they are greatest at the edges of the region. They are also larger for the rural areas where the data are most sparse, than for the urban areas where the wards are smallest and where the data are most dense.

The pattern in the risk of childhood cancer for the WMHAR is similar to that observed elsewhere in Great Britain by, for instance Alexander *et al.* (1990) who found that the risk of children's developing lymphoblastic leukaemia was largest in wards farthest from large urban centres, and Greaves (1988) who observed more childhood lymphoblastic leukaemia in rural areas than in towns.

It now seems worth searching for possible environmental causes where the risk is large, and the present author plans to do that in the future. Environmental properties can be analysed by disjunctive kriging, as described earlier, to provide both optimal estimates and estimates of the probabilities associated with certain critical levels of concentrations in the environment. Such maps should provide epidemiologists with the kind of spatial information that would enable them to focus attention on certain areas for further investigation.

For this study 5 years of the 20 years of data available have been analysed, taking all children less than 15 years of age and all forms of the disease to ensure enough data to estimate the variogram reliably. More detailed analyses will be possible and are planned for the full dataset using the major diagnostic subgroups and other subsets related to age, sex, ethnicity, social class and so on. The West Midlands conurbation will also be analysed as a separate unit to examine the spatial variation of the risk in a large urban area.

Conclusions

The case study of salinity in Israel shows how the conditional probabilities can be used for assesssing whether or not to ameliorate the land. Geostatistics is already a well established technology in the Earth sciences, and our adaptation of it for analysing childhood cancer shows how it could also be applied to estimating the risk of diseases. It seems possible that both of the geostatistical approaches described here could assist epidemiologists to focus further attention on certain areas in their search for possible causal links between some diseases and the environment. This is the approach that our team intends to pursue in the West Midlands

Health Authority Region to determine whether environmental factors contribute to the risk of childhood cancer.

I thank the West Midlands Regional Children's Tumour Research Group and Dr G. Wood for the use of their data, Professor R. Webster for his help and the use of his software, Dr K. R. Muir for his suggestions, S. Parker for help with graphics and two referees for helpful comments.

References

ALEXANDER, F. E., RICKETTS, T. J., MCKINNEY, P. A. & CARTWRIGHT, R. A. (1990) Community lifestyle characteristics and risk of acute lymphoblastic leukaemia in children. *The Lancet*, **336**, 1461–1465.
ARMSTRONG, M. (1989) *Geostatistics*, Kluwer, Dordrecht.
BERNARDINELLI, L. & MONTOMOLI, C. 1992. Empirical Bayes versus fully Bayesian analysis of geographical variation in disease risk. *Statistics in Medicine*, **11**, 983–1007.
BESAG, J., YORK, J. & MOLLIE, A. (1991) Bayesian image restoration with two applications in spatial statistics. *Annals of the Institute of Statistical Mathematics*, **43**, 1–59.
BITHELL, J. F. 1990 An application of density estimation to geographical epidemiology. *Statistics in Medicine*, **9**, 691–701.
BURKITT, D. P. & WRIGHT, H. D. 1970 *Burkitt's Lymphoma.* Livingstone, Edinburgh.
CARRAT, F. & VALLERON, A.-J. 1992. Epidemiologic mapping using the "kriging" method: application to an influenza-like illness epidemic in France. *American Journal of Epdemiology*, **135**, 1293–1300.
CLAYTON, D. & KALDOR, J. 1987. Empirical Bayes estimates of age-standardized relative risks for use in disease mapping. *Biometrics*, **43**, 671–681.
CRESSIE, N. A. C. 1993. *Statistics for Spatial Data.* Wiley, New York.
GREAVES, M. F. 1988. Speculations on the cause of childhood acute lymphoblastic leukaemia. *Leukaemia*, **2**, 120–125.
LAJAUNIE, C. 1991. *Local Risk Estimation for a Rare Noncontagious Disease based on Observed Frequencies.* Note N-36/91/G. Centre de Géostatistique, Ecole des Mines de Paris, Fontainebleau.
LASLETT, G. M., MCBRATNEY, M. B., PAHL, P. J. & HUTCHINSON, M. F. 1987. Comparison of several spatial prediction methods for soil pH. *Journal of Soil Science*, **38**, 325–341.
LOVETT, A. A. & GATTRELL, A. C. 1988. The geography of spina bifida in England and Wales. *Transactions of the Institute of British Geographers New Series*, **13**, 288–302.
MARSHALL, R. J. 1991. A review of methods for the statistical analysis of spatial patterns of disease. *Journal of the Royal Statistical Society, Series A*, **154**, 421–441.
MATHERON, G. 1965 *Les Variables Régionalisées et leur Estimation*, Masson, Paris.
——— 1971 *The Theory of Regionalized Variables and its Applications.* Cahiers du Centre de Morphologie, Mathématique de Fontainebleau, **5**, Fontainebleau, Ecole des Mines de Paris.
——— 1976. A simple substitute for the conditional expectation: the disjunctive kriging. In: GUARASCIO, M., DAVID, M. & HUIJBREGTS, C. (eds) *Advanced Geostatistics in the Mining Industry*, Reidel, Dordrecht, 221–236.
OLIVER, M. A. & WEBSTER, R. 1990. Kriging: A method of interpolation for geographical information systems. *International Journal of Geographical Information Systems*, **4**, 313–332.
———, LAJAUNIE, C., WEBSTER, R. et al. 1993a Estimating the risk of childhood cancer. In: SOARES, A. (ed.) *Geostatistics Tróia '92, Vol. 2.* Kluwer, Dordrecht, 899–910.
———, ———, ———, ——— 1993b. Binomial kriging the risk of a rare disease. *Cahiers de Géostatistique, Fascicule, 3*, Ecole des Mines de Paris, 159–165.
———, MUIR, K. R., WEBSTER, R., PARKES, S. E., CAMERON, A. H., STEVENS, M. C. G. & MANN, J. R. 1992. A geostatistical approach to the analysis of pattern in rare disease. *Journal of Public Health Medicine*, **14**, 280–289.
RIVOIRARD, J. 1994. *Introduction to Disjunctive Kriging and Non-linear Geostatistics.* Clarendon, Oxford.
ROSS, G. J. S. 1987. *Maximum Likelihood Program.* Numerical Algorithms Group, Oxford.
SNOW, J. 1855. *On the Mode of Communication of Cholera.* 2nd Edition, Churchill, London.
SOARES, A. 1993. *Geostatistics Tróia '92.* Kluwer, Dordrecht.
WALTER, S. D. 1993. Visual and statistical assessment of spatial clustering in mapped data. *Statistics in Medicine*, **12**, 1275–1291.
WEBSTER, R. & OLIVER, M. A. 1989. Optimal interpolation and isarithmic mapping of soil properties. IV. Disjunctive kriging and mapping the conditional probability. *Journal of Soil Science*, **40**, 497–512.
——— & ——— 1990. *Statistical Methods for Soil and Land Resource Survey.* Oxford University Press.
WOOD, G., OLIVER, M. A. & WEBSTER, R. 1990. Estimating soil salinity by disjunctive kriging. *Soil Use and Management*, **6**, 97–104.
WHITTLE, P. 1954. On stationary processes in the plane. *Biometrika*, **41**, 434–449.
YATES, S. R. & YATES, M. V. 1988 Disjunctive kriging as an approach to management decision making. *Soil Science Society of America Journal*, **52**, 1554–1558.

A review of the geochemical factors linked to podoconiosis

R. HARVEY, J. J. POWELL & R. P. H. THOMPSON

The Gastrointestinal Laboratory, The Rayne Institute, St Thomas' Hospital, London SE1 7EH, UK

Abstract: Podoconiosis, or non-filarial elephantiasis, was named and characterized by the late Ernest Price (Price 1988, 1990). He described the clinical features of swelling and deformity of the legs associated with enlargement of the draining lymph nodes. Histopathological examination of these nodes showed them to contain bi-refringent particles. He identified the epidemiological association between the local type of soil and the disease, and finally he confirmed this association by microanalysis of the particles in diseased tissue, showing them to be sub-micron aluminosilicate and silica, which are characteristic of the clay fractions of the local soil. It remains unclear what it is within the heterogeneous range of microparticles identified in tissue, and soil factors, that is responsible for the toxicity, and what are the host factors that determine certain individuals to be sensitive to them. Prevention of progression of the disease could be achieved by preventing further uptake of particles by using adequate footwear.

The local fine reddish-brown soils are similar in all areas of tropical Africa where non-filarial elephantiasis is prevalent. These soils are in areas of high altitude (> 1250 m), modest average temperature (20°C), and high, hot season rainfall (> 1000 mm annually) (Price & Bailey 1984). This precise epidemiological relationship has been shown in the Wollamu district of Ethiopia (Price 1974), the Nyambene range of Kenya (Crivelli 1986), and the Cameroon highlands (Price & Henderson 1981). A study (Kloos *et al.* 1992) in the Illubabor region of western Ethiopia of the incidence of disease in migrant and indigenous populations in an area of podoconiosis, showed that prevalence in migrants resident for more than six years was comparable to that of the indigenous population (*c.* 8%). This migrant effect emphasizes the environmental causation of podoconiosis and its geographical specificity. Superimposing the prevalence data for podoconiosis on a geological map of Africa reveals a correlation between alkali basalt rocks and podoconiosis. Weathering of these rocks under the above climatic conditions produces the characteristic clay soil. A study (Price & Plant 1990) of particle size in soils from endemic (upland) and non-endemic (lowland) areas using a disc centrifuge and a Multisizer to measure the size range, found significantly more particles of less than 5 μm in the upland podoconiosis areas. Analysis of particles from a digest of a diseased femoral lymph node showed a size range of 0.3–6 μm, suggesting either toxic particles are within this size range or a smaller size facilitates uptake into the tissues.

Evidence for the pathogenic role of local soils in podoconiosis came from Price's microanalysis studies (Price & Henderson 1979; Blundell *et al.* 1989) on diseased tissue, namely the skin, lymphatics and lymph nodes of the legs. Birefringent particles were initially identified by light and polarizing microscopy, and further analysis was then performed with transmission electron microscopy and X-ray microanalysis. This showed accumulations of sub-micron microparticles within phagolysosomes of macrophages. Microanalysis showed these microparticles to contain mainly silicon and aluminium with lesser amounts of titanium and iron. Analyses of the morphology of particles and elemental ratios identified the aluminosilicate kaolinite, silica, as mainly amorphous silica, with some quartz and some iron oxides. Electron diffraction studies confirmed the presence of kaolinite and small amounts of quartz, in addition to small amounts of haematite, goethite and gibbsite. These microparticles make up the clay fraction of the local soils, but they are environmentally ubiquitous, and, apart from increased proportions of the microparticles in soils of endemic areas, it is not clear why they should become toxic.

The toxicity of a mineral (Hochella 1993) is related to the following. (a) Mechanical or dimensional properties: the size of particles in diseased tissue has been emphasized. Smaller particles are more easily absorbed through abrasions in the feet and transported in the lymphatics. In other particle related diseases a specific fibrous shape is the most important toxicological determinant (Stanton *et al.* 1981), but in podoconiosis tissue there is no preponderance of fibrous minerals. (b) With decreasing size the surface area to volume ratio of particles increases and different chemical properties assume importance. These include surface and near surface composition, surface atomic structure, surface microtopography, surface charge and its pH and dependence on surrounding solutions, durability (dissolution rate), associated minor or trace elements and associated minor and trace phases. Such factors determine the ability of particles to interact in biological systems and the nature of that interaction. Some work has been performed to assess these factors in soils taken from either podoconiosis endemic or non-endemic areas. Blanke et al. (1983) used the thermoluminescence to discriminate between both groups. The volcanic rock from which the soils form has undergone rapid cooling. At high temperatures there are many imperfections in the crystal lattice in addition to extended defects as phase transitions, which are fixed by rapid cooling, forming a metastable system at ambient temperature. The process by which the bedrock is weathered to form the particulates in the soil will further alter these imperfections. Thermoluminescence operates by trapping electrons and holes at the imperfections at ambient temperature, electrons and holes are generated by ionizing radiation, and subsequent heating liberates characteristic emission spectra. Analyses were performed on the soil fraction of size less than 10 μm. Thermoluminescent curves resembled those of pure quartz for both soil groups, but soils from endemic areas had a characteristic second peak which, interestingly, disappeared on heating. It was postulated that these characteristic lattice defects related to the different biological behaviour of these soil fractions. Differences in the surface are likely to be of more importance to the biological behaviour of particles, however, with decreasing particle size and increasing surface area to volume ratio, lattice defects will increasingly affect surface properties.

Soil particle surface has been directly assessed (Davies & Townsend 1990) using thermostimulated exoelectronic emission. This technique is similar to thermoluminescence but exoemission involves analyzing electron emission from the surface of the material under study, rather than electromagnetic spectra from the crystal lattice. A characteristic emission was identified from podoconiosis soils, with a peak at 75°C. However, neither technique is able to specify the cause of the altered reactivity of the podoconiosis endemic soils.

Price did not identify any association between specific trace elements in soil and areas of disease, which contrasts with a recent study (Frommel et al. 1993). Soils were collected from an area of podoconiosis in the Ethiopian Rift valley and from other nearby areas of low incidence, and thirty-five elements were analysed by X-ray fluorescence spectroscopy and inductively coupled plasma mass spectroscopy. The soil was described as consisting of weathered rock fragments with limited argillaceous and organic components. High values of sulphur, cerium, lanthanum, neodymium, vanadium, beryllium and zirconium were found in soils from endemic areas, and concentrations of the latter two elements were double those in non-endemic areas at 4.6 v. 2.6 ppm and 618 v. 323 ppm respectively. Beryllium is associated with a well described disease whose microscopic pathology has a superficial resemblance to tissue changes in podoconiosis (Seaton 1987; Aller 1990), and zirconium minerals have been suggested as a cause of chronic lung inflammation (Bartter et al. 1991). The authors suggest these trace elements are directly involved in the pathogenesis of podoconiosis. However, beryllium has not been identified in podoconiosis tissue, and zirconium only at low levels (Price & Henderson 1978). Microanalysis of particles in tissues may not be sensitive enough to detect adsorbed trace elements, while lighter elements, such as beryllium require special techniques for detection. However, zirconium minerals are extremely durable and would be expected to be present as easily identified particulates, not as diffusely adsorbed ions, and the low level in diseased tissue make them an unlikely aetiological factor. There is good evidence that trace amounts of iron on mineral surfaces increases their toxicity (Ghio et al. 1992; Keeling et al. 1994). The surface of silicates can complex iron, both in soils and within the body, basic rocks have a high iron content, which, mobilized by weathering can interact with clay minerals. Iron is often identified in particles from podoconiosis tissue. As yet no studies have specifically examined the role of iron in podoconiosis.

Price identified cases of podoconiosis in Ethiopia, Kenya, Tanzania, Rwanda, Burundi, Cameroon and the Cape Verde Islands (Price 1990) and areas of prevalence within these areas were consistently associated with the red clay soils described above. One case report (Corachan et al. 1988) describes two cases of podoconiosis from Equatorial Guinea in areas of two distinct soil types, from Rio Mumi a coastal enclave with a red clay soil, and Biolo, a volcanic island with a black lava soil and high annual rainfall. Analysis of lymph nodes from

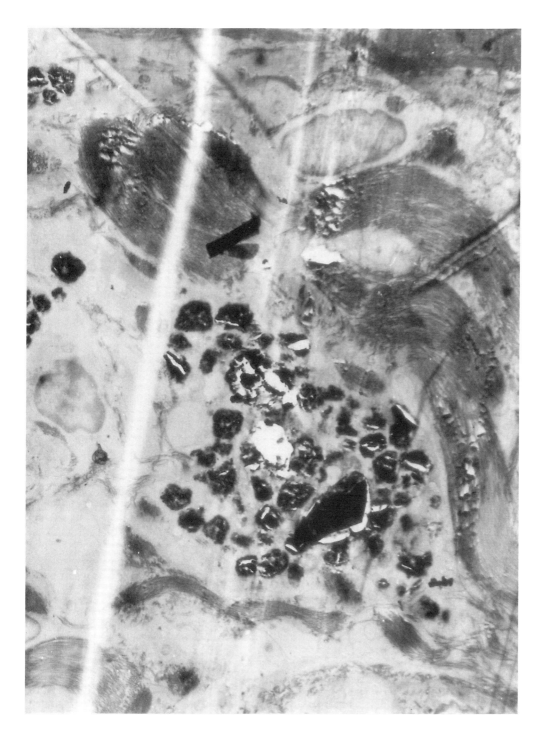

Fig. 1. Transmission electron micrograph of a macrophage engorged with microparticles from a lymph node of an Ethiopian patient of E. W. Price with podoconiosis. X-ray microanalysis shows microparticles to contain aluminium, silicon, titanium and iron. Micrograph provided by E. W. Price.

diseased tissues showed the presence of microparticles, which were further analysed by electron microscopy, energy dispersive X-ray analysis and X-ray diffraction (Fig. 1). In the Rio Mundi case particles ranged in size from 1–7 μm, were of predominantly silicon and aluminium, while X-ray diffraction studies gave an amorphous pattern. In the Bioko case particles were bigger (range 2–25 μm), predominant elements were aluminium and silicon with a significant peak for titanium and X-ray diffraction studies were compatible with the amphibole, eckermanite. The authors suggest this mineral to be the pathological agent in this case, and that a weathered volcanic bedrock is the more general prerequisite for the disease.

The ability of a given mineral to produce disease (i.e. its pathogenicity) is finally determined by the nature of its interaction with effector cells of the immune system, particularly the macrophage. This cell has the ability to take up (phagocytose) particles, and subsequent processing may activate the cell to produce a range of chemical species including cytokines (Driscoll & Mauer 1991), chemical messengers modulating the local and systemic immune response, and reactive oxygen species (Gabor et al. 1975; Kennedy et al. 1989), which are highly reactive molecules capable of reacting with and damaging a range of biomolecules. The final result is the tissue changes recognized in disease, namely inflammation and fibrosis. The extreme consequence of particle uptake and macrophage activation is death of the cell, which liberates insoluble particles that are then available for re-uptake and perpetuation of this process.

A study by Spooner & Davies (1986) assessed the toxicity to murine macrophages in culture of soil fractions from podoconiosis prevalent areas. Particles of less than 2 μm were separated out and added to a culture of peritoneal macrophages at a concentration of 50 μg ml^{-1}, with a standard microparticle quartz (a quartz with defined toxicity to macrophages, used in pneumoconiosis research) as a positive control. In contrast to the quartz, which produced 100% cell death in 24 hours, the soil fraction from endemic and non-endemic areas produced 50% cell death in 96 hours. Soil analysis on the specimens used in this study showed a higher proportion of smaller particles in the specimen from the endemic area and a significant difference in aluminium/silicon ratio, namely 0.48 and 0.63 for the non-endemic and endemic areas respectively. The authors suggested that the inability to discriminate between soil types in their study is related to the bioavailability of the small particle fraction in vivo. Thus the greater number of small particles, whose different Al/Si ratios may reflect different mineral species in the endemic soil, increases the number that are taken up which may interact with the immune system.

Crystalline silica has well established biological properties. Its effects at a cellular level on macrophages have been described above, but in the lungs, dusts produce a chronic inflammation, silicosis, in proportion to their quartz content. Lung inflammation and neoplasia can also be produced in animal models of pneumoconiosis, by inhalation of microparticles of crystalline silica. Fyfe & Price (1985) demonstrated that crystalline silica injected into a lymphatic of the leg of a rabbit damaged and blocked the vessel. Quartz particles of 0.1–4.0 μm were injected in suspension directly into an isolated limb lymphatic, and histopathological examination showed a vigorous macrophage response with marked inflammation, while lymphangiograms demonstrated complete obstruction of the injected vessel extending to the draining lymph node. These changes were progressive following a single injection and closely resemble those seen in podoconiosis. However, most silica in podoconiosis tissue is amorphous, and so it cannot be assumed that this has the same effect as crystalline quartz, and is responsible for the pathological changes.

Price & Henderson (1978) also analysed the elemental content of lymphatic tissue in subjects with and without podoconiosis. Light microscopy showed birefringence at the site of accumulation of particles; this was not observed in non-elephantiasis tissue. Electron microscopy, however, revealed that there were accumulations of similar microparticles in both groups. X-ray microanalysis showed an identical range of elements, but a difference in the Al/Si ratio, with particles in elephantiasis tissues tending to have higher concentrations of silicon. Price's hypotheses were either that this difference reflected a quantitative difference in uptake of a toxic fraction or that in elephantiasics the nature of the interaction of the host with particles determined their toxicity. The mechanism for this could be differences in particle absorption and dissolution, neutralization of the active surface of particles by interaction with chelating proteins, or involvement of specific immune effector mechanism that recognize particles to be antigenic and trigger a hypersensitivity response. It has been suggested by Ziegler (1994) that the interaction between soil derived inorganic microparticles and the immune system, in addition to causing podoconiosis, predisposes to Kaposi's sarcoma. This tumour is associated with immune deficiency, which can be generalized, as in for example AIDS, or localized to the limbs, as occurs in podoconiosis areas.

Price's studies have therefore shown that podoconiosis is related to certain soil types, and that minerals from the clay fraction of this soil taken up into the lymphatics of the lower limb can be toxic and cause a destructive inflammatory process responsible for the signs and symptoms of this disease. It remains unclear, however, what it is within the heterogeneous range of microparticles identified in tissue and soil fractions that is responsible for the toxicity, and what are the host factors that determine certain individuals to be sensitive to them.

References

ALLER, A. J. 1990. The clinical significance of beryllium. *Journal of Trace Elements and Electrolytes in Health and Disease*, **4**, 1–6.

BARTTER, T., IRWIN, R., DASEAL, A., NASH, G., HIMELSTEIN, J. & JEDERLINIC, J. 1991. Zirconium compound induced pulmonary fibrosis. *Archives of Internal Medicine*, **151**, 1197–1201.

BLANKE, J. H., PRICE, E. W., RENDELL, H. M., TERRY, J. & TOWNSEND, P. D. 1983. Correlations between elephantiasis and thermoluminescence of volcanic soil. *Radiation Effects*, **73**, 103–113.

BLUNDELL, G., HENDERSON, W. J. & PRICE, E. W. 1989. Soil particles in the tissues of the foot in endemic elephantiasis of the lower legs. *Annals of Tropical Medicine and Parasitology*, **83**, 381–385.

CORACHAN, M., TURA, J. M., CAMPO, E., SOLEY, M. & TRAVERIA, A. 1988. Podoconiosis in Aequatorial Guinea. Report of two cases from different geological environments. *Tropical and Geographical Medicine*, **40**, 359–364.

CRIVELLI, P. E. 1986. Non-filarial elephantiasis in Nyambene range: a geochemical disease. *East African Medical Journal*, **63**, 191–194.

DAVIES, J. E. & TOWNSEND, P. D. 1990. Exoemission of Ethiopian soils and the endemicity of non-filarial elephantiasis. *Radiation Protection Dosimetry*, **4**, 185–188.

DRISCOLL, K. E. & MAURER, J. K. 1991. Cytokine and growth factor release by alveolar macrophages: potential biomarkers of pulmonary toxicity. *Toxicologic Pathology*, **19**, 398–405.

FROMMEL, D., AYRANCI, B., PFEIFER, H. R., SANCHEZ, A., FROMMEL, A. & MENGISTU, G. 1993. Podoconiosis in the Ethiopian Rift Valley. Role of beryllium and zirconium. *Tropical & Geographical Medicine*, **45**, 165–167.

FYFE, N. C. & PRICE, E. W. 1985. The effects of silica on lymph nodes and vessels – a possible mechanism in the pathogenesis of non-filarial endemic elephantiasis. *Transactions of the Royal Society of Tropical Medicine and Hygiene*, **79**, 645–651.

GABOR, S., ANCA, Z. & ZUGRAVU, E. 1975. In vitro action of quartz on alveolar macrophage lipid peroxides. *Archives of Environmental Health*, **30**, 499–501.

GHIO, A., KENNEDY, T. P., WHORTON, A., CRUMBLESS, G., HATCH, G. & HOIDAL, J. 1992. Role of surface complexed iron in oxidant generation and lung inflammation induced by silicates. *American Journal of Physiology*, **263**, L511–L518.

HOCHELLA, M. F. 1993. Surface chemistry, structure, and reactivity of hazardous mineral dust. *In:* GUTHERIE, G. D. & MOSSMAN, B. T. (eds) *Health effects of mineral dusts*. Mineralogical Society of America, Washington, DC, 275–308.

KEELING, B., LI, K. & CHURG, A. 1994. Iron enhances uptake of mineral particles and increases lipid peroxidation in tracheal epithelial cells. *American Journal of Respiratory Cell and Molecular Biology*, **10**, 683–688.

KENNEDY, T. P., DODSON, R., RAO, N. V., BASER, M., TOLLEY, E. & HOIDAL, J. R. 1989. Dusts causing pneumoconiosis generate OH and produce haemolysis by acting as Fenton Catalysts. *Archives of Biochemistry and Biophysics*, **269**, 359–364.

KLOOS, H., BEDRI, K. A. & ADDUS, 1992. Podoconiosis (endemic non-filarial elephantiasis) in two resettlement schemes in western Ethiopia. *Tropical Doctor*, **22**, 109–112.

PRICE, E. W. 1974. The relationship between endemic elephantiasis of the lower legs and the local soils and climate. *Tropical and Geographical Medicine*, **26**, 225–230.

—— 1988. Non-filarial elephantiasis – confirmed as a geochemical disease, and renamed podoconiosis. *Ethiopian Medical Journal*, **26**, 151–153.

—— 1990. *Podoconiosis, non-filarial elephantiasis*. Oxford University Press.

—— & BAILEY, D. 1984. Environmental factors in the etiology of endemic elephantiasis of the lower legs in tropical Africa. *Tropical and Geographical Medicine*, **36**, 1–5.

—— & HENDERSON, W. J. 1978. The elemental content of lymphatic tissues of barefooted people in Ethiopia, with reference to endemic elephantiasis of the lower legs. *Transactions of the Royal Society of Tropical Medicine and Hygiene*, **72**, 132–136.

—— & —— 1979. Silica and silicates in femoral lymph nodes of barefooted people in Ethiopia with special reference to elephantiasis of the lower legs. *Transactions of the Royal Society of Tropical Medicine and Hygiene*, **73**, 640–647.

—— & —— 1981. Endemic elephantiasis of the lower legs in the United Cameroon Republic. *Tropical and Geographical Medicine*, **33**, 23–29.

—— & PLANT, D. A. 1990. The significance of particle size of soils as a risk factor in the etiology of podoconiosis. *Transactions of the Royal Society of Tropical Medicine and Hygiene*, **84**, 885–886.

SEATON, A. 1987. Pneumoconiosis. *In:* WEATHERALL, D. J., LEDINGHAM, J. G. G. & WARREL, D. A. (eds) *Oxford Textbook of Medicine*. Oxford University Press, 15–16.

SPOONER, N. T. & DAVIES, J. E. 1986. The possible role of soil particles in the aetiology of non-filarial (endemic) elephantiasis: a macrophage cytotoxicity assay. *Transactions of the Royal Society of Tropical Medicine and Hygiene*, **80**, 222–225.

STANTON, M. F., LAYARD, M., TEGERIS, A., MILLER, E., MAY, M., MORGAN, E. & SMITH, A. 1981. Relation of particle dimension to carcinogenicity in amphibole asbestos and other fibrous minerals. *Journal of the National Cancer Institute*, **67**, 965–975.

ZIEGLER, J. L. 1994. Endemic Kaposi's sarcoma in Africa and local volcanic soils. *The Lancet*, **342**, 1348–1351.

Index

Aberdares 49, 52–3, 54–60
absorption 13–4, 31, 98, 118
adenylate markers 109–10, 122–3, 125–8
AFB_1 see aflatoxin B_1
aflatoxin B_1 231–7
aluminium 10–9, 50, 69–71, 73–8
   groundwater 91–5, 110–20, 124–8, 141–51
   macrophages 255, 258
   mining 156, 190
   silica 10, 13, 95, 147, 258
   solubility 2–3, 82–3, 86, 95, 100, 102
Alzheimer's disease 95
Amboseli 49, 51–2, 54–60
ammonia 69, 77–8, 93, 108–9, 127, 169
analyses, see mineral, rock, soil and water analyses
andosols 49–54, 56, 111
antagonist reactions 1, 30–36, 50–60, 94–5
antimony 12–14, 17, 91–3, 101, 168–79
aquatic roots and mosses 82–4, 86–8
arsenates 18, 96–7, 154–5, 164–5
arsenic 2–4, 7–8, 11–9, 83, 208
   groundwater 91–4, 96–7, 102, 163–79
   sources 153–61, 163–179, 192
Aspergillus spp. 231
atmospheric transport 13, 17, 83, 86–7, 90
   mining 156–9, 165, 167, 190
   see also iodine, volatilization

bacteria 11, 96, 108–10, 165, 169–71
   concentrates 4, 122–8, 156
barium 11, 83, 91–4, 100–1, 112–7
   mining 155, 168–77, 192
baseline inputs 11–15, 83, 100, 195, 199
bauxite 12, 154–5
beryllium 12, 14, 17, 256
   groundwater 91–5, 100, 102
Bet Shean 248–9, 253
binomial kriging 251–54
bioavailability 1–5, 8–19, 26–36, 43, 56–60
   iodine 208, 220
   metals 83–4, 95, 158
biomass 108–10, 118, 120–8
   water quality 4, 18, 71, 73–8, 169–71
bismuth 14
Black Triangle 190
Blackfoot disease 153, 163–4
blood pressure 94
boron 3–4, 100, 118, 225
   toxicity 169–71, 177, 225
Bowen's disease 163
bromine 169–70
Burkitt's lymphona 24

cadmium 2, 7, 11, 13–4, 17
   groundwater 91–4, 101–2, 168–70
   interaction 26–36, 58, 101–2
   mining 185–92
calcareous buffering 1–2, 17, 87, 94
calcium 10–1, 68–78, 144–5, 168–71
   forage 26–36, 39–45

   groundwater 91–5, 111–22, 133, 136
   halogens 97–100, 102, 205–8, 215, 225
   soils 1–5, 49–60, 82, 86–72
   weathering 110–9, 121–2, 126–8
cambisols 49
capillary effects 55, 216
carbon 13, 98, 127, 141–4
carbonates 10–1, 13, 55, 71, 73–8
   groundwater 91–4, 101, 113–6, 144–5, 168–72
   water quality 122, 133, 168–79, 189
carcinogens 7, 101, 259–54
   arsenic 96–7, 153, 157–8, 163–4
   mycotoxins 231, 233, 236–7
cardiomyopathy, endemic see Keshan disease
cardiovascular disease 19, 94, 98, 101, 153, 157, 163
cattle 15–17, 23–35, 58–9, 157
cereal crops 3–4, 11, 43, 56–8, 83–4
childhood cancer 247–254
China 2, 4, 17, 19, 97, 201, 233–235
chloride 10–1, 242
chromium 2, 18, 69–70, 74, 112–4
   pollution 83, 86–7, 101, 168–71
clay 4, 163, 196, 255–9
   particulate 141–2, 148, 255–9
   tropical weathering 4–6, 10–11, 14, 55–6
coal 8, 16–7, 159, 183–6, 186–92, 96, 154–6
cobalt 7, 11–5, 155, 168–72, 175
   forage 1, 27–36, 33–6, 39–45, 83, 86
   soils 1–3, 5, 50–60, 69–71, 74
colloids 13, 95, 127–8, 141–9
copper 1–2, 4–5, 7, 11–17
   antagonists 30–5, 82–3, 102
   arsenic 155, 168–79
   forage 26–36, 39–45, 58, 82–3, 86–8
   free radicals 232–6
   pollution 183–92, 197–9
   soils 50–60, 68–71, 74
   water 91–5, 112–5
Cornwall 155–7, 164
cretinism 19, 207, 223, 228, 235
crystal lattice 256
cyanide 8, 18, 101, 166, 169, 177

databanks 3, 12, 83, 87
deficiencies, see also fluorine, iodine, magnesium,
   selenium
deficiencies 1–5, 11–9, 44–5, 97–102
   plants 25–36, 50–60, 87
defluorator 139–40
dental health 94, 99–100, 131–40, 177
depression cones 184
Derbyshire Neck 225
desertification 8
dewatering 183–4, 190
disjunctive kriging 245, 247–9
Dodoma region 107–28
dry season, deficiencies 44–5, 243
Dry Zone, Sri Lanka 100, 131–40
dump water 165, 169, 175, 187–90
duricrust 10–14

dust inhalation 156–9, 164–5, 190

eckermanite 258–9
Eh 11, 14–7
  aluminium 143–5
  arsenic 155, 164–5, 170–5
  cadmium 82
  groundwater 55–60, 91–7, 102, 126–8, 148–51
elephantiasis, non-filarial *see* podoconiosis
elephants 63–78
enzymes 232–6
evaporation 10–1, 55, 75–6, 131, 136, 165

fericrete 10–1, 110–1, 113, 118
ferralsols 10, 55–6, 110–1
ferrisols 55
fersiallitic soil 24–27, 35
fibrocystic desease 223
fish 17, 177–8, 195
fluorine 1–4, 14–7, 116, 126, 233
  fluorosis 2–4, 11, 99–102, 136–40, 225
  groundwater 91–102, 131–40, 190
  volcanic 242–3
forests 14, 86, 167, 190
free radicals 232–236

gases 13–4, 16–7, 165, 186–7, 190–2, 239–43
  see also, iodine, volatilization
genu vulgans 2
geophagia 4–5, 32, 59, 157, 159
geostatistics 245–54
Ghana 56, 95, 155, 163–79
glaciation 226–7
goitre 5, 17, 97–9, 201–28, 235
goitrogens 201, 207–8, 218, 220, 224–5, 231
gold 3, 8, 12, 14, 154–5, 164–179
groundwater 13, 17–9, 65, 82–8, 91–125
  caves 71, 75–8
  contamination 108–10, 115, 117, 122–8
  Poland 183–92

hafnium 11
halogens 11, 115–7, 123, 126–8, 177
  see also, bromine, chloride, fluorine *and* iodine
heavy metals 8, 17–9, 81–8, 96–7, 101–2, 184–92
hepatoma 233
herbage, and soils 39–45, 49, 56–60
Hermite polynomials 248
hydromorphic salt crust 91–102

IGCP Project 259 (Darnley UNESCO 1995) 12
Illubabor region 255–60
immunosuppressants 231–236
ingestion 14, 17
  arsenic 96, 156–7, 159, 163–5, 177–9
  see also fluorine, salt, selenium, toxic
inhibition 16, 31
interaction 1, 4–5, 13–9, 27–36, 56–60, 94, 102
iodine 2–5, 7–8, 15–9, 63, 73–8
  air *and* sea 202–9, 213–7, 219–20, 227–8, 242
  free radicals 232–3, 235
  geology 97–9, 201–8, 215–20, 226–8
  groundwater 11, 91–2, 97–9, 102, 169–71
  Iodine deficiency disorders 2, 11, 19, 201–11, 213–21, 223–30
  Iodine fixation potential 227
iron 14–8
  ferricrete 10–13, 111, 113, 118
  forage 26–36, 39–45
  groundwater 69–71, 73–8, 91–6, 100–2, 190
  interaction 1–5, 14–9, 30–5, 50–60, 163–79
  mobility 108–24, 126–8, 143–51
  particles 255–7
  protective 101–2, 233–5
irrigation 1–5, 8, 101, 157
isarithmic maps 247–8

kaolinite 10, 126–8, 143, 149, 167
Kaposi's sarcoma 259
Kashin–Beck disease 233, 235
Kenya 39–45, 47–62, 63–79, 100, 255, 258
keratosis 153, 164
Keshan disease 2, 97, 233
Kitum cave 63–78
kriging 245–54
kwashiorkor 235

lactation requirements 59
lake sediments 195–200
laterite 4–5, 10–12, 127, 143, 226–7
  arsenic 157, 164–5, 167
  see also bauxite, duricrust, ferricrete
leaching 2–5, 10–14, 59, 107, 164
  alkalis 115, 127
  eluvation 55–6, 58, 60, 82
  halogens 100, 136, 227
  Kitum 69–71, 75–6
  mining 17–9, 164, 189
lead 1–7, 11–5, 82–3, 86–7, 225
  groundwater 91–5, 101–2
  mining 153–6, 168–78, 184–92, 199
  Sweden 82–3, 86–7
leukaemia 94, 253
Lewa Downs 49, 53–4
limestone 17, 87, 94, 100
lithium 17, 114–6, 225
lithosol 11, 25

macrophages 255, 258
magnesium 10, 69–71, 73–8
  cattle 82–3, 86–7
  forage 26–35, 39–41
  groundwater 91–6, 144–5, 168–75
  soils 50–60
  weathering 110–19, 121, 126–8
Main Karakorum Thrust 225–6
Makutuapora aquifer 107–28
management 1–2, 4, 14–5, 102, 165
manganese 10–1, 86–7, 92–3, 96
  antagonist 32–3
  forage 26–36, 39–45
  groundwater 82, 102, 111–20, 127–8
  pollution 199
  protective 232–6
  soil 2, 10–19, 50
mass transfer 10–9, 83
mercury 7–8, 11, 13–5, 81–3, 191
  groundwater 92–3, 101–2

methane 8
methylation 96–7, 155, 157–8, 164–5
methyliodide 98, 202
microbial effects, groundwater 71, 73–8, 109, 116–28
   soil 155, 169–79
migration 8, 47, 60, 63
mineral analyses 69–72, 74
mining 3, 8, 12, 14–9, 203
   elephants 63–78
   pollution 97, 153–9, 163–79, 183–92
mobility 10–9, 82, 94–102, 111–8
   arsenic 153–61, 163–79
molybdenum 1–4, 11–7
   forage 39–45, 82–3, 87
   soils 50–60, 225
   water 92–3, 168–78
monitors 82–7, 98
montmorillonite 56, 126, 220
moose 59, 85–8
mountain areas 97–9, 102, 201, 207, 223–6
Mukono 143–51
myotoxins 231–6
myxoedema madness 224

Nakuru Lake 49–51, 54–60
Nawaikoke 143–151, 143–51
neurological effects 5, 153, 163
   see also cretinism
Neutral Red Retention 18
nickel 11–14, 69–71, 74, 82–3, 86
   Poland 185, 191
   soils 49–60, 92–3, 101
niobium 83, 91–2, 111, 113
nitosols 52–3, 56, 60, 111
nitrogen 10–11, 86–7, 168–7, 185, 191
nitrates 10, 13, 82–7
   bacteria 109, 122, 127–8
   groundwater 91–4, 96, 114–7, 144–5, 215
   salt caves 69–71, 76–8

Obuasi 155–6, 163–181
oceanic effect 97–8, 202–7, 209, 213–20, 227–8
Ok Tedi gold–copper mine 195–6
Ol Tukai salt pan 56, 195
organic, compounds 7, 118, 122–8, 184
   matter 13–4, 114–20
   metal 43, 96, 108, 149, 164–5
   leaching 2, 11, 51, 164
oxidation 10–11, 17–8, 95–9, 102
   arsenic 4, 154–5, 157–9, 164–5
   metalloids 97, 154, 168–78
   states 11, 13, 83, 86, 101, 206–7

Papua New Guinea 195–200
particle surfaces 118, 256–9
peat 98, 204–5, 208
pH 1–3, 11–4, 17, 60, 91–2
   forage 39–45, 50–60
   groundwater 91–5, 101–2, 112–8, 120–1
   Kitum 71, 73, 76–7
phaezemes 49
phosphorus 3, 10–3, 16–8, 154
   bacteria 112–8, 122–5
   fertilizers 82–4, 175

forage 26–36, 39–45, 58–9, 83
Kashin–Beck disease 235
Kitum caves 69–71, 73–8
   pollution 191, 199
   soil 50–60, 78, 110–1
plant uptake 1–4, 7–19, 157, 179, 242–3
   forage 23–35, 39–45, 49, 56–60, 82–8
Poas volcano 239–244
podoconiosis 255–59
Poland 156, 183–93
pollution 14–7, 95–7, 101–2
pollution see also mining, toxic levels
potassium 10–1, 50–60, 69–71, 73–8, 83, 110–1
   groundwater 91–4, 113–22, 126, 133, 144–5
precipitation sequence 77
protists 73, 108, 110, 118, 123, 126

quartz 11, 18, 169, 255–9
radioactive elements 8, 11–4, 16–7, 101–2, 184
   iodine 200, 234
   see also analyses and elements
radium 11, 14, 184
rain, acid 82–8, 94–5
   fall 49, 119–21, 131, 143
   heavy 10, 165, 196
   seasonal 115, 242–3
rare earths 11, 69, 256
Raynaud's syndrome 163
Regionalized Variables, Theory of 246–7
regosol 49
respiration 14, 83, 92, 112–3, 153, 168–70
rock analyses 28, 41–2, 69, 71, 74, 112–3, 202
rubidium 83, 92, 110–3, 168–70, 177

salination 11, 55–6, 60, 168–71
salinity 131, 136, 190, 248–9, 253
salt intake 138–40, 149
   animal 59–60, 63–78
   iodide 98, 216–7, 220, 224
   seasonal, changes 56–8, 83, 115–6, 195–6, 240–3
   uptake 56–8, 83, 228
seaweed 98, 144–5, 224, 226
sediment polluted 184–6, 195–9
selenium 1–2, 7–19
   forage 39–45
   free radicals 232–6
   groundwater 3–5, 91–7, 101–2
   selenosis 2, 19, 97, 233
   soils 47–59, 82–3
self-swallowing processes 56, 60
serum, 15–7, 23–35, 39–45, 58–9
shallow groundwater 16, 95, 114–28, 142–51
silcrete 11
silica, colloidal aluminium 10, 13, 95, 143–51
   groundwater 83, 91–4, 100, 111–20, 168–71
   particle 255–9
   see also quartz
silver 168–79
smelting 101, 153–6, 164–5, 169, 183–92
sodication 55, 59
sodium 10–1, 63–7, 144–5
   groundwater 91–4, 111–22, 126–8, 168–71, 178
   selenite 232–6
   soils 50–60, 215

soil 2–5, 10–1, 24–35, 98, 110–1, 167
   analyses 27–8, 33, 50–9, 112–3, 181–7
   degradation 8, 14–9, 55, 185–92
   Kenya 39–45, 47–60, 167, 255–9
soil analyses, iodine 206, 227
solonchaks 52, 56, 58
solonetz 50–2, 54–60
sorption 13–4, 18–9, 28–34, 97
sorption arsenic, 64–5 179
spatial distribution 58, 245–54
speciation 3, 8, 12–9, 96, 148–50, 228
Sri Lanka 94, 96, 99–100, 131–40, 213–220, 225
stochastic approach 245–54
stream sediments 12, 15–7, 23–36, 155–6
strontium 11, 113, 168–9, 177
sulphur 205, 256
   bacteria 109–10, 114–7, 121–2, 127–8
   cave salts 69–71, 73–8
   dioxide 17, 19, 178, 186, 190, 240–3
   forage 40–1
   groundwater 17, 91–4, 100–2, 133, 144–5
   metalloids 96–7, 154–6, 163–79
   pollution 186–7, 190, 192
supergene alteration 68–9, 77–8
surface properties 13, 213, 256
surface water, 116, 143 155, 167–77
Sweden 81–9

Taiwan 153, 156–9, 163, 225, 227
Tanzania, Makutapora aquifer 107–28
tellurium 13
Thailand 18, 96, 155–6
thallium 14
thorium 17, 101
thyroid 218, 220, 223–5, 249
tin 8, 10–1, 17, 92, 164
titanium 10, 14, 18, 83, 258
toxic levels 1–3, 11–9, 91–7, 128, 136–8
   aluminum 95, 120, 141, 149–51, 175
   metalloids 93–7, 153–4, 158–9, 163–5, 175–9
   metals 83–4, 87–8, 101–2, 255–9

   see also halogens, Poland
trace elements 92, 168–71, 177
   aflatoxin B_1 232–6
   tropical weathering 111–6
   see also analyses *and* elements
transition elements 1–2, 11–5, 232–6
   see also analyses *and* elements
tungsten 17, 69–71, 169–70

Uganda 100, 141–52
uranium 11–3, 16–7, 83, 91–7, 101–2
urban pollution 2, 8, 14, 19, 190–2, 198–9

vaccination 30, 69–70, 74
vanandium 11, 16–7, 59, 83, 86, 92, 112–3, 168
variogram 246–7, 250–2
vertisol 52–6, 60, 111
volatilization, iodine 202–7, 228
volcanic gases 99–101, 239–243

water, analyses 76, 93, 99, 114–26, 144–9, 168–74, 191
   iodine deficiencies 213, 215
   quality 131–40
   uptake 4, 76, 136–8
waterborne diseases 167
weathering 10–1, 17–8, 49–58, 110–4, 126–8
   see also laterites
wildlife 47–59, 63–79, 85–8

yttrium 83

zeolite veins 64, 68–70, 77–8
Zimbabwe 12, 15–8, 23–37, 164
zinc 1–5, 7, 11, 14–7, 92–3, 153
   forage 2, 23–36, 39–45
   free radicals 232–6
   immobile 11, 111
   pollution 183–92, 199
   sediment 112, 208
   soil 50, 58, 69–70
zirconium 11, 83, 112, 256